Smart Water Grids

A Cyber-Physical
Systems Approach

Smart Water Grids

A Cyber-Physical Systems Approach

Edited by
Panagiotis Tsakalides
Athanasia Panousopoulou
Grigorios Tsagkatakis
Luis Montestruque

CRC Press is an imprint of the
Taylor & Francis Group, an **informa** business

CRC Press
Taylor & Francis Group
6000 Broken Sound Parkway NW, Suite 300
Boca Raton, FL 33487-2742

© 2018 by Taylor & Francis Group, LLC
CRC Press is an imprint of Taylor & Francis Group, an Informa business

No claim to original U.S. Government works

Printed on acid-free paper
Version Date: 20180308

International Standard Book Number-13: 978-1-138-19793-0 (Hardback)

This book contains information obtained from authentic and highly regarded sources. Reasonable efforts have been made to publish reliable data and information, but the author and publisher cannot assume responsibility for the validity of all materials or the consequences of their use. The authors and publishers have attempted to trace the copyright holders of all material reproduced in this publication and apologize to copyright holders if permission to publish in this form has not been obtained. If any copyright material has not been acknowledged please write and let us know so we may rectify in any future reprint.

Except as permitted under U.S. Copyright Law, no part of this book may be reprinted, reproduced, transmitted, or utilized in any form by any electronic, mechanical, or other means, now known or hereafter invented, including photocopying, microfilming, and recording, or in any information storage or retrieval system, without written permission from the publishers.

For permission to photocopy or use material electronically from this work, please access www.copyright.com (http://www.copyright.com/) or contact the Copyright Clearance Center, Inc. (CCC), 222 Rosewood Drive, Danvers, MA 01923, 978-750-8400. CCC is a not-for-profit organization that provides licenses and registration for a variety of users. For organizations that have been granted a photocopy license by the CCC, a separate system of payment has been arranged.

Trademark Notice: Product or corporate names may be trademarks or registered trademarks, and are used only for identification and explanation without intent to infringe.

Library of Congress Cataloging-in-Publication Data

Names: Tsakalides, Panagiotis, editor. | Panousopoulou, Athanasia, editor. | Tsagkatakis, Grigorios, editor. | Montestruque, Luis A., editor.
Title: Smart water grids : a cyber-physical systems approach / [edited by] Panagiotis Tsakalides, Athanasia Panousopoulou, Grigorios Tsagkatakis, and Luis Montestruque.
Description: Boca Raton, FL : CRC Press/Taylor & Francis Group, 2018. | "A CRC title, part of the Taylor & Francis imprint, a member of the Taylor & Francis Group, the academic division of T&F Informa plc." | Includes bibliographical references and index.
Identifiers: LCCN 2017055149| ISBN 9781138197930 (hardback : acid-free paper) | ISBN 9781315271804 (ebook)
Subjects: LCSH: Smart water grids. | Water--Distribution.
Classification: LCC TD484 .S62 2018 | DDC 628.1--dc23
LC record available at https://lccn.loc.gov/2017055149

Visit the Taylor & Francis Web site at
http://www.taylorandfrancis.com

and the CRC Press Web site at
http://www.crcpress.com

Contents

Preface xi

About the Editors xiii

Contributors xv

SECTION I   Theory and Design

Chapter 1   Deployment and Scheduling of Wireless Sensor Network for
Monitoring Water Grids 3
RONG DU, CARLO FISCHIONE

    1.1   Introduction and Motivation 4
          1.1.1   Sensor nodes for water grids 5
          1.1.2   Current WSN systems for water grids 7
    1.2   Deployment of WSN for Monitoring Water Grids 8
          1.2.1   Deployment of static WSNs 10
          1.2.2   Deployment of mobile WSNs 21
    1.3   Transmission Scheduling of Wireless Sensor Networks
for Water Grids 25
          1.3.1   Routing 26
          1.3.2   Sleep/awake scheduling 27
          1.3.3   Transmission time scheduling with wireless en-
ergy transmission 31
          1.3.4   Scheduling of WET 33
    1.4   Concluding Remarks and Future Directions 36
    1.5   Acknowledgments 38
    1.6   References 38

Chapter 2   Consensus-Based Distributed Detection and State Estimation of
Biofilm in Reverse Osmosis Membranes by WSNs 41
DANIEL ALONSO-ROMÁN, CÉSAR ASENSIO-MARCO, BALTA-
SAR BEFERULL-LOZANO

    2.1   Introduction and Motivation 42
    2.2   Formation and Evolution of the Biofilm 45
          2.2.1   Assumptions and models 46
    2.3   Biofilm Monitoring by a Wireless Sensor Network 48

v

vi Contents

	2.3.1	Autonomous control of biofilm evolution.................50
	2.3.2	Optimal detection of biofilm formation 50
	2.3.3	Optimal state estimation of biofilm evolution..........52
2.4	Consensus-Based Distributed Detection and Tracking........53	
	2.4.1	Iterative consensus under unreliable links 55
	2.4.2	Consensus-based distributed detection 57
	2.4.3	Consensus-based distributed state estimation.........60
2.5	Numerical Results.................................64	
2.6	Concluding Remarks and Future Directions........................68	
2.7	References...69	

Chapter 3 Communication Optimization and Edge Analytics for Smart Water Grids 71

SOKRATIS KARTAKIS, JULIE A. MCCANN

3.1	Introduction and Motivation 71	
3.2	Low Power Wide Area Communication Technologies.........74	
	3.2.1	Theory behind LPWA technologies 76
	3.2.2	Evaluation platform 78
	3.2.3	Experimental methodology and environments 80
	3.2.4	Evaluation results..82
3.3	Lossless Compression and Compressive Sensing 87	
	3.3.1	Lossless and lossy compression...........................88
	3.3.2	Compressive sensing......................................92
	3.3.3	Edge adaptive compression................................96
3.4	Edge Processing...98	
	3.4.1	Adaptive thresholding-based real-time anomaly detection exploiting compression rates...................98
	3.4.2	Incorporating edge analytics and offline graph-based burst localization algorithm 101
3.5	Concluding Remarks and Future Directions......................103	
3.6	References...104	

Chapter 4 Matrix and Tensor Signal Modelling in Cyber Physical Systems 107

GRIGORIOS TSAGKATAKIS, KONSTANTINA FOTIADOU, MICHALIS GIANNOPOULOS, ANASTASIA AIDINI, ATHANASIA PANOUSOPOULOU, PANAGIOTIS TSAKALIDES

4.1	Introduction and Motivation 107	
4.2	Sparse Signal Coding..108	
	4.2.1	Dictionary learning 110
4.3	Low Rank Modelling of Matrices................................111	
	4.3.1	Matrix recovery...113
4.4	From Matrices to Tensors 115	
	4.4.1	Tensor decompositions...................................115
	4.4.2	Tensor completion.......................................118

Contents **vii**

4.5 Applications in CPS systems 120
    4.5.1 State-of-the-art in CPS monitoring 120
    4.5.2 Recovery from compressed measurements 120
    4.5.3 Recovery from noisy measurements 121
    4.5.4 Recovery from missing measurements 122
4.6 Concluding Remarks and Future Directions 124
4.7 Acknowledgments 125
4.8 References 125

Chapter 5 Sensing-Based Leak Quantification Techniques in Water Distribution Systems 129

ARY MAZHARUDDIN SHIDDIQI, RACHEL CARDELL-OLIVER, AMITAVA DATTA

5.1 Introduction and Motivation 129
5.2 Mathematical Modelling Background 131
5.3 Simulating a Water Distribution System 132
5.4 Machine Learning Algorithms for Leak Quantification 133
    5.4.1 Optimization and classification algorithms 133
    5.4.2 Single and multi-leaks quantification 134
5.5 Sensor Placement Strategies 137
    5.5.1 State space search 137
    5.5.2 Minimum set cover 138
    5.5.3 Genetic algorithms 139
    5.5.4 One sensor to find all leaks 140
    5.5.5 Small leak quantification 142
5.6 Leak Quantification in Large-Scale Water Distribution Systems 143
5.7 Concluding Remarks and Future Directions 145
5.8 References 146

SECTION II Applications and Emerging Topics: Water Infrastructures and Urban Environments

Chapter 6 An Agent-Based Storm Water Management System 151
LUIS A MONTESTRUQUE

6.1 Introduction and Motivation 151
6.2 Approach 153
6.3 Architecture 154
    6.3.1 Monitoring hardware 154
    6.3.2 Communication subsystem 156
    6.3.3 Distributed control subsystem 158

viii   Contents

6.4   Implementation 163
6.5   Results.. 166
6.6   Concluding Remarks and Future Directions...................... 166

Chapter 7   Gamified Approaches for Water Management Systems: An Overview 169
ANDREA CASTELLETTI, ANDREA COMINOLA, ALESSANDRO FACCHINI, MATTEO GIULIANI, PIERO FRATERNALI, SERGIO HERRERA, MARK MELENHORST, ISABEL MICHEEL, JASMINKO NOVAK, CHIARA PASINI, ANDREA-EMILIO RIZZOLI, CRISTINA ROTTONDI

7.1   Introduction and Motivation 169
7.2   Technical Concepts and Methodologies 171
        7.2.1   Smart metering technologies.................................. 172
        7.2.2   Big data analytics for end use disaggregation and user profiling 173
        7.2.3   Persuasive technologies 174
        7.2.4   Visualization interfaces... 175
        7.2.5   Gamification.. 175
7.3   An Overview on Gamified Water Management Systems ... 178
        7.3.1   Urban water project... 178
        7.3.2   WatERP project.. 179
        7.3.3   Waternomics 180
        7.3.4   SmartH2O 183
        7.3.5   A brief review of some commercial products 186
        7.3.6   Gamified applications for resource efficiency 187
7.4   Some Experiences from the Design and Deployment of the SmartH2O System 191
        7.4.1   Design guidelines for visualization and gamification in water saving applications 191
        7.4.2   Evaluating user responses and first effects on water consumption 194
7.5   Concluding Remarks and Future Directions...................... 195
7.6   References... 196

Chapter 8   Integrated Brine Management: A Circular Economy Approach. 203
DIMITRIS XEVGENOS, DESPINA BAKOGIANNI, KATHERINE-JOANNE HARALAMBOUS, MARIA LOIZIDOU

8.1   Introduction and Motivation 203
8.2   Brine Management.. 204
        8.2.1   Sources and nature of brines 204
        8.2.2   Current concentrate management options.............. 206
8.3   Desalination: Technologies, Challenges, and Opportunities 208
        8.3.1   Desalination technologies 208

Contents ix

	8.3.2	Energy considerations and renewable driven desalination	212
	8.3.3	Valorization of seawater desalination brine	215
8.4	Industrial Water Use and Desalination		216
	8.4.1	Industrial water use	216
	8.4.2	Industrial use of desalinated water	219
8.5	Integrated Brine Management: the Mediterranean Case		220
	8.5.1	Mediterranean desalination market	221
	8.5.2	Industrial users of desalination in Mediterranean countries	224
8.6	Concluding Remarks and Future Directions		224
8.7	References		227

Chapter 9 Model-Assisted, Real-Time Monitoring and Control of Water Resource Recovery Facilities 231

EVANGELIA BELIA, JOHN B. COPP

9.1	Introduction and Motivation		231
9.2	Driving Forces in Real-Time Monitoring and Control of WRRF		232
9.3	Essential Modules for an Integrated Control Software Tool		233
9.4	Challenges of Integrated Control Software Tools		234
9.5	Examples of MA-RT-PW-MC Software Platforms		235
	9.5.1	The Integrated Computer Control System (IC^2S)	235
	9.5.2	STAR Utility Solutions^{TM}	235
	9.5.3	IBM advanced simulation, forecasting, and optimization platform	236
	9.5.4	inCTRL Solutions Monitoring and Control Platform	236
9.6	Case study: Real-Time Control and Optimization of the Grand Rapids WRRF Targeting Energy Reduction		238
	9.6.1	The plant	238
	9.6.2	Grand rapids WRRF journey to real-time monitoring and control	239
	9.6.3	Project objective and tasks	240
	9.6.4	Process model development	241
	9.6.5	Data quality of online sensors	242
	9.6.6	The ABAC-SRT controller	244
	9.6.7	Controller testing in the process model: ABAC-SRT controller	244
	9.6.8	Controller implementation and full-scale testing: ABAC-SRT controller	248
9.7	Concluding Remarks and Future Directions		250
9.8	References		251

SECTION III Applications and Emerging Topics: Water Natural Resources

Chapter 10 Methods for Estimating the Optical Flow on Fluids and Deformable River Streams: A Critical Survey 255
KONSTANTINOS BACHARIDIS, KONSTANTIA MOIROGIORGOU, GEORGE LIVANOS, ANDREAS SAVAKIS, MICHALIS ZERVAKIS

10.1 Introduction and Motivation 255
10.2 Methodological Analysis of Algorithms 257
    10.2.1 Particle-based methods 257
    10.2.2 Probabilistic methods 267
    10.2.3 Differential methods guided by fluid properties 271
10.3 Comparison and Experimental Results 274
    10.3.1 Methodology description and evaluation 274
    10.3.2 Experimental Results 276
10.4 Challenges in Real-World Applications 286
10.5 Concluding Remarks and Future Directions 289
10.6 References 290

Chapter 11 Computational Approaches for Incorporating Short- and Long-Term Dynamics in Smart Water Networks 297
KARTIKEYA BHARDWAJ, DIMITRIOS STAMOULIS, RUIZHOU DING, DIANA MARCULESCU, RADU MARCULESCU

11.1 Introduction and Motivation 298
    11.1.1 Work contributions 300
11.2 Related Work 301
    11.2.1 Multivariate and multiscale climate dynamics 301
    11.2.2 River flow forecasting 302
11.3 Multivariate and Multiscale Climate Dynamics 303
    11.3.1 Methodology 303
    11.3.2 Experimental results 304
11.4 River Flow Forecasting: K-Hop Learning 309
    11.4.1 Problem formulation 309
    11.4.2 Experimental results 315
11.5 Concluding Remarks and Future Directions 319
11.6 References 321

Biographical Notes of Contributors 325

Index **339**

Preface

The last few decades have witnessed a rapidly expanding water crisis, which, according to the UN-Water, by 2025 could affect the daily lives of 67% of the world population. Triggered by these alarming figures, the modernization of water management has become a key policy for the 21st century, setting out the framework for sustainable solutions on water resources utilization. As a result, novel directions for the water industry have emerged, featuring the most recent advances in sensing, networking, processing, and control. The term *Smart Water Grid* was coined and is gaining an increasing role in the water industry lexicon, expressing the data-driven systems for the intelligent management of the water life cycle.

By definition, Smart Water Grids have an inherent dependence on Cyber-Physical Systems (CPS), since they provide the technological suite for the transition from episodic sampling to fast, responsive, and scalable architectures. Indeed, Cyber-Physical Systems are the driving force for automating and smartening the complex interaction of cyber components with physical processes, by combining computing, communication, and control. Smart Water Grids exploit such characteristics for establishing three pillars in the water arena:

- A global consciousness on water consumption and its impact on natural systems;
- New pathways on treatment, contaminations detection, and reusability;
- Responding to the impact of climate change on water natural sources and infrastructures.

Driven by such tight alignments, this book provides a thorough description on the best practices for designing, implementing, and deploying Cyber-Physical Systems that are tailored to different aspects of Smart Water Grids. This book is organized in three parts, each containing a collection of chapters contributed by leading experts in the field, who share their expertise on modelling, designing, and implementing Smart Water Grids, going well beyond smart metering aspects. Each part covers distinct and self-contained, yet complementary topics on the theory behind water-oriented CPS, existing case studies, and emerging topics.

In a nutshell, the contents of this book are organized as follows:

Part I: Theory and design, covering aspects related to the front-end networked sensors deployment and signal processing, state-of-art communication technologies and platforms, and algorithms for processing, learning over urban water infrastructures.

Part II: Applications and emerging topics of CPS over urban water infrastructures, including real-life deployments, user empowerment and consumer awareness, economical aspects, and modern control tools for industrial water facilities.

Part III: Applications and emerging topics associated to monitoring natural water resources, with emphasis on modern estimation and learning techniques for understanding the impact of climate change on fresh water sources and coastal/inland hydraulic basins.

The resulting structured discussion aims at elevating the design trends on CPS for Smart Water Grids to a sophisticated paradigm of science of systems integration, thereby going well beyond summarizing the current state of the art. As the phenomenon of water-stress intensifies on a global scale, the lack of any other edition on Smart Water Grids, makes this book a must-have both for professionals and researchers who are already active or newcomers in this new and exciting area, as well as all academic libraries.

This book would not have been possible without the contribution of several people; the authors and their willingness to share their knowledge and expertise in this area, and the referees and their commitment to ensuring high-quality of the contributed chapters.

The Editors

About the Editors

PROF. PANAGIOTIS TSAKALIDES
DEPARTMENT OF COMPUTER SCIENCE, UNIVERSITY OF CRETE, GREECE
INSTITUTE OF COMPUTER SCIENCE, FOUNDATION FOR RESEARCH AND TECHNOLOGY-HELLAS, GREECE

Professor Tsakalides received his diploma in Electrical Engineering from Aristotle University of Thessaloniki, Greece, in 1990. The same year, he moved to Los Angeles for graduate studies with scholarships from the Fulbright and Bodossakis Foundations. He received an M.Sc. degree in computer engineering and Ph.D. degree in electrical engineering from the University of Southern California (USC) in 1991 and 1995, respectively. From 1996 to 1999, he was a research assistant professor with the Signal & Image Processing Institute at USC and he consulted for the US Navy and Air Force. From 1999 to 2002 he was with the Department of Electrical Engineering, University of Patras, Greece. In September 2002, he joined the Computer Science Department at the University of Crete as an associate professor and FORTH-ICS as a researcher. Currently, he is a professor of computer science, the Vice Rector for Finance at the University of Crete, and the Head of the Signal Processing Laboratory at FORTH-ICS. His research interests lie in the field of statistical signal processing with emphasis in non-Gaussian estimation and detection theory, sparse representations, and applications in sensor networks, audio, imaging, and multimedia systems. He has co-authored over 150 technical publications in these areas, including 35+ journal papers. Prof. Tsakalides has extended experience in transferring research and interacting with the industry. During the last 10 years, he has been the project coordinator on 7 European Commission and 9 national projects totaling more than 4 M Euros in actual funding for FORTH-ICS and the University of Crete.

DR. ATHANASIA PANOUSOPOULOU
FOUNDATION FOR RESEARCH AND TECHNOLOGY-HELLAS, GREECE

Dr. Panousopoulou has been a research scientist with FORTH since 2013 and a visiting lecturer at the Computer Science Department at the University of Crete since 2015. She received her diploma and Ph.D. in electrical and computer engineering from the University of Patras, Greece in 2004 and 2009, respectively. Athanasia has 10 years of expertise on the design and deployment of wireless sensor networks and CPS in hostile and RF-harsh environments, and she was involved in several large- and small-scale research projects funded by the European Union, Greece, and the United Kingdom, including the FP7 STREP Hydrobionets Project on smart water treatment. Her research interests include wireless sensor/actuator networks, distributed networked architectures for machine learning and signal processing, and she has co-authored more than 30 peer reviewed conference papers, including journal articles and book chapters. In 2015 she established the International Workshop series on Cyber-Physical Systems for Smart Water Networks, and is a member of the organizing committee since then. She is a chartered engineer from the Technical Chamber of Greece since 2004, a member of IEEE, and she is serving as a reviewer on numerous research forums, edited by IEEE and Springer.

DR. GRIGORIOS TSAGKATAKIS
INSTITUTE OF COMPUTER SCIENCE, FOUNDATION FOR RESEARCH AND TECHNOLOGY-HELLAS, GREECE

Dr. Tsagkatakis received his diploma and M.Sc. degrees in electronics and computer engineering from the Technical University of Crete, Greece in 2005 and 2007, respectively, and his Ph.D. in imaging science from the Rochester Institute of Technology, New York, in 2011. During the period 2011-2013 he was a Marie Curie post-doctoral fellow at the Institute of Computer Science (ICS), FORTH and he is currently working as a research associate with the Signal Processing Laboratory at FORTH-ICS. He has been involved in various European projects, including H2020 PHySIS and DEDALE, FP7 CS-ORION & HYDROBIONETS, FP6 OPTAG, and in industry-funded projects by leading companies including Kodak and Cisco. He has co-authored more than 30 peer-reviewed conferences, journals, and book chapters in the areas of signal and image processing, computer vision and machine learning. He was awarded the best paper award in the Western New York Image Processing Workshop in 2010 for his paper "A Framework for Object Class Recognition with No Visual Examples."

DR. LUIS A. MONTESTRUQUE
EMNET LLC, USA

Luis A. Montestruque received a Ph.D. degree in control systems from the electrical engineering department at the University of Notre Dame in 2004. His research on model-based networked control systems concentrates on the use of real-time data and models to control networked infrastructure. He founded EmNet in 2004 to focus on the application of Internet of Things, Big Data, and Artificial Intelligence to solve problems in stormwater and wastewater sewers. Its first pilot in South Bend, Indiana, deployed a wireless sensor network of over 150 sensors, making it the most densely monitored sewer system in the country. The South Bend project eliminated over a billion gallons of combined sewer overflows per year and decreased E.coli concentrations by half in the Saint Joseph River. Today EmNet is helping over 20 cities across the U.S. reduce flooding and overflows through smart sewer applications.

Contributors

Anastasia Aidini
Institute of Computer Science, Foundation for Research and Technology-Hellas and Department of Computer Science, University of Crete, GR

Daniel Alonso Román
University of Agder, NO

César Asensio Marco
University of Agder, NO

Konstantinos Bacharidis
Technical University of Crete, School of ECE, GR

Despina Bakogianni
National Technical University of Athens, GR

Baltasar Beferull-Lozano
University of Agder, NO

Evaggelina Belia
Primodal US Inc., US

Kartikeya Bhardwaj
Department of Electrical and Computer Engineering, Carnegie Mellon University, US

Rachel Cardell-Oliver
CRC for Water Sensitive Cities, University of Western Australia, AU

Andrea Castelletti
Politecnico di Milano, IT

Andrea Cominola
Politecnico di Milano, IT

John B. Copp
Primodal Inc.

Amitava Datta
University of Western Australia, AU

Ruizhou Ding
Department of Electrical and Computer Engineering, Carnegie Mellon University, US

Rong Du
KTH Royal Institute of Technology, Stockholm, SE

Alessandro Facchini
Dalle Molle Institute for Artificial Intelligence (IDSIA), CH

Carlo Fischione
KTH Royal Institute of Technology, Stockholm, SE

Konstantina Fotiadou
Institute of Computer Science, Foundation for Research and Technology-Hellas, and Department of Computer Science, University of Crete, GR

Piero Fraternali
Politecnico di Milano, IT

Michalis Giannopoulos
Institute of Computer Science, Foundation for Research and Technology-Hellas, and Department of Computer Science, University of Crete, GR

Matteo Giuliani
Politecnico di Milano, IT

Katherine - Joanne Haralambous
National Technical University of Athens, GR

Sergio Herrera
Politecnico di Milano, IT

Sokratis Kartakis
Imperial College London and Intel Labs Europe, UK

George Livanos
Technical University of Crete, School of ECE, GR

Maria Loizidou
National Technical University of Athens, GR

Diana Marculescu
Department of Electrical and Computer Engineering, Carnegie Mellon University, US

Radu Marculescu
Department of Electrical and Computer Engineering, Carnegie Mellon University, US

Julie A. McCann
Imperial College London, UK

Mark Melenhorst
European Institute for Participatory Media, Berlin, DE

Isabel Micheel-Mester
European Institute for Participatory Media, Berlin, DE

Konstantia Moirogiorgou
Technical University of Crete, School of ECE, GR

Luis A. Montestruque
EmNet, LLC, US

Jasminko Novak
European Institute for Participatory Media, Berlin, DE

Athanasia Panousopoulou
Foundation for Research and Technology-Hellas, GR

Chiara Pasini
Politecnico di Milano, IT

Andrea-Emilio Rizzoli
Dalle Molle Institute for Artificial intelligence (IDSIA), CH

Cristina Rottondi
Dalle Molle Institute for Artificial Intelligence (IDSIA), CH

Andreas Savakis
Rochester Institute of Technology, Department of Computer Engineering, US

Ary Shiddiqi
CRC for Water Sensitive Cities, University of Western Australia, AU, and Institut Teknologi Sepuluh Nopember, ID

Dimitrios Stamoulis
Department of Electrical and Computer Engineering, Carnegie Mellon University, US

Grigorios Tsagkatakis
Institute of Computer Science, Foundation for Research and Technology-Hellas, GR

Panagiotis Tsakalides
Institute of Computer Science, Foundation for Research and Technology-Hellas, and Department of Computer Science, University of Crete, GR

Dimitris Xevgenos
SEALEAU B.V., NL

Michalis Zervakis
Technical University of Crete, School of ECE, GR

Part I

Theory and Design

1 Deployment and Scheduling of Wireless Sensor Networks for Monitoring Water Grids

RONG DU, CARLO FISCHIONE

CHAPTER HIGHLIGHTS

This chapter focuses on the deployment of wireless sensor networks for water grid monitoring, as well as the scheduling of the sensor nodes, in terms of routings, duty cycles, to have a good monitoring performance. The major contribution of the chapter is as follows:

- We study different node deployment problems, such as using minimum nodes to cover the whole water grid and using a given number of nodes to maximize the coverage of the monitoring. Due to the difficulties of the problems, we provide approximation algorithms based on the idea of submodularity to achieve the suboptimal solutions with low complexity. We also consider the use of mobile sensor nodes for water grid monitoring to improve the coverage of the sensor network;
- After the deployment, we also consider managing the running of the sensor network systems. More specifically, we analyze how the data of the sensor nodes should be routed such that the sensor network lifetime is maximized. Such a problem is formulated as a linear optimization, thus it is easy to solve. Then, we further consider the sleep/awake scheduling of the sensor node, and the resulting problem is an integer optimization, which is difficult to solve. Therefore, we propose an algorithm based on the idea of energy balancing to achieve a near optimal solution;
- We also consider using wireless energy transmission to prolong the lifetime of the sensor network. In this scenario, we use a dedicated base station to power up the sensor nodes by transmitting electromagnetic waves. The sensor nodes will use the harvested energy for sensing and data transmission, such that the sensor network can have a higher data throughput or a longer lifetime.

R Du ⊠ • C Fiscione

KTH Royal Institute of Technology, Stockholm, SE. E-mail: rongd@kth.se

1.1 INTRODUCTION AND MOTIVATION

A wireless sensor network (WSN) is an essential infrastructure in smart water grids. It provides information on the running states of the whole water smart grid system, such that one can make proper decisions to ensure that the water smart grid works in a desired state. As the deployment of the sensor nodes and the working pattern of the nodes greatly determine the performance of the WSN in terms of detection time, accuracy, coverage, and durations, these two issues should be carefully analyzed. In this chapter, we give a brief introduction to WSNs for water smart grids, and discuss the nodes deployment problems and the network scheduling problems.

Water supply network relates to the daily water usage of citizens and factories. Thus, it is a essential infrastructure in a smart city. However, it runs on aging pipelines that are buried underground, and it faces at least two major threats, i.e., contaminations and leakages, as described in the following:

1. Contaminations: Water in the pipelines can be easily polluted by chemical or biological contaminants. Such contaminants can enter a water supply network by accident or by malicious actions, and then spread all over the water network by flowing and affecting larger areas. Thus, the water may become harmful and threaten the public health.
2. Leakages: The water pipelines are often underground, thus the pipelines are easily eroded by the moist environment. Moreover, the pressure variations create strong solicitations to the pipes. These factors can cause leakages. When a leakage is small, it is difficult to detect, and is called background leakage; when the leakage becomes large, it is called a burst. A huge amount of water is wasted due to background leakages or bursts. Even a small leak could result in a huge loss if undetected over time. It is reported that, in a developing country, an average 45 million cubic meters are lost every day due to leakages [2]. This translates to 2.9 billion dollars per year. Similar problems hold for developed countries. For instance, London loses 0.6 million cubic meters of water per day, which is about 1/4 of its overall daily water supply [1], which translates to about 35 million Euro loss per year.

To resolve such threats, we need to have real-time measurements of some key parameters of the water in pipelines, such as flow rates, pressures, water levels, pH levels, and oxygen levels. However, as the pipelines are mostly buried underground and the scale of water grids is extremely large, it is almost impossible to use human labour to measure the necessary real-time data. Consequently, we employ a smart WSN in the pipeline network to measure and to provide remote monitoring.

A WSN consists of multiple sensor nodes that make measurements and communicate with each other wirelessly. The nodes are generally of small size and with limited resources on energy, computation and storage. Therefore, they are easy to deploy underground, but the performance of the WSN greatly depends on where they are deployed and how they utilise their resources. Consequently, in this chapter, we are going to discuss how to deploy and schedule the WSNs operation for water grids monitoring, such that the WSNs will operate in a cost and energy efficient way.

Deployment and Scheduling of WSN for Monitoring Water Grids   5

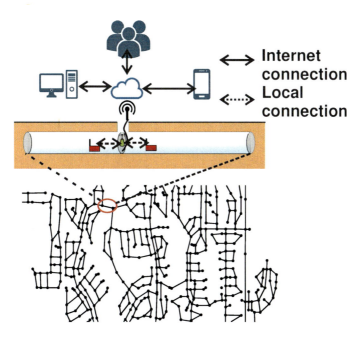

Figure 1.1: A typical wireless sensor network for water grid monitoring, where the lower part is the graph of the water grid, $\mathcal{G}_g$, with vertices and edges representing junctions and pipelines, respectively. The upper part shows a wireless sensor network $\mathcal{G}_s$ in the pipelines. The sensor network measures the data of the water grid, and transmits the measurements to a remote monitoring center, where decisions are made.

In this section, we will provide a basic introduction of WSN for water grids. Then, in Section 1.2, we will focus on the WSN configuration aspect and study the node placement problems. In Section 1.3, we will move to the WSN running management perspective and discuss the transmission scheduling problems. Finally, we will discuss the future research directions of designing WSNs for water grids, and summarize the chapter.

1.1.1 SENSOR NODES FOR WATER GRIDS

Sensor nodes are the essential units in water grid monitoring systems, which automatically provide real-time measurements. A sensor node consists of the following three components according to the functionality: sensing units, data processing and storage units, and data transmission units.

Regarding the sensing units, a sensor node may contain several types of sensors for different monitoring purposes. For example, to detect leaks and bursts, we can use sensors that measure pressure, flow velocity and acoustic signals; whilst to monitor water quality, we can use sensors that measure flow velocity, pH value, chlorine residual, and conductivity. For the data transmission part, wired cables are commonly used to connect the sensor units to the processing unit. Besides wired communication, a sensor node also has the capability to transmit data wirelessly. Generally, a node will use Bluetooth, WiFi, or Zigbee protocols for short-range communication, and 3G/4G modem for long-range communication. In the following, we will briefly describe some sensor nodes that are used in water grid monitoring systems.

The sensor nodes used in the PipeNet project [27] are based on the Intel Mote platform [19], which has an ARM7 core, a 64kB RAM, a 512kB Flash. We can see that the sensor nodes do not have a very large storage space. Thus, the nodes have to either transmit their measurements frequently to avoid data buffer overflows, or compress the measurements to save space. The sensor board contains various analog sensors, such as pressure, flow velocity, vibration, which can be used for leakage detection. Regarding the data transmission unit, the node uses Bluetooth protocol, which supports a short-range, one-to-one transmission with low energy consumptions.

The nodes used in Singapore's Wireless Water Sentinel project [29] have a processing unit running on a Linux operating system. The node has 2GB storage space, which is much larger than the one in the PipeNet project, and thus can buffer data for several days. However, it can support simultaneously up to three types of sensors to measure pressure, acoustics, and flow velocity. For data transmission units, it uses a 3G modem for long-range communication and a WiFi radio for short-range communication. Besides that, a GPS unit is used for time synchronization. The node is contained in a water proof plastic tube to protect its electrical circuit.

The deployment of the previous sensor nodes is fixed. Thus, we call it static sensor nodes. Besides such static nodes, we have mobile sensor nodes that can move around in the water grid to make measurements. Mobile sensor nodes for water grid monitoring are working inside pipelines. Thus, they must be water proof to protect the circuits, and be small enough to move inside the pipelines. Therefore, the prototypes of the mobile sensor nodes for water grid monitoring are specially designed. In [14], a prototype called Triopus is designed similar to an octopus, i.e., each node is equipped with a motor that drives three arms that can attach the sensor node onto the inner surface of pipelines. Consequently, we can control the placement of the sensor nodes inside the pipelines. The same group of researchers developed another kind of mobile sensor node called PipeProbe [15] to detect the spatial topology of hidden water pipelines. The node has no motor unit, and its movement is determined by the water flows inside the pipelines. As a result, we cannot control the movement of the mobile nodes. The work in [22] introduces a mobile sensor node to conduct continuous measurement as it moves in the flow stream to monitor water quality inside pipelines. The sensor node also has no motor unit for space and power saving. Commercial mobile sensor nodes are also available, such as SmartBall [18]. The SmartBall uses acoustic sensors to detect pipeline leakages, and it can run for approximately 16 hours with its own battery, which is long enough for an inspection.

1.1.2 CURRENT WSN SYSTEMS FOR WATER GRIDS

Based on the sensor nodes mentioned above, researchers and engineers have built or are building water grid monitoring systems in different cities, to reduce water leakage and/or contamination. In this subsection, we will describe some representative monitoring systems.

One of the earliest projects to monitor a water grid was conducted in Ann Arbor, MI, USA [25]. Its major goal was to minimize the public exposure to chemical or biological contaminants that are released into the water grid. To simulate how the contaminants spread with water flows, the research team developed a hydraulic model of the city water grid using a software called EPANET [23]. Then, the researchers simulated the population that is affected by contaminations under various contaminant-release scenarios. Such Monte Carlo-type simulation results are used as the input to identify vulnerable areas of the water grid. To determine the placement of the sensor nodes, the researchers used a toolkit called TEVA-SPOT. They formulated an optimization problem to minimize average public health effects resulting from contamination, and used a heuristic based approach to achieve a good solution. The team also made some field tests in Ann Arbor with four locations for security monitoring and four locations for water quality monitoring.

PipeNet [27] is a monitoring system to detect and locate leaks in a water grid, and it was tested in Boston, MA, USA. The researchers first conducted a field test to collect data and examined the preliminary design in terms of hardware durability, communication reliability, and system costs. To monitor water grids, sensors that measure pressure, pH level, water level, etc., are used. Different sensors are deployed at different locations according to the requirements. Basically, the WSN consists of three tiers (three different kinds of devices). The first tier is the sensor nodes that are deployed at the junctions, and battery powered. These sensor nodes have limited storage capability and short communication range for energy saving. They sample the raw data periodically, make some local data processing and compression, and then transmit the data to the second tier, i.e., gateways. The gateways are powered by electrical grid, and thus have better capability in communication and storage. These gateways are deployed lampposts near the junctions. The gateways are responsible for forming and configuring the sensor networks, collecting data from the sensor nodes, and relaying data to the back-end server which is the third tier, with long-range communications. The third tier is the back-end, which provides the server of applications and database, and also the tools for data visualization and network management. The system was tested in Boston for 22 months, and the collected data were used for testing the algorithms on leak detection and localization. The researchers used the data on pressure and flow velocity to detect large bursts and leakage events, with the use of an integer Haar wavelet transform [26]. Such a transformation detects the pressure pulses, and thus, will trigger the transmission of the recent sampled data to the back-end. The back-end server will further analyse to model pressure wave propagation using data from different locations, and to estimate the location of the leakage. For detecting and localizing small leaks, the vibration data are used. Each sensor node transforms the time-series of data it measured into frequency domains, and looks for an increased energy in the frequency bands that are characteristics of leaks. If the increment exists, it transmits a short period of data to the back-end server through a gateway, and the server performs cross correlation operation to localize the leakage.

In Singapore, a project called Wireless Water Sentinel (WaterWise@SG) [29] aims at building a real-time monitoring system of water grids for the city. The goals cover both detecting pipeline leakages and monitoring water quality. The initial system consists of 8 nodes, whose deployment is determined by experts, and the later deployment of the nodes is determined by the result of an optimal sensor placement problem formulated according to the system hydraulics in Singapore. The sensing unit measures data such as flow velocity, vibrations, acoustic signals underground, whilst the processing units are above ground. The node uses a WiFi radio for short-range communication, and a USB 3G modem for long-range communication. Similar to PipeNet, the researchers use both time-domain and spectrum-domain methods to detect pipeline leakages.

As mentioned by the projects above, node deployment problems are a common issue in current water grid monitoring systems, and these projects are also trying to find a proper deployment of the nodes. Therefore, we are going to study such deployment problems in the next section. Besides, we can also see that the current water grid monitoring systems are mainly based on static sensor nodes, and the use of mobile sensor nodes is seldom considered. Consequently, we will also discuss the deployment of the mobile sensor nodes to further improve the monitoring of water grids.

Before we discuss the deployment and scheduling problems, we summarize in Table 1.1 the major notations that are used in this chapter.

1.2 DEPLOYMENT OF WSN FOR MONITORING WATER GRIDS

When we are planning to use WSN to monitor a water grid, one of the first problems we must consider is where to deploy the sensor nodes. This is a typical node deployment problem. Different from the traditional node deployment problems for open space monitoring, for water grids, the deployment of the nodes is constrained by the topology of the water grids. More specifically, we generally can only deploy sensor nodes in the the junctions and inside the pipelines of the water grid. Thus, the placement problems receive lots of research interest.

In this chapter, we use $\mathcal{G}_g = \{\mathcal{V}_g, \mathcal{E}_g\}$ to represent the water grid, where $\mathcal{V}_g$ is the set of junctions, and $\mathcal{E}_g$ is the set of pipelines. If without further specification, we assume that water grid $\mathcal{G}_g$ is fixed and given as problem input throughout the chapter. The sensor network is denoted by $\mathcal{G}_s = \{\mathcal{V}_s, \mathcal{E}_s\}$, where $\mathcal{V}_s$ is the set of the sensor nodes, and $\mathcal{E}_s$ is the edge set representing the wireless communication among the sensor nodes. The common objectives of the node deployment problems, denoted by $f(\mathcal{G}_s)$, include maximizing the covered area of the WSN, maximizing the covered population of the WSN, maximizing the detection probability of the WSN, and minimizing the detection time of the WSN [21]. Thus, we formulate the node deployment

Deployment and Scheduling of WSN for Monitoring Water Grids

Table 1.1
Major notations used in this chapter

Symbols	Meanings
$\mathcal{C}(v_i)$	set of junctions that can be monitored by a sensor deployed at v_i
$\mathcal{E}_g$	set of pipelines in water grid
$\mathcal{E}_s$	wireless links among sensor nodes
$\mathcal{G}_g$	water grid
$\mathcal{G}_s$	sensor network
$\mathcal{N}(v_i)$	neighbor set of junction v_i
$\mathcal{V}_g$	set of junctions in water grid
$\mathcal{V}_s$	set of sensor nodes
A	adjacency matrix
B	deployment budget
C_i	achievable uplink data throughput of node i
E_i	initial energy of sensor node i
M_{cs}	cardinality constraint
P_i^r	received power of node i
W_i	importance of monitoring junction v_i
$c(v_i)$	cost to deploy a sensor at junction v_i
c_i	energy consumption rate of node i
$e_{i,j}^r$	energy consumption of node j to receive a unit data from i
$e_{i,j}^t$	energy consumption of node i to transmit a unit data to j
e_{ij}	pipeline connecting junction i and j
$f_{i,j}$	amount of data flow from node i to j
p_{ij}	probability of a mobile node moving from node v_i to v_j
q_{ij}	water flow rate from junction v_i to v_j
r	approximation ratio
v_r	root vertex
α_i	energy harvesting efficiency of node i
$\emptyset$	empty set

problem as follows:

$$\max_{\mathcal{G}_s} \quad f(\mathcal{G}_s) \tag{1.1a}$$

$$\text{s.t.} \quad \mathcal{G}_s \in \mathbb{G}, \tag{1.1b}$$

where $\mathbb{G}$ is the set of all possible WSN deployment, which depends on requirements from reality including cardinality constraint (the number of sensor nodes that can be deployed) and connectivity constraint (the sensor nodes should form a connected communication network). We should mention that the problems discussed in this section are difficult to solve; some are even NP hard. This means that the optimal solution of the problem may not be achievable when the size of the problem is large. Consequently, researchers are using approximation solutions with low complexity to find some suboptimal solutions. To measure how good the approximation solutions are, we use the metric called approximation ratio as defined below:

Definition 1 (Approximation ratio). *Consider a feasible maximization problem, with optimal solution x^* and optimal value $f(x^*)$. An algorithm for this maximization problem is said to have approximation ratio $r \geq 1$ if and only if the algorithm guarantees to achieve a solution x^g that satisfies $rf(x^g) \geq f(x^*)$. On the other hand, for a minimization problem, with optimal solution x^* and optimal value $f(x^*)$, an algorithm for this minimization problem is said to have approximation ratio $r \geq 1$ if and only if the algorithm guarantees to achieve a solution x^g that satisfies $rf(x^*) \geq f(x^g)$.*

The approximation ratio is used to describe how close the approximation solution is to the optimal in the worst case. The closer the approximation ratio to 1, the better performance the algorithm has.

In this chapter, we will consider the deployment of both static WSNs and mobile WSNs. To begin with, we consider the problems for static WSNs, i.e., the sensor nodes in the WSNs are statically deployed.

1.2.1 DEPLOYMENT OF STATIC WSNS

In the static WSNs case, the sensor nodes are deployed at some fixed points in $\mathcal{G}_g$. Once the nodes are deployed, their locations cannot be changed. Thus, the monitoring performance $f(\mathcal{G}_s)$ depends on where the nodes are deployed. In the prior state of arts on node deployment for water grid monitoring, the sensor nodes can only be deployed at the junctions of the water grid. One reason is that, in so doing, a node at a junction can measure the parameters of the water in the pipelines that are connected with that junction. Therefore, the coverage performance of deploying a node at a junction is better than that of deploying it in the middle of a pipeline, especially for contamination detections. As a result, the variable $\mathcal{G}_s$ is a subset of the junctions V_g. Furthermore, we assume that the sensor nodes are the same in terms of prices, sensing capability, communication capability, etc., as default, if without further specifications. Based on this setting, we start to consider several representative node deployment problems of WSNs for monitoring a water grid.

MIN Node Coverage Problem

Consider the case where we want to deploy sensor nodes to cover the whole water grid. Then, the problem arises, what is the minimum number of the sensor nodes and the corresponding deployment, such that the whole water grid is covered. Here are some necessary definitions:

Definition 2 (Node Cover). *If a sensor node is deployed at $v_i \in \mathcal{V}_g$, then we define the junctions that can be monitored by v_i by a set $C(v_i)$. That is, for all junctions $v_j \in C(v_i)$, we say v_j is covered by v_i.*

Then, the MIN Node Cover problem can be formulated as follows:

$$\min_{\mathcal{V}_s} \quad |\mathcal{V}_s| \tag{1.2a}$$

$$\text{s.t.} \quad \bigcup_{v_i \in \mathcal{V}_s} C(v_i) = \mathcal{V}_g, \tag{1.2b}$$

where $|\mathcal{V}|$ denotes the cardinality (the number of elements) of set $\mathcal{V}$. The problem is to find the minimum number of nodes to deploy at the junctions, such that every junction in the water grid is covered by at least one sensor node. Recall that $C(v_i)$ is the set of junctions that is covered by v_i. In practice, $C(v_i)$ can be the junctions that are one-hop or two-hops from v_i in $\mathcal{G}_g$. If it is defined as one-hop neighbors, then the MIN-Node Cover problem is to find the minimum dominating set of $\mathcal{G}_g$, which is an NP-hard problem. Therefore, the optimal solution of our Problem (1.2) is difficult to find in polynomial time. However, due to a special proposition called submodularity [20] in the coverage, we can get an approximate solution to the problem. The submodularity is defined as follows:

Definition 3 (Submodularity [20]). *Consider a finite set V and a function $f : 2^V \to \mathbb{R}$, where 2^V denote the power set of V (the family of all subsets of V). The function f is said to be submodular if and only if for any two sets $\mathcal{A}, \mathcal{B} \in 2^V$,*

$$f(\mathcal{A} \cup \mathcal{B}) + f(\mathcal{A} \cap \mathcal{B}) \geq f(\mathcal{A}) + f(\mathcal{B}). \tag{1.3}$$

Remark 1. *An alternative definition of submodularity is, for any two sets $\mathcal{A} \subseteq \mathcal{B} \in 2^V$, and element $a \in V \backslash \mathcal{B}$, it holds that*

$$f(\mathcal{A} \cup \{a\}) - f(\mathcal{A}) \geq f(\mathcal{B} \cup \{a\}) - f(\mathcal{B}).$$

In practice, it is easier to use this definition to check if a function is submodular.

For the MIN Node Cover problem, if we define a potential function $f(\mathcal{V}) = \left| \bigcup_{v_i \in \mathcal{V}} C(v_i) \right|$, i.e., the number of junctions that are covered by $\mathcal{V}$, then it is easy to achieve that such a potential function is submodular. Also, we know that f is monotone increasing, i.e., $\forall \mathcal{A} \subseteq \mathcal{B} \in 2^{\mathcal{V}_g}$, we have that $f(\mathcal{A}) \leq f(\mathcal{B})$. Based on these propositions, we can use the approach in Algorithm 1 to find an approximation solution for the MIN Node Cover problem (1.2).

The idea is that we start from an empty deployment $\mathcal{V}_s = \emptyset$. Then, as long as $\mathcal{G}_g$ is not covered, we deploy an additional sensor node into $\mathcal{G}_g$ at the junction that maximizes the marginal of coverage, where the marginal of coverage to deploy a node at v_i given the deployment of sensor network $\mathcal{V}_s$ is defined as: $\Delta_{v_i} f(\mathcal{V}_s) \triangleq f(\mathcal{V}_s \cup \{v_i\}) - f(\mathcal{V}_s)$. We have the following proposition of greedy MIN Node Cover Algorithm:

Algorithm 1 greedy MIN Node Cover

Require: $\mathcal{G}_g, \mathcal{C}(\cdot)$
Ensure: Deployment $\mathcal{V}_s$
1: Set $\mathcal{V}_s \leftarrow \emptyset, \mathcal{U} \leftarrow \mathcal{V}_g$
2: **while** $f(\mathcal{V}) \neq \mathcal{V}_g$ **do**
3:    $v_k \leftarrow \arg\max_{v_i \in \mathcal{U}} \{\Delta_{v_i} f(\mathcal{V}_s)\}$.
4:    Set $\mathcal{V}_s \leftarrow \mathcal{V}_s \cup \{v_k\}, \mathcal{U} \leftarrow \mathcal{U} \backslash \{v_k\}$
5: **end while**
6: **return** $\mathcal{V}_s$

Proposition 1 ([5]). *Consider a feasible MIN Node Cover Problem (1.2). Denote $\mathcal{V}_s^g$ the solution achieved by the greedy MIN Node Cover Algorithm 1, and $\mathcal{V}_s^*$ the optimal solution. Then,*

$$|\mathcal{V}_s^g| \leq (1 + \ln|\mathcal{V}_g|)|\mathcal{V}_s^*| .$$

It means that, the approximation ratio of greedy MIN Node Cover Algorithm is $(1 + \ln|\mathcal{V}_g|)$.

In some cases where the costs of deploying a node at different junctions are different, we can take the total cost of deploying the WSN into account. Assume that the cost to deploy a node at junction v_i is $c(v_i) > 0$. The objective is to minimize the total cost of deployment such that the whole $\mathcal{G}_g$ is covered. Then, we can formulate the MIN Cost Node Cover Problem as follows:

$$\min_{\mathcal{V}_s} \quad \sum_{v_i \in \mathcal{V}_s} c(v_i) \tag{1.4a}$$

$$\text{s.t.} \quad \bigcup_{v_i \in \mathcal{V}_s} \mathcal{C}(v_i) = \mathcal{V}_g . \tag{1.4b}$$

Note that, if the deployment cost is location independent, i.e., $c(v_i) = c, \forall v_i \in \mathcal{V}_g$, then the MIN Cost Node Cover Problem (1.4) is equivalent to the MIN Node Cover Problem (1.2). Therefore, the optimal solution of Problem (1.4) is hard to achieve, but we can use a greedy algorithm to get an approximate solution, by revising Algorithm 1, as shown in Algorithm 2. The only difference is in Step 3.

Similarly, the performance of Algorithm 2 is given as follows:

Proposition 2 ([5]). *Consider a feasible MIN Node Cover Problem (1.4). Denote $\mathcal{V}_s^g$ the solution achieved by the greedy MIN Node Cover Algorithm 2, and $\mathcal{V}_s^*$ the optimal solution. Then,*

$$\sum_{v_i \in \mathcal{V}_s^g} c(v_i) \leq (1 + \ln|\mathcal{V}_g|) \sum_{v_i \in \mathcal{V}_s^*} c(v_i) .$$

That is, the approximation ratio of the algorithm is also $1 + \ln|\mathcal{V}_g|$.

Algorithm 2 greedy MIN Cost Node Cover

Require: $\mathcal{G}_g, \mathcal{C}(\cdot)$
Ensure: Deployment $\mathcal{V}_s$
 1: Set $\mathcal{V}_s \leftarrow \emptyset, \mathcal{U} \leftarrow \mathcal{V}_g$
 2: **while** $f(\mathcal{V}) \neq \mathcal{V}_g$ **do**
 3:    $v_k \leftarrow \arg\max_{v_i \in \mathcal{U}} \{\Delta_{v_i} f(\mathcal{V}_s)/c(v_i)\}$.
 4:    Set $\mathcal{V}_s \leftarrow \mathcal{V}_s \cup \{v_k\}, \mathcal{U} \leftarrow \mathcal{U} \backslash \{v_k\}$
 5: **end while**
 6: **return** $\mathcal{V}_s$

We can see that, when costs are considered, the approximation ratio of the greedy approach based on the idea of submodular is still $1 + \ln|\mathcal{V}_g|$. We can also consider Algorithm 2 to be the general version of Algorithm 1, where the cost to deploy at any junction is the same.

MIN Node Cover Problem with Connectivity Constraint

In the previous cases, we do not take the connectivity issue into account. However, it is also an important requirement to consider. The reason is that, if the WSN $\mathcal{G}_s$ is disconnected, it may require the sensor nodes to use long-range communication to transmit their measurement to the monitoring center; whereas if the WSN is connected, all the sensor nodes can route the measurements to a sink node first with short-range communication, and then the sink node transmits all the measurements to the monitoring center. We also consider the connectivity requirement here, and we formulate the MIN Node Cover Problem with Connectivity Constraint as follows:

$$\min_{\mathcal{V}_s} \quad |\mathcal{V}_s| \tag{1.5a}$$

$$\text{s.t.} \quad \bigcup_{v_i \in \mathcal{V}_s} \mathcal{C}(v_i) = \mathcal{V}_g \tag{1.5b}$$

$$\mathcal{V}_s \text{ is connected}. \tag{1.5c}$$

Recall that the sensor nodes are deployed in the pipelines and their transmission ranges are usually short. Consequently, we say two nodes can communicate with each other directly if and only if they are in the same pipeline, i.e., $\forall v_i, v_j \in \mathcal{V}_s, \langle v_i, v_j \rangle \in \mathcal{E}_s$ if and only if $\langle v_i, v_j \rangle \in \mathcal{E}_g$. Furthermore, if we let $\mathcal{C}(v_i) = \{v_j | \langle v_j, v_i \rangle \in \mathcal{E}_g\}$, i.e., a sensor node at v_i covers junctions v_j that are the neighbors of v_i in $\mathcal{G}_g$, then Problem (1.5) is a MIN Connected Dominating Set Problem [5]. We will again use a greedy based algorithm [5] to achieve the solution. The key step is to find a proper potential function f. In the solution algorithm, the potential function consists of two parts, i.e., $f(\mathcal{V}_s) = -p(\mathcal{V}_s) - q(\mathcal{V}_s)$. The function $p(\mathcal{V}_s)$ denotes the number of connected components in the induced graph of $\mathcal{V}_s$, and $q(\mathcal{V}_s)$ denotes the number of connected components of the subgraph with vertex set $\mathcal{V}_g$ and edge set $\mathcal{D}(\mathcal{V}_s)$, where $\mathcal{D}(\mathcal{V}_s)$ is the set of all edges incident on some vertices in $\mathcal{V}_s$.

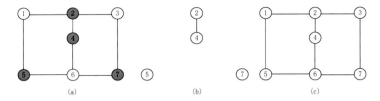

Figure 1.2: An example for the potential function f for the MIN Node Cover Problem with Connectivity Constraint [5]. (a) the water grid $\mathcal{G}_g$, with the sensor nodes are deployed at the gray points, i.e., the sensor network is $\mathcal{V}_s = \{v_2, v_4, v_5, v_7\}$. (b) the induced graph of $\mathcal{V}_s$, and thus $p(\mathcal{V}_s) = 3$. (c) the subgraph with vertex set $\mathcal{V}_g$ and edge sets $\mathcal{D}(\mathcal{V}_s)$. Thus, $q(\mathcal{V}_s) = 1$

We give an example in Fig. 1.2. Consider a water grid $\mathcal{G}_g$ as shown in Fig. 1.2 (a), and the sensor nodes are deployed at v_2, v_4, v_5, v_7, i.e., $\mathcal{V}_s = \{v_2, v_4, v_5, v_7\}$. Then, Fig. 1.2 (b) is the induced graph by $\mathcal{V}_s$, and it has three connected components $(v_2, v_4), (v_5), (v_7)$. Therefore, $p(\mathcal{V}_s) = 3$. Fig. 1.2 (c) is the subgraph with vertex set $\mathcal{V}_g$ and edge set $\mathcal{D}(\mathcal{V}_s)$. It is connected and thus $q(\mathcal{V}_s) = 1$.

It can be known that, $f(\mathcal{V}_s)$ is monotone increasing. However, since $q(\mathcal{V}_s)$ is not submodular, $f(\mathcal{V}_s)$ is not submodular. However, when $\mathcal{V}_s$ is connected and is a dominating set, we have that $f(\mathcal{V}_s) = -2$ and $\Delta_{v_i} f(\mathcal{V}_s) = 0, \forall v_i \in \mathcal{V}_g$. Based on this observation, the greedy MIN Connected Node Cover algorithm is given by Algorithm 3. Note that the major difference between Algorithm 3 and the greedy MIN Node Cover Algorithm 1 is the potential function $f(\mathcal{V}_s)$. Notice that the potential function is not submodular. Consequently, Algorithm 3 does not have an approximation ratio $1 + \ln |\mathcal{V}_g|$. However, the performance of Algorithm 3 is given by the following proposition:

Proposition 3 ([5]). *Consider a feasible MIN Node Cover Problem (1.5). Denote $\mathcal{V}_s^g$ the solution achieved by the greedy MIN Node Cover Algorithm 2, and $\mathcal{V}_s^*$ the optimal solution. Then,*

$$|\mathcal{V}_s^g| \le (2 + \ln |\mathcal{V}_g|) |\mathcal{V}_s^*|.$$

That is, the approximation ratio of Algorithm 3 is $2 + \ln |\mathcal{V}_g|$.

We can see that the approximation ratio is $2 + \ln |\mathcal{V}_g|$ instead of $1 + \ln |\mathcal{V}_g|$, and the major reason is that f is not submodular.

MAX Coverage Problem

Now we consider the problem from another perspective, where the number of sensor nodes to deploy, N, is limited, and we want to maximize the coverage of these N

Algorithm 3 greedy MIN Connected Node Cover

Require: $\mathcal{G}_g, \mathcal{C}(\cdot)$
Ensure: Deployment $\mathcal{V}_s$
  1: Set $\mathcal{V}_s \leftarrow \emptyset, \mathcal{U} \leftarrow \mathcal{V}_g$
  2: **while** $f(\mathcal{V}) \neq -2$ **do**
  3:      $v_k \leftarrow \arg\max_{v_i \in \mathcal{U}} \{ \Delta_{v_i} f(\mathcal{V}_s) \}$.
  4:      Set $\mathcal{V}_s \leftarrow \mathcal{V}_s \cup \{v_k\}, \mathcal{U} \leftarrow \mathcal{U} \backslash \{v_k\}$
  5: **end while**
  6: **return** $\mathcal{V}_s$
---

nodes. In this case, the MAX Coverage Problem can be formulated as follows:

$$\max_{\mathcal{V}_s} \quad f(\mathcal{V}_s) \tag{1.6a}$$

$$\text{s.t.} \quad |\mathcal{V}_s| \leq N, \tag{1.6b}$$

where Constraint (1.6b) is the cardinality constraint. $f(\mathcal{V}_s)$ depends on the objectives of monitoring. For instance, if the objective is maximizing the covered junctions, then we can set $f(\mathcal{V}_s) = |\bigcup_{v_i \in \mathcal{V}_s} \mathcal{C}(v_i)|$; if the objective is maximizing the monitored population, then we can set $f(\mathcal{V}_s) = \sum_{v_i \in \bigcup_{v_j \in \mathcal{V}_s} \mathcal{C}(v_j)} W_i$, where W_i is the population of people who use the water from v_i. In most of the cases that are considered in the deployment problems in water grid, including the three cases above, $f(\mathcal{V}_s)$ is a monotone increasing and submodular function. Thus, we can use the following greedy MAX Cover Algorithm 4 to find an approximation solution. The idea is similar to the one used for the MIN Node Cover Problem (1). That is, we begin with an empty set (no nodes are deployed in $\mathcal{V}_g$), and we deploy the sensor nodes one after another such that the node deployed in each step maximizes the marginal of coverage. The performance of the algorithm is given by the following proposition:

Proposition 4 ([5]). *Consider a feasible MAX Cover Problem (1.6). Denote $\mathcal{V}_s^g$ the solution achieved by the greedy MAX Cover Algorithm, and $\mathcal{V}_s^*$ the optimal solution. If the objective function $f : 2^{\mathcal{V}} \rightarrow \mathbb{R}$ is monotone increasing and submodular, then the approximation ratio is $e/(e-1)$, i.e.,*

$$f(\mathcal{V}_s^*) - f(\emptyset) \leq \frac{e}{e-1} \left(f(\mathcal{V}_s^g) - f(\emptyset) \right). \tag{1.7}$$

Similar to the case in the MIN Node Coverage Problem, we can consider also a budget constraint instead of the Cardinality Constraint (1.6b). Then, the MAX-Cover Problem with Budget Constraint is formulated as follows:

$$\max_{\mathcal{V}_s} \quad f(\mathcal{V}_s) \tag{1.8a}$$

$$\text{s.t.} \quad \sum_{v_i \in \mathcal{V}_s} c(v_i) \leq B. \tag{1.8b}$$

Algorithm 4 greedy MAX Cover

Require: $\mathcal{G}_g, f(\cdot), N$
Ensure: Deployment $\mathcal{V}_s$
  1: Set $\mathcal{V}_s \leftarrow \emptyset, \mathcal{U} \leftarrow \mathcal{V}_g$
  2: **while** $|\mathcal{V}_s| \leq N$ **do**
  3:    $v_k \leftarrow \arg\max_{v_i \in \mathcal{U}} \{\Delta_{v_i} f(\mathcal{V}_s)\}$.
  4:    Set $\mathcal{V}_s \leftarrow \mathcal{V}_s \cup \{v_k\}, \mathcal{U} \leftarrow \mathcal{U} \backslash \{v_k\}$
  5: **end while**
  6: **return** $\mathcal{V}_s$

Note that if $c(v_i) = 1, \forall i$, then Problem (1.8) becomes a MAX Cover Problem (1.6). Therefore, the complexity of Problem (1.8) is no less than that of Problem (1.6). Note that the approximation ratio of the greedy approach for Problem (1.6) is $e/(e-1)$. Thus, an intuitive question is whether we have a greedy approach to achieve the same approximation ratio $e/(e-1)$. The answer is positive, according to the approach developed in [28], and is shown in Algorithm 5. The approach is based on a heuristic enumeration of all subsets with 2 elements, and then uses the greedy approach to find the approximation result. More specifically, in Line 1 of Algorithm 5, the enumeration of all subsets of 2 elements is to solve the following problem:

$$\mathcal{V}_1 = \arg\max_{\mathcal{V}_s} \quad f(\mathcal{V}_s) \tag{1.9a}$$

$$\text{s.t.} \quad \sum_{v_i \in \mathcal{V}_s} c(v_i) \leq B \tag{1.9b}$$

$$|\mathcal{V}_s| \leq 2, \tag{1.9c}$$

The time complexity to find $\mathcal{V}_1$ is $O(|\mathcal{V}_g|^2)$. Then, in the second part from Line 2 to 16, we need to enumerate with initial set of size 3. For each initial set, we keep deploying the node that maximizes the marginal coverage per unit cost into the water grid (Line 6) one after another, if the cost constraint is still satisfied (Line 7). Finally, in the third part we compare the performance of the result achieved in the first part and the second part, and take the better one as solution deployment. The performance of greedy MAX Cover with Budget Constraint Algorithm is given as follows:

Proposition 5 ([28]). *Consider a feasible MAX Cover Problem (1.8). Denote $\mathcal{V}_s^g$ the solution achieved by the greedy MAX Cover with Budget Constraint Algorithm 5, and $\mathcal{V}_s^*$ the optimal solution. If the objective function $f: 2^\mathcal{V} \rightarrow \mathbb{R}$ is monotone increasing and submodular, then the approximation ratio is $e/(e-1)$, i.e.,*

$$f(\mathcal{V}_s^*) - f(\emptyset) \leq \frac{e}{e-1} \left(f(\mathcal{V}_s^g - f(\emptyset)) \right).$$

The total complexity of Algorithm 5 is $O(|\mathcal{V}_g|^5)$, which comes from finding $\mathcal{V}_2$. We can see that, for both MIN Node Cover Problem and MAX Cover Problem, the

---
Algorithm 5 greedy MAX Cover with Budget Constraint

Require: $\mathcal{G}, f(\cdot), B$
Ensure: Deployment $\mathcal{V}_s$
    // Enumerate the case with cardinality less than or equal to two
1: Find $\mathcal{V}_1$ by solving Problem (1.9)
    // Enumerate the case with initial sets of size 3
2: Set $\mathcal{V}_2 \leftarrow \emptyset$
3: **for all** $\mathcal{V} \subseteq \mathcal{V}_g, |\mathcal{V}| = 3, \sum_{v_i \in \mathcal{V}} c(v_i) \le B$ **do**
       // greedy approach
4:    Set $\mathcal{U} = \mathcal{V}_g \backslash \mathcal{V}$
5:    **while** $\mathcal{U} \ne \emptyset$ **do**
6:        $v_k \leftarrow \arg\max_{v_i \in \mathcal{U}} \{\Delta_{v_i} f(\mathcal{V}_s)/c(v_i)\}$.
7:        **if** $\sum_{v_i \in \mathcal{V} \cup \{v_k\}} \le B$ **then**
8:            Set $\mathcal{V} \leftarrow \mathcal{V} \cup \{v_k\}, \mathcal{U} \leftarrow \mathcal{U} \backslash \{v_k\}$
9:        **else**
10:           Set $\mathcal{U} \leftarrow \mathcal{U} \backslash \{v_k\}$
11:        **end if**
12:    **end while**
13:    **if** $f(\mathcal{V}) > f(\mathcal{V}_2)$ **then**
14:        Set $\mathcal{V}_2 \leftarrow \mathcal{V}$
15:    **end if**
16: **end for**
    // Find the best among $\mathcal{V}_1$ and $\mathcal{V}_2$
17: Set $\mathcal{V}_s \leftarrow \arg\max_{\mathcal{V} \in \{\mathcal{V}_1, \mathcal{V}_2\}} f(\mathcal{V})$
18: **return** $\mathcal{V}_s$

---

approximation ratio does not change if we replace the cardinality part by the budget part. However, the additional complexity of doing so for the MAX Cover Problem is much higher.

MAX Coverage with Connectivity Constraint

Similar to the case of MIN Node Cover Problems, we also consider the case of the connectivity requirement for the MAX Coverage Problem. When connectivity of the WSN is also required, the MAX Coverage Problem becomes more difficult. The MAX Cover Problem with Connectivity Constraint is formulated as follows:

$$\max_{\mathcal{V}_s} \quad f(\mathcal{V}_s) \tag{1.10a}$$

$$\text{s.t.} \quad |\mathcal{V}_s| \le N, \tag{1.10b}$$

$$\mathcal{V}_s \text{ is connected.} \tag{1.10c}$$

Note that, if Connectivity Constraint (1.10c) is relaxed, then Problem (1.10) becomes a MAX Cover Problem (1.6). Therefore, the complexity of Problem (1.10) is

Algorithm 6 greedy MAX Rooted Cover ($MRC(\mathcal{G}, f, N, v_r)$)

Require: $\mathcal{G}, f(\cdot), N, v_r$
Ensure: Deployment $\mathcal{V}_s$
1: Set $\mathcal{V}_s \leftarrow \emptyset, \mathcal{U} \leftarrow \mathcal{V}_g \backslash \{v_r\}$
2: **while** $|\mathcal{V}_s| \leq N - 1$ **do**
3:     $v_k \leftarrow \arg\max_{v_i \in \mathcal{U}} \{\Delta_{v_i} f'_{v_r}(\mathcal{V}_s)\}.$
4:     Set $\mathcal{V}_s \leftarrow \mathcal{V}_s \cup \{v_k\}, \mathcal{U} \leftarrow \mathcal{U} \backslash \{v_k\}$
5: **end while**
6: **return** $\mathcal{V}_s \leftarrow \mathcal{V}_s \cup \{v_r\}$

---

no less than Problem (1.6). However, the approach to solve Problem (1.6) is a key step for the approach to solve Problem (1.10), as we will describe next.

The approach to achieve an approximate solution for MAX Coverage with Connectivity Constraint Problem (1.10) consists of two parts: (i) First, we relax the connectivity constraint and achieve a disconnected deployment with part of the nodes; (ii) Then, we deploy the rest of the nodes to make the WSN connected. This approach is proposed in [13], and we call it greedy MAX Connected Cover algorithm.

To better describe the algorithm, we first formulate a subproblem called MAX Rooted Cover. The problem is, given a graph $\mathcal{G}$, a positive integer N, a root vertex v_r, and the cover function $f()$, find a MAX coverage with N nodes that include v_r:

$$\max_{\mathcal{V}_s} \quad f(\mathcal{V}_s) \tag{1.11a}$$

$$\text{s.t.} \quad |\mathcal{V}_s| \leq N, \tag{1.11b}$$

$$v_r \in \mathcal{V}_s \tag{1.11c}$$

For MAX Rooted Cover Problem (1.11), due to that v_r is in the solution set, we only need to find the remaining $N - 1$ elements. Based on this observation, one can modify the submodular function $f(\mathcal{V})$ to be $f'_{v_r}(\mathcal{V}) \triangleq f(\mathcal{V} \cup \{v_r\}) - f(\{v_r\})$, which is still submodular and monotone increasing. Therefore, we can use Algorithm 4 to achieve a solution, denoted by Sol_r. The greedy MAX Rooted Cover algorithm is given by Algorithm 6. From the proposition of submodularity, we have that the approximation ratio of Sol_r is $e/(e-1)$.

Now we are ready to present the complete greedy MAX Connected Cover algorithm in Algorithm 7. In the first part, we enumerate every junction $v_i \in \mathcal{V}_g$ as the root, to find the MAX Rooted Cover with cardinality $\lfloor \sqrt{N} \rfloor$ in a subgraph whose vertices have distance to v_i with at most $\lfloor \sqrt{N} \rfloor$, as shown in Lines 1 to 8. The result till Line 8 is a disconnected subgraph with $\lfloor \sqrt{N} \rfloor$ nodes, each of which has distance to v_r^* at most $\lfloor \sqrt{N} \rfloor$. Then, in the second part (Lines 9 to 11), we deploy nodes to connect every node $v_i \in Sol_r^*$ to v_r^*. Due to that we have at most $\lfloor \sqrt{N} \rfloor - 1$ disconnected nodes that are not v_r^* in Sol_r^* and for each node we need to deploy additionally at most $\lfloor \sqrt{N} \rfloor - 1$ nodes to make it connected v_r^*, we have that we need to deploy extra $(\lfloor N \rfloor - 1)^2 \leq N - \sqrt{N}$ nodes. Thus, the resulting WSN is connected and it satisfies Cardinality Constraint (1.10b). The performance of the algorithm is given as

Deployment and Scheduling of WSN for Monitoring Water Grids

Algorithm 7 greedy MAX Connected Cover [13]

Require: $\mathcal{G}_g, f(\cdot), N$
Ensure: Deployment $\mathcal{V}_s$
     // Enumerate Maximum Cover with $\lfloor \sqrt{N} \rfloor$ nodes in a subgraph
1:   Set $\mathcal{V}_s \leftarrow \emptyset$
2:   **for all** $v_r \in \mathcal{V}_g$ **do**
3:      Set $Sol_r = \{v | dist(v,r) \leq \lfloor \sqrt{N} \rfloor\}$, and find the induced subgraph $\mathcal{G}_r$
        // Solve a greedy MAX Rooted Cover problem using Algorithm 6
4:      Set $Sol_r \leftarrow MRC(\mathcal{G}_r, f, \lfloor \sqrt{N} \rfloor, v_r)$
5:      **if** $f(Sol_r) > f(\mathcal{V}_s)$ **then**
6:         Set $v_r^* \leftarrow v_r$, $\mathcal{V}_s \leftarrow Sol_r$
7:      **end if**
8:   **end for**
     // Make the WSN connected
9:   **for all** $v_i \in \mathcal{V}_s$ **do**
10:     Find the shortest path from v_i to v_r^* in $\mathcal{G}_g$ and add the nodes along the path to $\mathcal{V}_s$
11:   **end for**
12:   **return** $\mathcal{V}_s$

---

follows:

Proposition 6 ([13]). *Consider a feasible MAX Cover Problem* (1.10). *Denote* $\mathcal{V}_s^g$ *the solution achieved by the greedy MAX Connected Algorithm 7, and* $\mathcal{V}_s^*$ *the optimal solution. If the objective function* $f : 2^\mathcal{V} \to \mathbb{R}$ *is monotone increasing and submodular, then the approximation ratio is* $O(\sqrt{N})$, *i.e.,*

$$f(\mathcal{V}_s^*) - f(\emptyset) \leq O(\sqrt{N})\big(f(\mathcal{V}_s^g - f(\emptyset)\big).$$

If Cardinality Constraint (1.10b) is replaced by a budget constraint B, one can use a similar approach, i.e., first find a MAX coverage given a root v_r with a budget constraint $\sqrt{B}$ using Algorithm 5, then deploy additional nodes to make the WSN connected. The detailed algorithm can be found in [13].

Now, we will compare the deployment algorithms for the four problems in a case study. Note that they have different objectives and constraints. We only show the deployment result achieved by the algorithm discussed here.

Example:

The topology of the water grid to monitor is shown in Fig. 1.3. First, we consider the case to cover the water grid with minimum number of sensor nodes, with and without the requirement of network connectivity. The set of the coverage junctions of a sensor node deployed in v_i is defined as $C(v_i) = \{v_j | \langle v_i, v_j \rangle \in \mathcal{E}_g\}$.

For the case without the connectivity requirement, we use Algorithm 1 to determine the deployment of the sensor nodes; and for the case with the connectivity requirement, we use Algorithm 3. The result is shown in Fig. 1.4 (a) and (b), where a

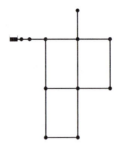

Figure 1.3: The network topology of the water grid to be monitored, where the vertices are the junctions to put sensor nodes at, and the edges are the pipelines

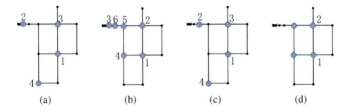

Figure 1.4: The deployment (blue circles) of the sensor nodes for different problems, where the number next to the circle is the sequence of the deployment according to corresponding algorithms. (a) The deployment achieved by Algorithm 1 for the minimum node cover problem; (b) The deployment achieved by Algorithm 3 for the minimum connected node cover problem; (c) The deployment achieved by Algorithm 4 for the maximum coverage problem; (d) The deployment achieved by Algorithm 7 for the maximum connected coverage problem.

blue circle on the vertex means the placement of a sensor node on that junction. The number next to the sensor node represents the sequence of the node that is deployed into the network according to the algorithm. For example, in Fig. 1.4 (a), the first node is deployed as it can cover in total 5 junctions. Then, nodes 2 and 3 are deployed because each can cover an additional 3 junctions. Finally, node 4 is deployed to cover the last uncovered junction. We should mention that, for the case without connectivity requirement, the result achieved by the greedy algorithm coincidentally is also the optimal solution. However, the algorithm does not guarantee to achieve the optimal solution. For example, in the second iteration, the number of the additional covered junction is the same for the case if we deploy the second node at its current location and the case to deploy it at the right one next to its current location, which are 3 for both cases. If we deploy the second sensor node as the second case does, then Algorithm 1 will give us a deployment to cover the water grid with 5 nodes in total. Besides, if we desire to construct a connected WSN, then additional nodes are required; compare to the case where connectivity is not required, as shown in Fig. 1.4 (b).

Next, we consider the case of using a fixed number of sensor nodes to cover the water grid. In the example, we assume to have 4 nodes in total. We again check the deployment of the nodes with and without the requirement of connectivity. The solutions are achieved by Algorithm 7 and 4, respectively.

Fig. 1.4 (c) shows the deployment achieved by Algorithm 4 for MAX Cover Problem (1.6). We have known from Fig. 1.4 (a) that using only four nodes can cover the whole water grid. Thus, the deployment in Fig. 1.4 (a) is an optimal solution for the MAX Cover problem. We have mentioned before that the greedy algorithm may not give us the optimal solution. We can see here that the four nodes deployed in Fig. 1.4 (c) fail to cover the whole water grid. However, such a deployment covers 11 out of 12 junctions, which satisfies Proposition 5.

In Fig. 1.4 (d), we show the deployment result of four sensor nodes that are connected. Node 1 in the figure corresponds to the v_r^* found in Line 6 of Algorithm 7. Then, V_s in Line 9 is the junctions where Nodes 1 and 2 are located. Due to the fact that Node 1 and Node 2 are connected and we still have two additional nodes, we can deploy the rest of the nodes at the junctions that provide additional coverage without breaking the connectivity of the WSN. In our case, the nodes without numbering are the locations of the additional nodes. We can see that the sensor nodes cover 10 out of 12 junctions, and it coincidentally is the optimal solution for Problem (1.10).

What we can conclude from the example is that, even though the algorithms only guarantee an approximation result, sometimes they can achieve an optimal solution. In general, if we require the connectivity of the sensor nodes, the coverage performance and the theoretical performance bounds of the algorithms are worse than the case where connectivity is not required.

1.2.2 DEPLOYMENT OF MOBILE WSNS

Now we have discussed the deployment of static sensor nodes. Notice that water grid $\mathcal{G}_g$ generally is huge for most cities. Therefore, full coverage by static sensor nodes might be impossible. In this case, mobile sensor nodes may greatly help in detecting or making measurements in the places not covered by the static nodes, as shown in Fig. 1.5.

Suppose we have deployed some static sensor nodes V_s in the water grid V_g to monitor events such as pollution. Once pollution in a pipeline is detected by the static sensor node that monitors the pipeline, the city government needs to take action, such as isolating the pipeline and dosing some anti-pollutant into the affected area. However, since the static sensor nodes cover only part of the water grid, the states of the pipelines that are not covered by V_s is still unknown. As a result, we can release some mobile sensor nodes from some junctions into the water grid and let the mobile ones move around in the pipeline and make measurements in the area that is not covered by the static ones. If the mobile ones have no motor units, their movement is carried by the water flow in the water grid. Thus, the coverage performance of the mobile sensor nodes greatly depends on their releasing locations. Based on such an observation, in this part, we discuss where the mobile nodes should be released.

For a water grid $\mathcal{G}_g$, we assume that we have the information of water flow rate in

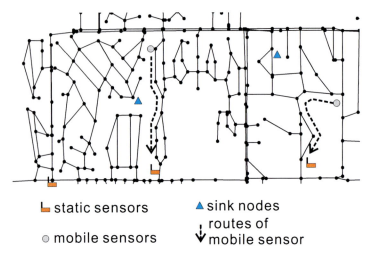

Figure 1.5: Monitoring water grid with mobile WSN, where the graph is a water grid. The mobile sensor nodes move with the flows in the water grid to make measurements and upload their measurement when they encounter static sensors or sink nodes.

the water grid. We denote $q_{ij} \geq 0$ the flow rate for pipeline e_{ij} if the water is flowing from junction v_i to v_j, and $q_{ji} = 0$. We assume that the mobile nodes are small enough such that they will not cause blockage in the pipeline, and their influence on the flow rate is negligible. Recall that the mobile ones have no motor unit. Therefore, we model the movement of a mobile node from a junction v_i to its neighboring node v_j with probability $p_{ij}^{(1)}$ that is proportional to the flow rate q_{ij}, i.e.,

$$p_{ij}^{(1)} = \frac{q_{ij}}{\sum_{k \in \mathcal{N}(v_i)} q_{ik}},$$

where $\mathcal{N}(v_i)$ is the neighbor set of junction v_i, and

$$p_{ii}^{(1)} = \begin{cases} 0 & \text{if } \sum_{k \in \mathcal{N}(v_i)} q_{ik} > 0 \\ 1 & \text{other wise,} \end{cases}$$

where $p_{ii}^{(1)} = 1$ corresponds to the case in which v_i is a sink junction (no outgoing flows from v_i). We use matrix $\boldsymbol{P}_1 = \{p_{ij}^{(1)}\}$ to denote the transition probability of a mobile node. Then, $\boldsymbol{P}_2 \triangleq \boldsymbol{P}_1 \boldsymbol{P}_1$ represents the two hops transition probability of a mobile node. Similarly, we define $\boldsymbol{P}_n \triangleq \boldsymbol{P}_1^n$ as the n hops transition probability. Then, the reachability matrix of mobile nodes is defined as $\boldsymbol{P} = \boldsymbol{I} + \boldsymbol{P}_1 + \boldsymbol{P}_2 + \ldots + \boldsymbol{P}_\infty = (\boldsymbol{I} - \boldsymbol{P}_1)^{-1}$, where its element p_{ij} denotes the probability of a mobile sensor at v_i will finally reach v_j at some time. Thus, for a mobile node released at v_i, the probability that it will pass by $v_j \in \mathcal{V}_g$ is p_{ij}.

Denote $\mathcal{V}_u$ as the set of junctions where a mobile node can upload its measurement to the monitoring center. $\mathcal{V}_u$ can be the set of the junctions where static sensors are deployed, if the mobile nodes can use static sensor nodes as relays to upload data. We can also deploy some sink nodes with long-range communication capability in the water grid such that the mobile ones can also upload data when they are close to the sink nodes. We denote the junctions that are deployed with sink nodes by $\mathcal{V}_k$. Then, $\mathcal{V}_u = \mathcal{V}_s \cup \mathcal{V}_k$. Thus, a junction v_i that is not monitored by static nodes can be monitored by a mobile node that is released from v_j with probability $1 - \prod_{v_k \in \mathcal{V}_u}(1 - p_{ji}p_{ik})$. If the releasing location of the mobile sensor nodes is $\mathcal{X}$, then the probability that a junction v_i is monitored by at least one of the mobile node is $1 - \prod_{v_k \in \mathcal{V}_u, v_j \in \mathcal{X}}(1 - p_{ji}p_{ik})$. Then, we formulate the MAX Mobile Coverage problem as follows:

$$\max_{\mathcal{X}} \quad \sum_{v_i \notin \mathcal{V}_s} \left(1 - \prod_{v_k \in \mathcal{V}_u, v_j \in \mathcal{X}} (1 - p_{ji}p_{ik})\right) W_i \tag{1.12a}$$

$$\text{s.t.} \quad |\mathcal{X}| \le N, \tag{1.12b}$$

where $W_i \ge 0$ is the importance of monitor v_i, which could be related to the population or area size. The objective function of Problem (1.12) is to maximize the coverage performance of the mobile nodes at the areas that are not monitored by the static ones, and Cardinality Constraint (1.12b) means that we have only N mobile nodes. The problem can be considered as a MAX Coverage Problem 1.6 and solved by Algorithm 4 if the objective function is monotone increasing and submodular. We have the following proposition:

Proposition 7 ([6]). *Consider set function* $f(\mathcal{X}) = \sum_{v_i \notin \mathcal{V}_s} \left(1 - \prod_{v_k \in \mathcal{V}_u, v_j \in \mathcal{X}}(1 - p_{ji}p_{ik})\right) W_i$. *If* $0 \le p_{i,j} \le 1, \forall i, j$ *and* $W_i \ge 0$, *then* f *is submodular and monotone increasing.*

Proof. We define $g_i(\mathcal{X}) = 1 - \prod_{v_k \in \mathcal{V}_u, v_j \in \mathcal{X}}(1 - p_{ji}p_{ik})$. Then, we show $g_i(\mathcal{X})$ is submodular as follows by definition. Consider set $\mathcal{A} \subseteq \mathcal{B}$ and $\{a\} \notin \mathcal{B}$. Then we have that

$$g_i(\mathcal{A} \cup \{a\}) - g_i(\mathcal{A}) = \left(1 - \prod_{v_k \in \mathcal{V}_u} (1 - p_{ai}p_{ik})\right) \prod_{v_k \in \mathcal{V}_u, v_j \in \mathcal{A}} (1 - p_{ji}p_{ik}) \ge 0,$$

where the inequality comes from $0 \le p_{ij} \le 1$. Therefore, g_i is monotone increase.

Similarly, for $\mathcal{B}$ we have that

$$g_i(\mathcal{B} \cup \{a\}) - g_i(\mathcal{B})$$

$$= \left(1 - \prod_{v_k \in \mathcal{V}_u} (1 - p_{ai}p_{ik})\right) \prod_{v_k \in \mathcal{V}_u, v_j \in \mathcal{B}} (1 - p_{ji}p_{ik})$$

$$= \left(1 - \prod_{v_k \in \mathcal{V}_u} (1 - p_{ai}p_{ik})\right) \prod_{v_k \in \mathcal{V}_u, v_j \in \mathcal{A}} (1 - p_{ji}p_{ik}) \prod_{v_k \in \mathcal{V}_u, v_j \in \mathcal{B} \setminus \mathcal{A}} (1 - p_{ji}p_{ik})$$

$$\leq \left(1 - \prod_{v_k \in \mathcal{V}_u} (1 - p_{ai}p_{ik})\right) \prod_{v_k \in \mathcal{V}_u, v_j \in \mathcal{A}} (1 - p_{ji}p_{ik}) = g_i(\mathcal{A} \cup \{a\}) - g_i(\mathcal{A}s).$$

Therefore, g_i is submodular. Note that f is a weighted sum of g_i, which preserves the submodularity and monotonicity. This gives us that f is submodular and monotone increasing. $\qquad\square$

According to Proposition 7, Problem (1.12) is a submodular maximization problem with cardinality constraint. Therefore, we can use Algorithm 4 to achieve an approximation solution, and the performance is given as follows:

Proposition 8 ([6]). *Consider a feasible MAX Mobile Cover Problem* (1.12). *Denote $\mathcal{V}_s^g$ the solution achieved by Algorithm 4, and $\mathcal{V}_s^*$ the optimal solution. Then, the approximation ratio is $e/(e-1)$, i.e.,*

$$f(\mathcal{V}_s^*) \leq \frac{e}{e-1} f(\mathcal{V}_s^g).$$

To compare the performance of mobile sensor nodes and that of static sensor nodes, we conduct the following simulation. The water grid to be monitored is given as Fig. 1.6, which is given by EPANET. More specifically, we use EPANET to achieve the water grid topology (junctions and pipelines) and the flow rates, as the input to MATLAB®. With the water grid topology, we are able to generate the graph $\mathcal{G}_g$; and with the flow rates, we can construct the matrix $\boldsymbol{P}$. Then, we run Algorithm 4 in MATLAB® to determine the releasing locations of the mobile sensors. Recall that W_i is the importance to monitor a junction v_i. In the simulation, we use W_i to represent the expectation of the population under the affection of junction v_i if it is polluted, and W_i is uniform randomly set from $(50, 150)$. We assume that there have been K number of static sensor nodes deployed in the water grid. Once the static sensor nodes detect a contamination, we can either deploy an additional sensor node in the water grid to monitor the water quality of that junction, or release a mobile sensor node into the water grid from a junction. This is to see if it is more beneficial to use mobile sensors than to use static sensors.

The releasing of the mobile sensor is determined by Algorithm 4. Noticing that we only need to determine the releasing location of one mobile node, the solution given by Algorithm 4 is optimal. Also, we will see how it performs if the mobile sensor node is released into the water grid in a location randomly picked. The simulation

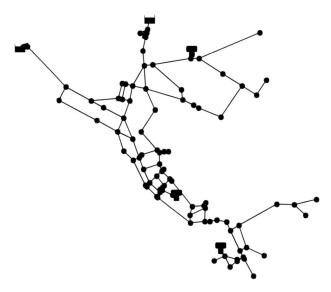

Figure 1.6: The water grid to be monitored with both static and mobile sensor nodes. It is achieved from the example water grid called 'Net3' in EPANET.

result is shown in Fig. 1.7. We can see that releasing an additional mobile node from the optimal junction leads to a much better coverage performance than using a static node. However, if the mobile node is released at a random junction, the improvement is not obvious. In the simulation, if the mobile nodes are released at random junctions, we need 5 mobile nodes to achieve a performance similar to the case where we release 1 mobile node at the optimal location. The reason is that, if the mobile node is released randomly, it may have a very low probability to encounter a sensor node to upload its measurement. That is why we need to optimize the releasing location of the mobile nodes.

1.3 TRANSMISSION SCHEDULING OF WIRELESS SENSOR NETWORKS FOR WATER GRIDS

In the previous section, we have discussed how to improve the coverage of WSN by deployment. Note that the pipelines are usually deployed underground and the sensor nodes are generally powered by battery. It is costly to replace the battery of the sensors in the pipeline. Thus, the lifetime of the sensor network greatly affects the long-term monitoring performance of the WSN for water grids. Consequently, in this section, we will discuss, given a proper deployment of WSN, how to extend the network lifetime by scheduling the transmission of the sensor nodes.

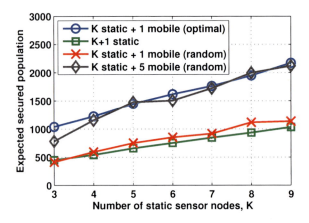

Figure 1.7: Comparison of using one additional mobile sensor node and one additional static sensor node in terms of coverage

1.3.1 ROUTING

In this part, we consider that the transmission range of every sensor node is not fixed. We assume that static sensor nodes have been deployed in every pipeline of the water grid. Sink nodes with long-range communication capability are deployed at the two ends of the pipeline, to collect the data from sensor nodes and to upload the data to the monitoring center. Here, we will try to extend the lifetime of the sensor nodes in every pipeline, such that the lifetime of the WSN is extended. Thus, in the following, we focus the sensor nodes in one pipeline.

Assume that there are N sensor nodes deployed in a pipeline, and the nodes are denoted by $n_1, n_2, \ldots, n_N$. Also, we denote the sink nodes at two ends to be n_0 and n_{N+1}. A node can transmit its data to any node it wants with proper transmission power. The further to the destination a node wants to transmit, the more power it needs to use. We denote $e^t_{i,j}$ the energy consumption that n_i spends to transmit to n_j data with unit size, and denote $e^r_{i,j}$ the energy consumption n_j spends to receive data with unit size from n_i. In each timeslot, every sensor node makes a measurement with a unit size, and no data compression is used during the data transmission. This can be formulated as a flow conservation constraint. The lifetime of the network is considered as the operation time until the first node that depletes its battery. Therefore, the nodes need to determine how much data it should transmit to other nodes, which is a routing problem. We denote $f_{i,j}$ the amount of data n_i transmits to n_j. Then the maximization lifetime routing problem [4] can be formulated as follows:

$$\max_{f_{i,j}, \forall i, j, T} \quad T \tag{1.13a}$$

$$\text{s.t.} \quad f_{i,j} \geq 0, \forall i, j \tag{1.13b}$$

$$\sum_{j:i \in \mathcal{N}(j)} f_{j,i} + T = \sum_{j \in \mathcal{N}(i)} f_{i,j}, \forall i \in \mathcal{V}_s \backslash \{n_0, n_{N+1}\} \tag{1.13c}$$

$$\sum_{j \in \mathcal{N}(0)} f_{j,0} + \sum_{j \in \mathcal{N}(N+1)} f_{j,N+1} = NT \tag{1.13d}$$

$$\sum_{j \in \mathcal{N}(i)} f_{i,j} e_{i,j}^t + \sum_{j:i \in \mathcal{N}(j)} f_{j,i} e_{j,i}^r \leq E_i, \forall i, \tag{1.13e}$$

where Constraint (1.13c) means that for every sensor node, it generates in total T unit size of measurements in T timeslots and the amount of data it transmits is the same as the amount of data it receives and measures; Constraint (1.13d) means that all the measured data are received by the sink nodes n_0 and n_{N+1}; Constraint (1.13e) means that the sensor nodes cannot use more energy than they have in data transmission. Note that the objective function and all the constraints are linear. Therefore, MAX Lifetime Routing Problem (1.13) is a linear programming, and such a problem is easy to solve by using optimization toolbox such as CVX [11].

1.3.2 SLEEP/AWAKE SCHEDULING

In this part, we will consider the case to determine when a sensor node should be activated to make a measurement and to transmit its data. We assume that N sensor nodes are densely deployed in a pipeline, denoted by $n_1, \ldots, n_N$. Two sink nodes, n_0 and n_{N+1} are at the two ends of the pipeline. To save energy in data transmission, the sensor nodes can use a fixed minimum power for data transmission. Thus, their data transmission ranges are the same and fixed. Then, the sensors use multi-hop data transmission to relay data to the sink at the end of the pipeline. Note that, once the sensors near the sink run out of battery, the network will become disconnected and the data of the sensors in the middle of the pipeline cannot reach the sink. As a result, we also need to balance the energy consumption of the sensors.

Recall that the sensor nodes are densely deployed. Consequently, their measurements have strong spatial correlation. Based on this observation, we can use Compressive Sensing [3, 12] to further reduce the total amount of energy consumption in data transmission and also balance the energy consumption of each node. We developed a data transmission scheme in [10] based on Compressive Sensing. More specifically, in each timeslot, we can activate a subset of the sensors that are connected with the sink, and set the rest of the sensors into sleep mode to save energy. The minimum number of the nodes that must be active is denoted by M_{cs}, which is related to the sparsity of the data, and we call such a requirement Cardinality Constraint. Besides, the active sensor nodes and the sink nodes must be connected, so that the measured data can reach the sink. We call such a requirement Connectivity Constraint. Then, as long as the active sensor nodes satisfy the Cardinality Constraint and the Connectivity Constraint, the monitoring center can achieve good estimation of the condition of the pipeline from the data of the active nodes.

According to Compressive Sensing, the sizes of the transmitted data of the active nodes are the same. Thus, the energy consumption of the active nodes are the same, and w.l.o.g, we normalize it to be 1. The energy consumption of a sleeping node is 0. Then, we can normalize the initial energy of node n_i to be E_i. The lifetime of the sensor networks in the pipeline is considered as the number of timeslots until either the Cardinality Constraint, or the Connectivity Constraint can not be satisfied any more. We denote the activation of the nodes at time slot t by $\boldsymbol{x}(t) = \{x_1(t), x_2(t), \ldots, x_N(t)\}^T$. Then the scheduling of the activation is $\boldsymbol{x}(1), \boldsymbol{x}(2), \ldots, \boldsymbol{x}(t), \ldots$. Then, the lifetime maximization problem is formulated as

$$\max_{\boldsymbol{x}(1),\ldots,\boldsymbol{x}(T)} \quad T \tag{1.14a}$$

$$\text{s.t.} \quad \sum_{i=1}^{N} x_i(t) \geq M_{cs}, \forall t \leq T \tag{1.14b}$$

$$\mathcal{G}_s(\boldsymbol{x}(t)) \text{ is connected}, \forall t \leq T \tag{1.14c}$$

$$\sum_{t=1}^{T} x_i(t) \leq E_i, \forall i \tag{1.14d}$$

$$x_i(t) \in \{0, 1\}, \forall i, t, \tag{1.14e}$$

where $\mathcal{G}_s(\boldsymbol{x}(t))$ is the induced graph of the active sensor nodes at t and the two sink, (1.14b) is the Cardinality Constraint, (1.14c) is the Connectivity Constraint, (1.14d) means that every sensor node cannot consume more energy than it has. The lifetime maximization Problem (1.14) is an integer optimization. The Cardinality Constraint (1.14b) and the Connectivity Constraint (1.14c) jointly make the problem not satisfy the totally unimodular proposition [24], and thus the problem is difficult to solve. However, the lifetime upper bound of the network is given by the following proposition:

Proposition 9 ([10]). *Consider a feasible network lifetime maximization Problem* (1.14). *Denote the optimal solution of the problem by* T^*. *Then,* T^* *is upper bounded by* $T^* \leq \sum_{i=1}^{N} E_i / M_{cs}$.

To find a good solution for Problem (1.14), we apply the idea of energy balancing. That is, we want to find the activation that balances the residual energy of the sensor nodes as much as possible, while guaranteeing the Cardinality Constraint (1.14b) and the Connectivity Constraint (1.14c). We define the residual energy of node n_i at timeslot t to be $p_i(t) = E_i(t)/E_i$, where $E_i(t) = E_i - \sum_{i=1}^{t} x_i(t)$. Then in each timeslot

Deployment and Scheduling of WSN for Monitoring Water Grids

Algorithm 8 Algorithm to Find M_c

Require: The adjacency matrix of nodes A.
Ensure: Minimum number of sensor nodes that ensures connectivity M_c.

1: Set $M_c \leftarrow 0, k \leftarrow 0$.
2: **while** $n_{N+1} \notin \mathcal{N}_-(n_k)$ **do**
3:    **if** $\mathcal{N}_-(v_k) \neq \emptyset$ **then**
4:       $k \leftarrow \max\{i | n_i \in \mathcal{N}_-(n_k), E_i(t) \geq 1\}, M_c \leftarrow M_c + 1$
5:    **else**
6:       $M_c \leftarrow +\infty$ // No feasible solution
7:       **return** M_c
8:    **end if**
9: **end while**
10: **return** M_c

---

t, we try to solve the following energy balancing problem:

$$\max_{\boldsymbol{x}(t)} \quad \sum_{i=1}^{N} p_i(t) \tag{1.15a}$$

$$\text{s.t.} \quad \sum_{i=1}^{N} x_i(t) = \max\{M_{cs}, M_c\} \tag{1.15b}$$

$$\mathcal{G}_s(\boldsymbol{x}(t)) \text{ is connected} \tag{1.15c}$$

$$x_i(t) \in \{0,1\}, \forall i, \tag{1.15d}$$

$$x_i(t) = 0 \text{ if } E_i(t) < 1, \tag{1.15e}$$

where M_c is the minimum number of nodes to be active such that Connectivity Constraint (1.14c) is satisfied. We have developed an algorithm based on dynamic programming to solve Problem (1.15) [10]. To explain the approach better, we define the down stream set (DNS) of a node n_i as $\mathcal{N}_-(n_i) = \{n_k | k > i, \langle n_i, n_k \rangle \in \mathcal{E}_s\}$, i.e., the set of the neighoring nodes of n_i with index larger than i. Now we are ready to describe the algorithm, which consists of two parts: 1) Finding M_c; 2) Determine $\boldsymbol{x}(t)$.

To calculate M_c, we follow the procedure as shown in Algorithm 8. In each iteration, the furthest node n_i that belongs to the DNS of the current node n_k is selected as current node (Line 4). If in any iteration, the DNS of current node is empty, the network is disconnected, which indicates that no feasible solution can be found and the WSN has expired.

To find $\boldsymbol{x}(t)$, we use a dynamic programming approach. We define $g(n_i, k)$ as the maximum sum of normalized residual energy of k sensor nodes among n_{i+1} to n_N and these k sensor nodes with the sink node n_{N+1} must be connected. Then, $g(n_i, k)$

30 Smart Water Grids: A Cyber-Physical Systems Approach

Algorithm 9 Optimal Activation Schedule for Each Timeslot

Require: Adjacency matrix $\boldsymbol{A}$, the minimum number of active node M_{cs}, and the normalized residual energy of the sensor nodes $\boldsymbol{p}$.
Ensure: A set of sensor nodes $\mathcal{V}_A$ that need to be activated.

1: Find M_c using Algorithm 8
2: **if** $M_c < +\infty$ **then**
　　// Find the minimum number of nodes that satisfy both connectivity constraint and cardinality constraint
3:　　$m = \max\{M_c, M_{cs}\}$
　　// Dynamic programming
4:　　Calculate $g(s_l, m)$ according to (1.16) and the corresponding $\boldsymbol{x}(t)$
5:　　**return** $\mathcal{V}_A$
6: **else**
7:　　**return** $\emptyset$
8: **end if**

---

can be calculated by the following recursive formula:

$$
g(n_i, k) = \begin{cases} \max\limits_{v_j \in \mathcal{N}'_-(n_i)} \{g(n_j, k-1) + p_j(t)\} & \text{if } k>0, \mathcal{N}'_-(n_i) \neq \emptyset \\ 0 & \text{if } k=0, n_{N+1} \in \mathcal{N}_-(n_i) \\ -\infty & \text{otherwise}, \end{cases} \tag{1.16}
$$

where $\mathcal{N}'_-(n_i) \triangleq \mathcal{N}_-(n_i) \backslash \{n_{N+1}\}$. Then, we activate the nodes that lead to $g(n_0, \max\{M_{cs}, M_c\})$ and achieve $\boldsymbol{x}(t)$. For timeslot t, we can use Algorithm 9 to determine which nodes to activate.

We have the following proposition for Algorithm 9.

Proposition 10 ([10]). *Consider a feasible energy balancing Problem (1.15). If the sensor nodes are deployed in a line and have the same transmission range, then Algorithm 9 achieves an optimal solution for Problem (1.15).*

Then, for any timeslot, we can update the residual energy of each sensor node, and apply Algorithm 9 to find the most energy balanced activation scheme for that timeslot, until the network expires. Consequently, we prolong the network lifetime by iteratively solving the energy balancing problem, as summarized by Algorithm 9. The performance of Algorithm 9 is given by the following two propositions:

Proposition 11 ([8]). *Consider a feasible network lifetime maximization Problem (1.14). Denote T^* the maximum lifetime of the WSN, and T^g the network lifetime achieved by Algorithm 10. If the sensor nodes are deployed in a line and have the same transmission range, then the approximation ratio of Algorithm 10 is 2, i.e., $T^* \leq 2T^g$.*

Deployment and Scheduling of WSN for Monitoring Water Grids   **31**

Algorithm 10 Node Activation for Network Lifetime Maximization

---

Require: Adjacency matrix A, the minimum number of active node M_{cs}, the battery
  of the nodes $E_i, \forall i$
Ensure: Network lifetime T
1: Set $t \leftarrow 1, Flag \leftarrow TRUE, E_i(t) = E_i$
2: **while** $Flag$ **do**
3:    Set $p_i \leftarrow E_i(t)/E_i, p = \{p_1, \ldots, p_N\}$
      // Node activation based on Algorithm 9
4:    $\mathcal{V}_A \leftarrow$ Call Algorithm 9 with input A, M_{cs}, p
5:    **if** $\mathcal{V}_A \neq \emptyset$ **then**
6:       Set $t \leftarrow t+1, E_j(t) \leftarrow E_j(t-1) - 1, \forall j \in \mathcal{V}_A$
7:    **else**
8:       Set $Flag \leftarrow FALSE$
9:    **end if**
10: **end while**
11: **return** $T \leftarrow t-1$

Proposition 12 ([8]). *Consider a feasible network lifetime maximization Problem (1.14), with initial energy of the sensor nodes to be E. Denote $T^*(k)$ and $T^*g(k)$ the maximum lifetime of the WSN and the lifetime achieved by Algorithm 10 with initial energy kE. Then,*

$$\lim_{k \to +\infty} \frac{T^g(k)}{T^*(k)} = 1.$$

Proposition 12 shows that the lifetime achieved by Algorithm 10 is asymptotic optimal. In other words, when the initial energy of the nodes is large enough, the lifetime gap is negligible, as is shown in Fig. 1.8. We set the initial normalized energy of sensor nodes to be 100, and $N = 100$ and $M_{cs} = 20$. We compare the lifetime achieved by Algorithm 10 to the benchmark algorithm from [9] and the lifetime upper bound. We can see that the lifetime achieved by Algorithm 10 is very close to the lifetime upper bound when the transmission range of the nodes is large enough, which corresponds to the case where the nodes are densely deployed and the WSN connectivity is good. It shows that, although Algorithm 10 may not achieve the optimal solution for Problem (1.14), its solution is close to the optimal one. Thus, it is efficient to use Algorithm 10 to determine the activation schedule of the WSN to prolong network lifetime.

1.3.3 TRANSMISSION TIME SCHEDULING WITH WIRELESS ENERGY TRANSMISSION

Note that the sensor nodes in pipelines are energy limited. An idea to prolong WSN lifetime is to provide additional energy to the sensor nodes wirelessly. This can be fulfilled by the idea of wireless energy transfer (WET), where the sensor nodes harvest the radio frequency (RF) energy that is transmitted from a dedicated energy

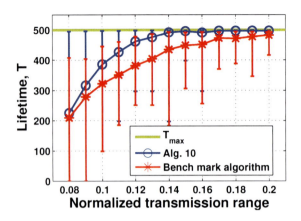

Figure 1.8: Comparison of the network lifetime achieved by Algorithm 10, and the lifetime upper bound.

source (we call it power base station). The sensor nodes can thus use the harvested RF energy to sense or transmit data, and therefore it is possible that the WSN has infinite network lifetime, i.e., none of the nodes will run out of energy. In this sense, how much data the sensor nodes can send to the sink is one performance metric of the WET-powered WSN.

Consider a WSN with N sensor nodes monitoring an underground pipeline [16]. A power base station transmits RF energy with power $P_{\max}$ in downlink with duration τ_0. Each sensor node uploads its measurements to the base station directly using the energy it receives, in a time division multiple access way. Then, the received power of n_i is

$$P_i^r = \frac{P_{\max} G_t G_r}{L_i^{DL}},$$

where G_t, G_r are the antenna gains at the transmitter and the receiver, L_i^{DL} accounts for the path loss in downlink. Let α_i be the energy harvesting efficiency of n_i, and P_{noise} be the power of environmental noise, the harvested energy of n_i is

$$E_i^r = \alpha_i (P_i^r + P_{\text{noise}}) \tau_0.$$

For sensor node n_i, it uses β_i portion of the energy to transmit its measurement, within duration τ_i. Then, its transmission power is

$$P_i^t = \frac{\beta_i E_i^r}{\tau_i}.$$

Then, the signal power received by the base station from n_i is

$$P_i^b = \frac{P_i^t G_t G_r}{L_i^{UL}} = \eta_i \frac{\tau_0}{\tau_1},$$

where L_i^{UL} is the pathloss in uplink from n_i to the base station, and $\eta_i = \alpha_i \beta_i (P_{\text{noise}} L_i^{\text{DL}} + P_{\text{max}} G_t G_r) G_t G_r / L_i^{\text{UL}}$. Thus, the achievable uplink throughput of n_i is given by

$$C_i = \tau_i \log_2 \left(1 + \frac{\eta_i \tau_0}{\tau_1 P_{\text{noise}}} \right),$$

where $\eta_i \tau_0 / (\tau_1 P_{\text{noise}})$ is the signal to noise ratio at the base station, and we denote it to be SNR_i. Define $\boldsymbol{\tau} = [\tau_0, \dots, \tau_N]^T$ the transmission scheduling of the nodes. Then, to maximize the throughput of WSN by scheduling the transmission duration of each node, we need to solve the following problem:

$$\max_{\boldsymbol{\tau}} \quad \sum_{i=1}^{N} C_i(\boldsymbol{\tau}) \tag{1.17a}$$

$$\text{s.t.} \quad SNR_i \geq \gamma_{\min}, \forall i \tag{1.17b}$$

$$C_i(\boldsymbol{\tau}) \geq C_{\min}^i \tag{1.17c}$$

$$\sum_{i=0}^{N} \tau_i = 1 \tag{1.17d}$$

$$\tau_i \geq 0, \forall i, \tag{1.17e}$$

where Constraint (1.17b) means that the SNR should be no less than the threshold $\gamma_{\min}$ such that the bit error rate at the base station is less than a desired value; Constraint (1.17c) means that the throughput for each sensor node should not be too small; Constraint (1.17d) guarantees that the transmission of the nodes and the base station will not be overlapped. We have the following proposition for the MAX throughput Problem (1.17):

Proposition 13 ([16]). *Problem* (1.17) *is a convex optimization.*

From this proposition, we know that Problem (1.17) is easy to solve. A more detailed approach and the closed form solution for the problem can be found in [16].

1.3.4 SCHEDULING OF WET

In the previous case, the base station broadcasts the wireless energy to the sensor nodes. To improve the energy that can be harvested by sensor node, we may use the idea of energy beamforming [17]. Energy beamforming concentrates the transmitted energy in one direction instead of spreading the energy to all directions. As a result, the scheduling of the energy beamforming (i.e., how long should the base station transmit energy to a sensor node), should also be considered to further prolong network lifetime or to improve the monitoring performance.

Based on the setting, we consider a WET-powered WSN for water grid monitoring systems described as follows. We use N sensor nodes, $n_1, \dots, n_N$, to monitor the water grid, and use a power base station to transfer energy to the nodes. The routing of the nodes is fixed and predetermined. Within a timeslot, the amount of data that

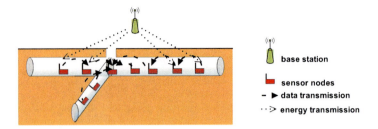

Figure 1.9: The wireless energy powered sensor network for water grid monitoring. The base station transmits energy using energy beamforming to charge sensor nodes. Then sensor nodes harvest the radio frequency energy from base station, make measurements, and relay data to the sinks.

node n_i needs to transmit is λ_i, which is determined by the routing and the required sampling rates of the nodes. Then, the energy consumption of n_i is a function of λ_i, and we denote it by $c_i(\lambda_i)$. Note that λ_i is predetermined. Therefore, we can simplify $c_i(\lambda_i)$ to be c_i.

The power base station forms an energy beam in each timeslot with power $P_{\max}$ to charge one node. Then, within a timeslot, the base station first transmits power to its target node for time duration τ_0, then the sensor nodes transmit and relay data to the sink in the remaining time one by one. We assume here that τ_0 and the transmission time of the sensor nodes are predetermined. Then, the transmitted energy from base station in a timeslot is $E_i^t = P_{\max}\tau_0$. Let $x_i(t)$ be the indicator if the base station is charging node n_i in timeslot t. Denote α_i the energy harvesting efficiency of n_i if the base station is charging n_i. Then, the harvested energy of n_i is

$$E_i^r(t) = \begin{cases} \alpha_i P_{\max}\tau_0 & \text{if } x_i(t) = 1 \\ 0 & \text{otherwise.} \end{cases}$$

Consequently, the energy dynamic of n_i is given by

$$E_i(t+1) = E_i(t) + \alpha_i P_{\max}\tau_0 x_i(t) - c_i. \tag{1.18}$$

We want to maximize the lifetime of the WSN by scheduling the WET of the base station, where the lifetime is defined as the time until the first node runs out of energy. Then, we can formulate the MAX lifetime by WET scheduling problem as follows:

$$\max_{\boldsymbol{x}(t)} \quad T \tag{1.19a}$$

$$\text{s.t.} \quad E_i(t) + \alpha_i P_{\max} \tau_0 x_i(t) \geq c_i, \forall i, \forall t \leq T \tag{1.19b}$$

$$E_i(t+1) = E_i(t) + \alpha_i P_{\max} \tau_0 x_i(t) - c_i \tag{1.19c}$$

$$\sum_{i=1}^{N} x_i(t) = 1, \forall t \leq T \tag{1.19d}$$

$$x_i(t) \in \{0,1\}, \forall i, \forall t \leq T, \tag{1.19e}$$

where T is the lifetime of the WSN, Constraint (1.19b) means that every node must have enough energy to transmit data in the first T timeslots, Constraint (1.19c) represents the energy dynamics of the sensor nodes, Constraints (1.19d) and (1.19e) mean that the base station can WET just one sensor node in a timeslot.

Notice that, if the energy consumption of the sensor nodes is small enough compared to the transmitted energy from the base station, the WSN can have infinite lifetime, which is characterized by the following proposition:

Proposition 14 ([7]). *Consider Problem* (1.19) *with* $c_i \ll E_i(0), \forall i$. *Suppose the following condition is satisfied:*

$$\sum_{i=1}^{N} \frac{c_i(\lambda_i)}{\alpha_i P_{\max} \tau_0} < 1. \tag{1.20}$$

Then, there exists a WET with high probability such that the network is immortal, i.e., $T \to +\infty$.

On the other hand, if Condition (1.20) is not satisfied, we have the following proposition on the upper bound of T:

Proposition 15 ([7]). *Consider Problem* (1.19). *Suppose that* $\sum_{i=1}^{N} c_i(\lambda_i)/(\alpha P_{\max} \tau_0) > 1$. *Then the network lifetime T is upper bounded by*

$$\bar{T} = \frac{\sum_{i=1}^{N} \frac{E_i(0)}{\alpha_i P_{\max} \tau_0}}{\sum_{i=1}^{N} \frac{c_i}{\alpha_i P_{\max} \tau_0} - 1}. \tag{1.21}$$

For the cases where the lifetime of the WSN is not infinity, we develop an approach as shown in Algorithm 11 to find the optimal WET scheduling to maximize the network lifetime. The basic idea of the algorithm is that, in each timeslot, the base station charges the node with the minimum residual lifetime $E_i(t)/c_i$. We have the following proposition to characterize the performance of the algorithm:

Proposition 16 ([7]). *Consider Problem* (1.19) *with its parameters satisfying* $\sum_{i=1}^{N} c_i(\lambda_i)/(\alpha P_{\max} \tau_0) > 1$. *Then, the WET scheduling $\boldsymbol{x}(t)$ that is achieved by Algorithm 11 results in the optimal solution for Problem* (1.19).

Algorithm 11 WET Scheduling for Maximizing WSN Lifetime

Require: $E_i(0), c_i, \alpha_i, P_{\max}, \tau_0$
Ensure: WSN lifetime T and WET schedule $x_i(t) \forall i, \forall t \leq T$
1: Set $x_i(t) \leftarrow 0, \forall i, t$ and set $t \leftarrow 0$
2: **while** $\min_i \{E_i(t)\} \geq 0$ **do**
3:   Set $k \leftarrow \arg\min_i \{E_i(0)/c_i\}$
4:   Set $x_k(t) \leftarrow 1$
5:   Update $E_i(t+1) = E_i(t) + \alpha_i P_{\max} \tau_0 x_i(t) - c_i$
6:   Set $t \leftarrow t + 1$
7: **end while**
8: Set $T \leftarrow t - 1$
9: **return** $T, x(t), \forall t \leq T$

The proposition shows that for the case where the power transmitted from the base station is not large enough compared to the consumption of the nodes, we can use Algorithm 11 to schedule the WET sequence of the base station and achieve the maximum lifetime of the WSN.

1.4 CONCLUDING REMARKS AND FUTURE DIRECTIONS

In this chapter, we have described the sensor nodes and the WSN that are used in monitoring water grids. Then, we have discussed the problems of node deployment. Such problems include MIN Node Cover (finding the minimum number of nodes that covers the water grids), MIN Cost Node Cover (finding the placement of nodes that covers the water grids with minimum cost), MIN Connected Node Cover (finding the placement of nodes that covers the water grid and ensures the connectivity of the sensor network), MAX Cover (finding the placement of a number of nodes that maximize the coverage of the sensor network), MAX Cover with Budget Constraint (finding the placement of sensor nodes that maximize the coverage with a certain budget), and MAX Connected Cover (finding the placement of sensor nodes that are connected and maximize the coverage). These problems are for static wireless sensor networks. Regarding mobile wireless sensor networks, we have considered a mobile node releasing problem. In general, such problems are combinatorial optimization problems, and are not easy to solve. Consequently, we have presented one algorithm for each problem, based on greedy approaches, and provided the performance bound of each algorithm. For the problems with connectivity requirements, i.e., the sensor nodes should be connected, the complexities of the algorithms are generally higher, and the performance bounds of the algorithms are worse.

Then, we have considered, given the deployed sensor networks, how to prolong the network lifetime or improve the network throughput by scheduling the sensing and transmission of the nodes. We have investigated a routing problem, where each node has to determine how much data to send to each neighboring node, to maximize the network lifetime. Such a problem is a convex optimization and thus easy to solve. Next, we have studied a sleep/awake scheduling problem to maximize the network lifetime. We have provided an approach based on the idea of balancing the energy of the nodes, and have shown that the result is asymptotically optimal. We have also taken wireless energy transfer into account to provide additional energy to the nodes. For the case of energy broadcasting from a base station, we need to schedule the time period for downlink energy broadcasting and uplink data transmission for each node, such that the network throughput is maximized. The formulated problem is convex. Thus, the optimal solution is achievable. We have also investigated a case with energy beamforming, where the power station can charge one node at a time. In this context, we formulated a network lifetime maximization problem, and provided a greedy algorithm that also achieves the optimal solution.

There are some research directions to improve the performance of water grid monitoring systems. For example, we believe that the usage of mobile sensor nodes and even the pervasive sensing devices could be widely used in water grid monitoring systems. Compared to the static sensor nodes, the mobile sensor nodes can monitor more areas, but also have higher costs; while the pervasive sensing devices may have lower costs but worse sensing accuracy. Then, a problem can be to determine how many static sensor nodes and mobile sensor nodes we should use to monitor the water grid, with a certain budget limit. Also, one can study how to provide incentives to the users who have pervasive sensing devices to participate in water grid monitoring, to improve the coverage of the monitoring systems. Regarding the usage of mobile sensor nodes, the problem discussed in the chapter requires/assumes that the mobile nodes are released into the water grid at the same time. It is also interesting to further study the case where we can release the mobile nodes at a different time, according to the measurements from the sensor nodes and the mobile nodes that have been released into the water grid.

Regarding the transmission scheduling, there are also some research efforts to improve the performance of monitoring. For instance, one could jointly consider the placement problem with the routing problem, such that the energy consumption issue also plays a role in the sensor network deployment. The resulting joint problem may be a mixed-integer optimization one, and thus more challenging. For the problem considering sleep/awake scheduling, the current solution is based on the assumption that the transmission ranges of the sensor nodes are fixed. Therefore, one can also consider the case where the sensor nodes can determine their power used for different transmission ranges. Regarding wireless energy transfer, we may also consider the cases with multiple power base stations. Then, one can study the power allocation of each power base station, i.e., what is the transmission power of each power base station to supports the whole wireless sensor network. Another interesting topic is to consider using mobile vehicles that move around in the city to charge the sensor nodes in the water grids, and also to collect information. Then, we can study how many vehicles we need, and what their travelling routes are. To provide energy to sensor nodes, we can also use energy harvesting techniques. In this case, the incoming energy may be unknown in advance, and the energy consumption on transmission scheduling should take this issue into account. Last, we can consider solving the problems mentioned above in a distributed manner, such that the sensor nodes can make decisions based on their local knowledge.

To summarize, to monitor water grids with sensor networks, we need to consider the placement of the nodes in the deployment phase, and also how to schedule the transmission of the devices during the network running phase. The performance of the water grid monitoring still has room to improve by using new types of devices and techniques. We hope that the solution approaches discussed in this chapter can provide some insights for the people who are designing monitoring systems for water grids.

1.5 ACKNOWLEDGMENTS

The authors of the chapter would like to thank the reviewers, the proofreaders and the editors for comments and proofreading. The work is partially supported by the Digital Demo Stockholm project iWater.

1.6 REFERENCES

1. Securing London's water future, 2011. [Online] `https://www.london.gov.uk/sites/default/files/gla_migrate_files_destination/water-strategy-oct11.pdf`.
2. Water leakage? Look to the clouds. *Water & Wastewater International*, 27(2):36–37, 2012. [Online] `https://library.e.abb.com/public/6a61049db9459d18c1257a10002e1aae/WWI_April_May_2012_TakaDu.pdf`.
3. R G Baraniuk. Compressive sensing. *IEEE Signal Processing Magazine*, 24(4), 2007.
4. J-H Chang and L Tassiulas. Maximum lifetime routing in wireless sensor networks. *IEEE/ACM Trans. on Networking*, 12(4):609–619, 2004.
5. D-Z Du, K-I Ko, and X Hu. *Design and analysis of approximation algorithms*. Springer Science & Business Media, 2011.
6. R Du, C Fischione, and M Xiao. Flowing with the water: On optimal monitoring of water distribution networks by mobile sensors. In *Proc. Annual IEEE International Conference on Computer Communications (INFOCOM)*, pages 1–9, April 2016.
7. R Du, C Fischione, and M Xiao. Lifetime maximization for sensor networks with wireless energy transfer. In *Proc. IEEE International Conference on Communications (ICC)*, pages 20–25, 2016.
8. R Du, L Gkatzikis, C Fischione, and M Xiao. On maximizing sensor network lifetime by energy balancing. *IEEE Trans. on Control of Network Systems*, 2017, `doi:10.1109/TCNS.2017.2696363`.
9. R Du, L Gkatzikis, C Fischione, and M Xiao. Energy efficient monitoring of water distribution networks via compressive sensing. In *Proc. of IEEE International Conference on Communications (ICC 2015)*, pages 8309–8314, 2015.
10. R Du, L Gkatzikis, C Fischione, and M Xiao. Energy efficient sensor activation for water distribution networks based on compressive sensing. *IEEE Journal on Selected Areas in Communications*, 33(12):2997–3010, 2015.
11. M Grant, S Boyd, and Y Ye. Cvx: MATLAB® software for disciplined convex programming, 2008.
12. C Karakus, A C Gurbuz, and B Tavli. Analysis of energy efficiency of compressive sensing in wireless sensor networks. *IEEE Sensors Journal*, 13(5):1999-2008, 2013.

13. T-W Kuo, K Ching-J L, and M-J Tsai. Maximizing submodular set function with connectivity constraint: Theory and application to networks. *IEEE/ACM Trans. on Networking*, 23(2):533–546, 2015.

14. T-T Lai, W-J Chen, K-H Li, P. Huang, and H-H Chu. Triopusnet: Automating wireless sensor network deployment and replacement in pipeline monitoring. In *Proc. ACM/IEEE International Conference on Information Processing in Sensor Networks*, pages 61–71, 2012.

15. T-T Lai, Y (Tiffany) Chen, P Huang, and H Chu. Pipeprobe: A mobile sensor droplet for mapping hidden pipeline. In *Proc. ACM Conference on Embedded Networked Sensor Systems*, pages 113–126, 2010.

16. G Liu, Z Wang, and T Jiang. Qos-aware throughput maximization in wireless powered underground sensor networks. *IEEE Transactions on Communications*, 99(99):1–13, 2016.

17. L Liu, R Zhang, and K-C Chua. Multi-antenna wireless powered communication with energy beamforming. *IEEE Transactions on Communications*, 62(12):4349–4361, 2014.

18. Pure Technologies Ltd. Smartball-water main leak detection and pipeline condition assessment. [Online] `https://puretechltd.com/technology/smartball-leak-detection/`

19. L Nachman, R Kling, R Adler, J Huang, and V Hummel. The Intel® mote platform: A Bluetooth-based sensor network for industrial monitoring. In *Proc. International Symposium on Information Processing in Sensor Networks*, pages 1–6, 2005.

20. G L Nemhauser, L A Wolsey, and M L Fisher. An analysis of approximations for maximizing submodular set functionsi. *Mathematical Programming*, 14(1):265–294, 1978.

21. A Ostfeld, J G Uber, E Salomons, J W Berry, W E Hart, C A Phillips, J-P Watson, G Dorini, P Jonkergouw, Z Kapelan, et al. The battle of the water sensor networks (bwsn): A design challenge for engineers and algorithms. *Journal of Water Resources Planning and Management*, 134(6):556–568, 2008.

22. L Perelman, W Salim, R Wu, J Park, A Ostfeld, M Banks, and D Porterfield. Enhancing water distribution system security through water quality mobile sensor operation. In *Proc. World Environmental & Water Resources Congress*, pages 663–674.

23. L A Rossman. Epanet 2: Users manual. 2000.

24. A Schrijver. *Theory of linear and integer programming*. John Wiley & Sons, 1998.

25. J Skadsen, R Janke, W Grayman, W Samuels, M Tenbroek, B Steglitz, and S Bahl. Distribution system on-line monitoring for detecting contamination and water quality changes. *Journal - American Water Works Association*, 100(7):81–94, 2008.

26. R S Stanković and B J Falkowski. The haar wavelet transform: Its status and achievements. *Computers & Electrical Engineering*, 29(1):25–44, 2003.

27. I Stoianov, L Nachman, S Madden, and T Tokmouline. PipeNet: A wireless sensor network for pipeline monitoring. In *Proc. IEEE International Symposium on Information Processing in Sensor Networks*, pages 264–273, 2007.

28. M Sviridenko. A note on maximizing a submodular set function subject to a knapsack constraint. *Operations Research Letters*, 32(1):41–43, 2004.

29. A J Whittle, L Girod, A Preis, M Allen, H B Lim, Mudasser Iqbal, Seshan Srirangarajan, Cheng Fu, Kai Juan Wong, and Daniel Goldsmith. WATERWISE@ SG: A testbed for continuous monitoring of the water distribution system in Singapore. In *Proc. 12th Annual Conference on Water Distribution Systems Analysis*, pages 1362-1378, 2010.

2 Consensus-Based Distributed Detection and State Estimation of Biofilm in Reverse Osmosis Membranes by WSNs

DANIEL ALONSO-ROMÁN, CÉSAR ASENSIO-MARCO, BALTASAR BEFERULL-LOZANO

CHAPTER HIGHLIGHTS

In this chapter, motivated by the operational problems caused by the formation of biofilm in RO membranes systems, we introduce a framework for the use of a WSN as a detector and an estimator of its evolution. The following related issues are tackled:

- We address our design as a complete distributed and decentralized scheme, while assuming a realistic scenario where cooperation between nodes is performed under unreliable links.
- We present the equations that, under several assumptions, model the evolution of the biofilm, and how, by applying some transformations, a linear model for both the evolution of the system to be controlled and the observation process are obtained.
- We show how the detection and the estimation of the biofilm can be implemented on a consensus-based basis, explaining how the failures of the links affect the performance of both processes. Then, we propose their redesign to adapt them to the probabilistic properties of the random consensus.
- We show how this adaptive scheme is able to compensate for the error introduced by the random consensus, and how its performance approximates that of a centralized approach.
- We show numerical results that provide a comparative study of the several mentioned designs and, offer a meaningful perspective of the different implications.

D Alonso-Román • C Asensio-Marco ⊠ • B Beferull-Lozano

University of Agder, NO. E-mail: cesar.asensio@uia.no

2.1 INTRODUCTION AND MOTIVATION

A biofilm is an accumulation of microorganisms, in which cells adhere to each other and to a certain surface. Inside the biofilm structure, the bacteria are protected from being washed away with the water flow, and can grow in locations where their food supply remains abundant [9]. Furthermore, the physical structure of the biofilm allows the growth and survival of microorganisms that otherwise could not compete successfully in a completely homogeneous system. Finally, microbial activity in biofilms can modify the internal environment to make itself more hospitable than the bulk liquid [15]. Thus, biofilm seems to constitute the preferred form of microbial life.

Formation of a biofilm begins with the attachment of free-floating microorganisms to a specific surface. If they are not immediately separated and removed, they might start taking some nutrients from the bulk liquid, with the subsequent growing process. Therefore, a gradient in the concentration of the nutrient substrates appears, which progressively transports this nutrient towards the surface, where a microbial synthesis takes place for its conversion into biofilm.

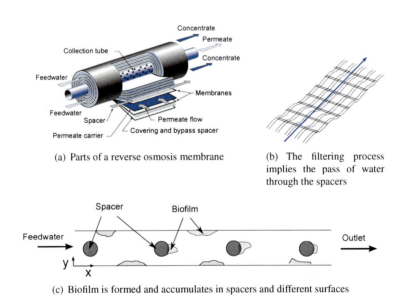

(a) Parts of a reverse osmosis membrane

(b) The filtering process implies the pass of water through the spacers

(c) Biofilm is formed and accumulates in spacers and different surfaces

Figure 2.1: Scheme of (a) a reverse osmosis membrane, (b) the flow of water through the spacer, and (c) how the biofilm accumulates in different surfaces.

While some biofilms can be considered beneficial and result in different useful processes (i.e., natural attenuation or controlled bioremediation of ground water and soils), most of them cause major problems for the systems where they are formed. For example, a biofilm is a problem for dental hygiene, since it causes infectious diseases and infections related to medical implants. In a larger scale, the appearance of biofilm in water treatment and distribution plants might result in a decrease of the quality of the water, and a reduction of the performance of the plants. More specifically, the biofilm formation remains a severe problem in water desalination plants whose technology is based on reverse osmosis (RO). The impact of biofilm formation can be deduced from Figure 2.1, where is shown a general scheme of a RO membrane and how the biofilm accumulates on its spacers. In the filtering process, a pressure greater than the osmotic pressure applied to the saline water causes fresh water to flow through the membrane while holding back the solutes. The higher the applied pressure above the osmotic pressure, the higher the rate of fresh water across the membranes. The two most important components in seawater RO systems are the high pressure feed pumps and the RO membranes, which must be properly optimized to ensure a successful operation. In this scenario, the biofilm formation affects the flux of water that is processed through the membranes, and implies an increase of the pressure applied by the pumps, which results in a higher energy consumption. Desalination plants currently avoid biofilm by doing periodic cleaning, which can vary depending on the type of biofilm. When the problem is severe, an autopsy of one of the membranes is required so that the type of biofilm can be identified. It has been shown that the application of pretreatment or dosing of biocides, such as with chlorine, is not an effective solution once the biofilm reaches an irreversible attachment phase. Its effect is often so severe that acceptable operation cannot be maintained and a RO membrane replacement is needed [14].

Given all the above, a continuous monitoring of the water in the plant may allow an early detection of the biofilm formation, together with an accurate estimation of its evolution, enabling the application of control measures before the biofilm reaches an irreversible attachment phase. The application of these tasks should result in a decrease of the costs of chemicals, membrane replacements and energy, and in a more correct operation of the plant. In this sense, wireless sensor networks (WSNs) have emerged as a powerful technology for the monitoring and control of natural and industrial environments. A key property of these networks is that the sensor nodes can be placed in areas that are inaccessible to wired and bulky devices. In addition, by deploying a large number of nodes, it is possible to establish correlation between different measurements taken in different parts of the system being controlled. Finally, the deployment as well as the management of the sensor network do not require disruption of the normal operation of the structure.

Although these networks are usually composed of very simple devices in terms of processing, storage, and communication capabilities, the performance of the whole network can be significantly improved over that of individual sensors by means of cooperation between them. Among the different possible cooperation schemes, those which rely on a fully distributed and complete decentralized scenario provide important advantages over centralized schemes, in terms of scalability, flexibility and robustness against failures and attacks. In those settings, where nodes only have access to local information and just communicate with one-hop neighbors, the achievement of the assigned task can be accomplished by iterative exchanges of information, until all nodes eventually agree on a common value. For example, most of the statistical signal processing tasks to be performed by WSNs can be divided into simpler local tasks, each one consisting of a first step that involves just local computation followed by a refinement through information exchanges between nodes, until the nodes reach a consensus in the final result. In this way, the decomposition of the main task in separable functions, computed locally by nodes and executed in parallel, has traditionally been a popular topic in the computer science community [6]. Under ideal conditions, these distributed designs are optimal, in the sense that they attain the same performance as a complete centralized system. However, since cooperation between nodes is performed by wireless communications, it is subject to several random phenomena such as interferences, packet losses, and fading, which implies that the different instantaneous topologies that arise at subsequent iterations of the process are completely random and, in general, nonsymmetric. Under these conditions, the consensus value becomes a random variable, which is different, in general, from the desired value [18]. Therefore, the previously mentioned distributed consensus-based approaches become suboptimal under these realistic constraints, since their performance is reduced due to the imperfection and randomness of the communications.

In this chapter, we present a framework for the use of a WSN as both a detector of the formation of the biofilm in a reverse osmosis membrane, and an estimator of its evolution. Given that the nodes of the network are equipped with sensors for the measurement of both the biofilm thickness and the flux of nutrient substrate into the biofilm, these can estimate the concentration of biomass, which is the solid phase of the biofilm and defines its morphologic and physical properties. Then, the nodes are able to detect when this concentration exceeds a specific threshold that can be considered dangerous for the reverse osmosis membrane or close to the irreversible attachment. After the detection, the nodes start tracking the concentration evolution, in order to activate the corresponding actuators when necessary. The design of both the detection and the state estimation tasks is addressed in a complete distributed and decentralized fashion, and subject to realistic constraints where cooperation between nodes is performed under unreliable links.

In this way, we first present the equations that, under several assumptions, model the evolution of the biofilm, and how, by applying some transformations, linear models for both the evolution of the system to be controlled and the observation process can be obtained [2]. Under the conditions of linearity and gaussianity, the optimal detector and estate estimator are the Neyman-Pearson detector and the Kalman filter, respectively. Then, we show how these tools can be decomposed to be implemented on a consensus-based basis. We explain how the failures of the links affect the performance of both processes, and how these can be redesigned to adapt to the probabilistic properties of the real communications [3] [4]. We show how this adaptive scheme is able to compensate the error introduced by the random consensus, and how its performance approximates the one of a centralized approach. We present numerical results that provide a comparative study of the mentioned designs, and offer a meaningful perspective of the different implications.

2.2 FORMATION AND EVOLUTION OF THE BIOFILM

A biofilm is a gel-like aggregation of microorganisms and other particles embedded in extracellular polymeric substances (EPS). The biofilm is attached to a solid surface called substratum on one side, and in contact with liquid (bulk liquid) on the other side, and is usually composed of two parts: the biofilm itself and the boundary layer (see Figure 2.2) [9].

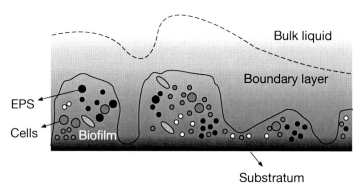

Figure 2.2: Typical parts of a biofilm: substratum, biofilm, boundary layer, and bulk liquid.

The biofilm contains diverse types of solid (particulate) and liquid (dissolved) components. The solid phase of the biofilm, expressed by the biomass concentration, gives the biofilm its reactive and structural properties. The formation of a biofilm begins with the attachment of free-floating microorganisms to a surface. In order to keep growing, the attached microorganisms take as a nutrient some substrate present in the bulk liquid, whose concentration controls the growth rate of the biofilm. This process causes a gradient in the concentration of this substrate, which results in its transport by diffusion towards the growing biofilm attached to the substratum, where it is consumed by the biomass. The concentration of the nutrient substrate is described by the mass balance equation:

$$\frac{\partial S}{\partial t} = -\nabla^T \mathbf{j}_S + r \qquad (2.1)$$

where S is the concentration of substrate, t is time, $\nabla^T = \left[\frac{\partial}{\partial x}, \frac{\partial}{\partial y}, \frac{\partial}{\partial z}\right]$ represents the gradient operator, $\mathbf{j}_S$ is the mass flux, and r is the net production rate. Usually the mass flux involves a diffusive flux $\mathbf{j}_d$ and a convective flux $\mathbf{j}_c$. The diffusive flux is given by Fick's first law:

$$\mathbf{j}_d = -D\nabla S$$

where D is the diffusion coefficient. The convective flux is expressed as:

$$\mathbf{j}_c = \mathbf{u}C$$

where $\mathbf{u}$ is the carrying fluid velocity field. By considering all the above, expression (2.1) can be written as follows:

$$\frac{\partial S}{\partial t} = -\nabla^T(\mathbf{u}C) + \nabla^T(D\nabla C) + r \tag{2.2}$$

The synthesis of microbial mass in the biofilm is given by the Monod equation:

$$r = q_{\max}\frac{S}{S+\Gamma}X_F \tag{2.3}$$

where $q_{\max}$ is the maximum specific substrate conversion rate, Γ is the affinity constant of the substrate, and X_F is the biomass concentration in the biofilm. This concentration of the solid component of the biofilm can be described by a mass balance equation similar to (2.1):

$$\frac{\partial X_F}{\partial t} = -\nabla^T \mathbf{j}_F + r \tag{2.4}$$

where $\mathbf{j}_F$ is the specific mass flux that describes the microbial growth and decay in the biofilm. This flux can be expressed by:

$$\mathbf{j}_F = \mathbf{u}_F X_F$$

where $\mathbf{u}_F$ are the distance by which the cells are displaced per unit time along both coordinates.

These differential equations express the behavior of both the substrate and the biomass concentration, and describe the structural and physical properties of the biofim. Unfortunately, due to their complexity, their exact solution can only be computed by numerical methods. However, by adopting several assumptions about the biofilm structure, a much simpler model can be obtained. This model captures the essentials of the biofilm behavior and is simple enough to be used by the nodes of the network as the state-space representation for the statistical inference tasks. In the next section, these assumptions are used to obtain linear expressions for the dependence of the substrate flux into the biofilm on the biomass concentration, and the evolution over time of this concentration.

2.2.1 ASSUMPTIONS AND MODELS

Since the characteristic times of the reaction and diffusion processes of soluble components are very small when compared to the biomass evolution, we assume a steady state for the substrate concentration. Besides, we consider a one-dimensional model and assume that the biofilm is flat and homogeneous. Finally, we assume no convection flux for the substrate nutrient. Under these assumptions (see Figure 2.3), equation (2.2) becomes:

$$D\frac{\partial^2 S(y)}{\partial y^2} + r = 0 \tag{2.5}$$

In the same way, the flux through the biofilm is given by the expression:

$$j_S = -D\frac{\partial S}{\partial y}$$

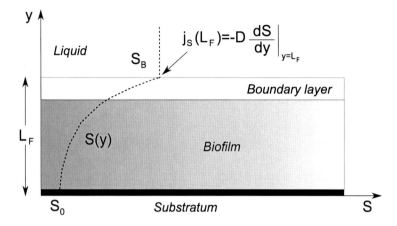

Figure 2.3: Profile of substrate concentration S(y) in a flat, homogeneous, and partially penetrated biofilm.

If a high substrate concentration is considered, we can assume $(S(y) \gg \Gamma)$, thus the reaction in (2.3) becomes the zero order expression:

$$r = q_{max}X_F \qquad (2.6)$$

At this stage, the solution depends on the penetration depth of the biofilm, which is defined as the thickness at which the substrate concentration becomes zero. We assume a fully penetrated biofilm, namely the penetration depth is greater than the actual thickness of the biofilm, as in Figure 2.3. If we integrate expression (2.5) by simultaneously considering (2.6), we obtain:

$$D\frac{\partial S}{\partial y} = -(q_{max}X_F)y$$

If we compute this expression at the biofilm surface $(y = L_F)$, we have the following:

$$j_S(L_F) = -D\frac{\partial S}{\partial y}\bigg|_{y=L_F} = q_{max}L_F X_F \qquad (2.7)$$

which gives the dependence between the value of the flux of nutrient into the biofilm $j_S(L_F)$, its thickness L_F, and the biomass concentration X_F. Similarly, if we apply the one-dimensional model, the flat and homogeneous condition of the biofilm and the high substrate concentration (zero-order kinetics), equation (2.4) becomes:

$$\frac{\partial X_F}{\partial t} = \left(q_{max} - \frac{\partial u_F}{\partial y}\right)X_F$$

which expresses the continuous evolution in time of the biomass concentration. If we discretize this equation for given intervals $\alpha = \Delta t$, we obtain the following:

$$X_F(k+1) = \left[1 + \alpha\left(q_{max} - \frac{\partial u_F}{\partial y}\right)\right] X_F(k) \qquad (2.8)$$

which gives the value of the biomass concentration at every time step k. Both equations (2.7) and (2.8) will be used in the following section for the state space representation of our system and considered by the nodes for their statistical inferences.

2.3 BIOFILM MONITORING BY A WIRELESS SENSOR NETWORK

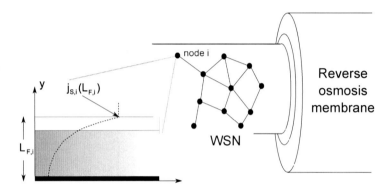

Figure 2.4: Deployment of a wireless sensor network with the objective of i) detecting the biofilm formation before it can damage the RO membrane, and ii) tracking its evolution. Each sensor i in the network obtains a local measurement of the biofilm thickness $L_{F,i}$ and a noisy observation of the flux of substrate in the biofilm surface $j_{S,i}(L_{F,i})$. We assume a multi one-dimension model, where this flux changes at each sensor but remains constant in its local domain.

The objective of this work is monitoring the formation and evolution of the biofilm, in order to detect when the biomass concentration reaches a specific level such that its state can be tracked by a state estimator. Eventually, a controller could actuate according to these estimations by applying some biocide in the system. To this end, a wireless sensor network is deployed inside the pipeline and before the RO membrane, as shown in Figure 2.4. We consider that the nutrient concentration changes along the pipeline and is different at each sensor location. However, as it has been previously exposed, we assume that in the local range of each node this concentration only changes along the y-coordinate and remains constant along the other two coordinates, that is, a multi one-dimensional model is assumed. In addition, since all the nodes are affected by the same nutrient under the same conditions, we assume that all the nodes have the same biomass concentration in the biofilm. Each sensor i

has a flat substratum where the biofilm may appear, and two sensors to measure the flux of substrate in the biofilm surface $j_{S,i}(L_{F,i})$ and the thickness of the biofilm L_i at its location. While the measurement of the flux involves an observation noise, for the sake of simplicity of our model we assume that the biofilm thickness is observed without error (an additional observation error could be introduced). Based on that, we consider that each node i obtains a noisy observation φ_i of the substrate flux into the biofilm at its location given by:

$$\varphi_i = j_{S,i}(L_{F,i}) + w_i$$

where the observation noise w_i is assumed to be zero mean Gaussian with variance σ_w^2 and spatially uncorrelated. By applying equation (2.7), this expression becomes:

$$\varphi_i = q_{max} L_{F,i} X_F + w_i$$

By taking $h_i = q_{max} L_{F,i}$, we have the following observation equation in matrix form:

$$\boldsymbol{\varphi} = \mathbf{h} X_F + \mathbf{w} \qquad (2.9)$$

which expresses the substrate flux into the biofilm in all the nodes of the network as a noisy and distorted version of the biomass concentration X_F. Similarly, we also assume that the time evolution of the biomass concentration in (2.8) is affected by a zero mean Gaussian noise $v(k)$ of variance σ_v^2 and uncorrelated in time. By doing $a = \left[1 + \alpha \left(q_{max} - \frac{\partial u_F}{\partial y}\right)\right]$, this expression becomes:

$$X_F(k+1) = a X_F(k) + v(k) \qquad (2.10)$$

Expressions (2.9) and (2.10) are the linear and Gaussian state-space representation of our biofilm system (see Figure 2.5), and are used by the nodes of the WSN to detect and track the biomass concentration in the biofilm. Each node i is assumed to know h_i, a, σ_w^2, and σ_v^2.

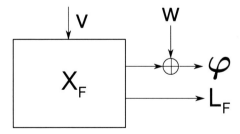

Figure 2.5: The state of the system is given by the biomass concentration X_F, whose time evolution is affected by noise v. The outputs of the system are the flux of the substrate in the surface of the biofilm, which is corrupted by noise w and whose observation results in $\boldsymbol{\varphi}$, and the biofilm thickness L_F.

2.3.1 AUTONOMOUS CONTROL OF BIOFILM EVOLUTION

Although it is out of the scope of this chapter, for the sake of completeness, we introduce how the detection and the state estimation of the biofilm evolution can be involved in a broader setting, which implies the autonomous control of the state of the system. In our specific case, once the biofilm formation has been detected, the final goal is to provide accurate information about the state of the biofilm, enabling a proper control of its evolution by automatically activating the biocide dosage [1]. Figure 2.6 shows the complete control loop and the role of the detector and the state estimator therein.

The specific threshold of biofilm detection from which the estimator and therefore the controller start working is a design parameter that must be chosen carefully based on the desired performance, biofilm properties, and system requirements. Roughly speaking, the threshold must be high enough so that the estimator works with enough signal-to-noise-ratio and the state evolution is properly tracked. On the other hand, it must be low enough so that it does not imply a high dose of biocide to keep it under control, and ultimately, that the biofilm does not reach a nonreversible state.

More specifically, if we denote by $\hat{X}_F(k)$ the estimation of the biomass concentration at time k, and by the M-dimensional vector $\mathbf{u}(k)$ the input to the biocide dosers or actuators (we assume that, in general, $M \neq N$), applying a linear control law, we can write the following [7]:

$$\mathbf{u}(k) = -\mathbf{l}(k)\hat{X}_F(k) \tag{2.11}$$

where the vector $\mathbf{l}(k)$ defines the control law at time k. Since the state of the system is affected by the input of the actuators, equation (2.10) should be rewritten as:

$$X_F(k+1) = aX_F(k) + \mathbf{b}^T(k)\mathbf{u}(k) + v(k) \tag{2.12}$$

where we have assumed a linear dependence, and where vector $\mathbf{b}(k)$ defines how each actuator impacts the state of the system.

The computation of the optimal control law in (2.11) and the study of the control loop with the mentioned feedback in (2.12) in terms of stability and optimality and under different constraints are tasks in the scope of automatic control. We refer the interested reader to [7], and the references therein. In this chapter, since we focus on statistical inference, namely detection and estimation, we will still rely on equation (2.10) to express the evolution of the system.

2.3.2 OPTIMAL DETECTION OF BIOFILM FORMATION

The first task of the deployed WSN is to detect the formation of biofilm, by considering when the biomass X_F in the biofilm reaches a specific and detectable concentration X_{th}, which can pose a threat to the reverse osmosis membrane, or approach dangerously near an irreversible attachment. Based on expression (2.9), this problem can be cast under the two following hypotheses:

$$\begin{aligned} \mathcal{H}_0 &: \boldsymbol{\varphi} = \mathbf{w} \\ \mathcal{H}_1 &: \boldsymbol{\varphi} = \mathbf{h}X_{th} + \mathbf{w} = \mathbf{s} + \mathbf{w} \end{aligned} \tag{2.13}$$

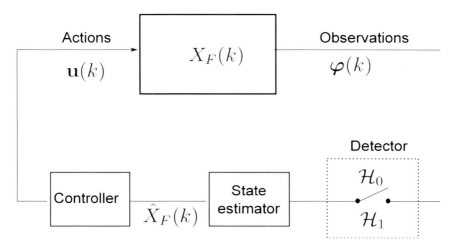

Figure 2.6: Diagram of the autonomous control loop, whose aim is to control the biofilm evolution. Based on the observations obtained by the sensors of the nodes, the detector triggers the controller once the biomass concentration reaches a specific threshold. Then, based on the estimation of the state of the biofilm, the actuators (biocide dosifiers) are activated if necessary.

where hypothesis $\mathcal{H}_0$ implies that no biofilm has still formed, and hypothesis $\mathcal{H}_1$ entails the appearance of biofilm. Since the signal $\mathbf{s}$ to be detected is known, the Neyman-Pearson theorem [12] provides the optimal decision criterion, where the likelihood ratio is compared with a specific value γ, previously computed to minimize a given false alarm probability P_{FA}. More precisely, the detection rule decides for $\mathcal{H}_1$, namely the biofilm formation is detected, if:

$$\Lambda(\boldsymbol{\varphi}) = \frac{p(\boldsymbol{\varphi}|\mathcal{H}_1)}{p(\boldsymbol{\varphi}|\mathcal{H}_0)} \geq \gamma \qquad (2.14)$$

where $p(\boldsymbol{\varphi}|\mathcal{H}_i)$ is the joint probability density function of the observed data under the hypothesis $\mathcal{H}_i$. Since in our case the measurements collected by the sensors are conditionally independent, and by applying the linear observation model in (2.13), it is straightforward to see that the rule in (2.14) becomes:

$$T(\boldsymbol{\varphi}) = \boldsymbol{\varphi}^T \mathbf{s} = \sum_{i=1}^{N} \varphi_i s_i \geq \gamma' \qquad (2.15)$$

where γ' is a new threshold defined as $\gamma' = \sigma^2 \log \gamma + \frac{\mathbf{s}^T \mathbf{s}}{2}$. We denote by $\mathcal{E}$ the energy of $\mathbf{s}$, such that $\mathcal{E} = \|\mathbf{s}\|_2^2 = \sum_{i=1}^{N} s_i^2$. Then, it can be shown that:

$$T(\boldsymbol{\varphi}) \sim \begin{cases} \mathcal{N}(0, \sigma^2) & \text{under } \mathcal{H}_0, \\ \mathcal{N}(\mathcal{E}, \sigma^2) & \text{under } \mathcal{H}_1. \end{cases} \qquad (2.16)$$

Considering the above, the resulting P_{FA} of the detector is given by:

$$P_{FA} = \Pr\left\{T(\mathbf{y}) > \gamma \mid \mathcal{H}_0\right\} = Q\left(\frac{\gamma}{\sqrt{\sigma^2}}\right) \tag{2.17}$$

where $Q(\cdot)$ is the tail probability of the standard normal distribution. Therefore, given a target false alarm probability, this expression allows to compute the optimal threshold to achieve it. Similarly, the detection probability of this detector can be computed as follows:

$$P_D = \Pr\left\{T(\mathbf{y}) > \gamma \mid \mathcal{H}_1\right\} = Q\left(\frac{\gamma - \mathcal{E}}{\sqrt{\sigma^2}}\right)$$

The detection performance can also be characterized by the deflection coefficient d^2, which measures the signal-to-noise ratio at the output of the detector. It is defined as the distance between the means of $T(\mathbf{y})$ under both hypotheses, measured in units of the variance under hypothesis $\mathcal{H}_0$ [13]:

$$d^2 = \frac{(\mathbb{E}[T;\mathcal{H}_1] - \mathbb{E}[T;\mathcal{H}_0])^2}{\mathrm{var}[T;\mathcal{H}_0]} = \frac{1}{\sigma^2}\mathbf{s}^T\mathbf{s} = \frac{\mathcal{E}}{\sigma^2}$$

The deflection coefficient gives an accurate insight about the performance of the detector in terms of the likelihood ratio even for non Gaussian scenarios [13]. Unlike P_A and P_D, d^2 can be computed without the knowledge of the probability distribution of the statistics $T(\mathbf{y})$ under both hypotheses.

2.3.3 OPTIMAL STATE ESTIMATION OF BIOFILM EVOLUTION

Once the biofilm formation has been detected, the next task is to track its evolution by means of the estimation of the state of the system (Figure 2.5), given by the biomass concentration, along time. In this way, given the observations (2.9) and the biomass concentration evolution (2.10), it is well known that the recursive scheme defined by the Kalman filter [19] provides the optimal estimate of the system state. Whenever a new set of observations $\boldsymbol{\varphi}(k)$ is obtained, the filter refines its *a priori* estimation $\hat{X}_F^-(k)$ to yield the *a posteriori* estimation $\hat{X}_F^+(k)$, as follows:

$$\hat{X}_F^+(k) = \hat{X}_F^-(k) + \mathbf{g}^T(k)\left[\boldsymbol{\varphi}(k) - \mathbf{h}\hat{X}_F^-(k)\right] \tag{2.18}$$

where $\mathbf{g}^T$ is the gain of the filter. Then, the filter projects this estimation in time to get the next *a priori* estimation:

$$\hat{X}_F^-(k+1) = a\hat{X}_F^+(k) \tag{2.19}$$

If we denote by $p_x^-(k)$ and $p_x^+(k)$ the variance of the estimation error of $\hat{X}_F^-(k)$ and $\hat{X}_F^+(k)$, respectively, the following holds:

$$p_x^+(k) = p_x^-(k)\left[1 - \mathbf{g}^T(k)\mathbf{h}\right]^2 + \sigma_w^2\mathbf{g}^T(k)\mathbf{g}(k)$$

Distributed Detection and State Estimation of Biofilm in RO Membranes **53**

and the next *a priori* variance is given by:

$$p_x^-(k+1) = a^2 p_x^+(k) + \sigma_w^2 \tag{2.20}$$

The optimal filter gain that minimizes the variance of the error at each time k is given by:

$$\mathbf{g}(k) = p_x^-(k)\mathbf{h}\left[p_x^-(k)\mathbf{h}^T\mathbf{h} + \sigma_v^2\right]^{-1} \tag{2.21}$$

Since $\mathbf{h}$, σ_v^2, and σ_w^2 are time-invariant, both the error variance and the filter gain can be pre-computed by running the filter off-line. Then, the filter must only compute expressions (2.18) and (2.19) during the estimation process.

2.4 CONSENSUS-BASED DISTRIBUTED DETECTION AND TRACKING

The control loop represented in Figure 2.6 can be designed in either of the two ways shown in Figure 2.7. The centralized scenario in 2.7(a) requires the nodes to send their observations to remote central entities, where those observations are processed and the different tasks such as the detection, estimation, and control are performed. Then, these entities send their decisions back to the nodes, which activate their actuators. In the distributed scenario shown in 2.7(b), no remote central entity is required. In this case the nodes, in a cooperative way and by exchanging observations with just one-hop neighbors, are able to carry out the different tasks and take their own decisions. As has been previously mentioned, a distributed approach to the design of the different involved tasks in the biofilm monitoring provides valuable advantages in terms of scalability, flexibility, and robustness. This approach additionally avoids delays from sending and receiving data between sensor nodes and the remote entities, hence ensuring more easily the stability of the control loop [10].

The distributed design of global inference tasks, when individual nodes only have access to their own local observations, is commonly addressed by iterative schemes that allow the involved nodes to reach a common value. More specifically, the case where the final agreement value is the average of the initial values of the nodes can be used to solve the distributed computation of many functions. In general, given a network of N nodes, where each one i has access to generic local information given by vector $\mathbf{y}_i$, a global objective function $f_0(\mathbf{y}_1,\dots,\mathbf{y}_N)$ can be decomposed in the following way:

$$f_0(\mathbf{y}_1,\dots,\mathbf{y}_N) = \sum_{i=1}^{N} f_i(\mathbf{y}_i)$$

which can be computed as the average of the initial values $x_i(0) = Nf_i(\mathbf{y}_i)$. Alternatively, if the objective function can be decomposed in a product of the form:

$$f_0(\mathbf{y}_1,\dots,\mathbf{y}_N) = \prod_{i=1}^{N} f_i(\mathbf{y}_i)$$

it can be computed as the average of the initial values $x_i(0) = N\log\left(f_i(\mathbf{y}_i)\right)$.

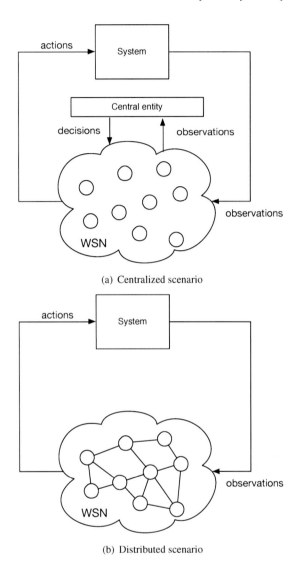

(a) Centralized scenario

(b) Distributed scenario

Figure 2.7: Two different designs of the control loop displayed in Figure 2.6. In (a), nodes must send the observations collected by their sensors to remote entities (namely detector, estimator, and controller), which perform the necessary computations and send their decisions back to the nodes. Based on that, the nodes activate their actuators. In (b), cooperation between neighboring nodes allows them to take their own decisions.

In the sequel we explain how the average of the initial values can be computed distributively by successive iterations on a consensus basis under ideal communications, and how random link failures affect the consensus process. Then, we show how the detection and the estate estimation tasks can be designed in distributed fashion by means of iterative consensus. Finally, we explain how these inference tasks can be redesigned in order to mitigate the error introduced by the deviation caused by the random failures to the consensus value.

2.4.1 ITERATIVE CONSENSUS UNDER UNRELIABLE LINKS

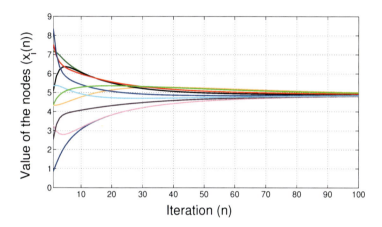

Figure 2.8: Example of the iterative consensus process, where $N = 9$ nodes iteratively combine their values until they reach the agreement of a value.

Since the purpose of this work is the consensus-based design of a distributed detector and a Kalman filter to monitoring the biofilm, we present below the basic principles that rule the consensus process, and how it is affected by random link failures. A network can be modeled as a random graph $\mathcal{G}(n) = (\mathcal{V}, \mathcal{E}(n))$ consisting of a set $\mathcal{V}$ of N nodes and the set $\mathbf{E}(n)$ of directed links existing at time n. A directed link from any node i to any other node j is denoted by e_{ij}. Then, $\mathbf{A}(n)$ is the $N \times N$ adjacency matrix, whose entry $[\mathbf{A}(n)]_{ij}$ is equal to 1 if $e_{ij} \in \mathbf{E}(n)$ and 0 otherwise. Thus, this matrix is random and, in general, not symmetric. The random set of neighbors of a node i at time n is defined as $\Omega_i(n) = \{j \in \mathcal{V} : e_{ij} \in \mathbf{E}(n)\}$. The degree matrix $\mathbf{D}(n)$ is a diagonal matrix whose entries are $[\mathbf{D}]_{ii} = |\Omega_i(n)|$, and the instantaneous Laplacian matrix is defined as $\mathbf{L}(n) = \mathbf{D}(n) - \mathbf{A}(n)$. Finally, we denote by $\mathbf{1}$ and $\mathbf{I}$ the all-one N-dimension vector and the $N \times N$ identity matrix, respectively.

Let us consider that each node i of the network takes a value $x_i(0)$ from a normal distribution with mean x_{avg} and variance σ_x^2. Then, the distributed consensus (or agreement) problem consists of a succession of iterations, at each of which every

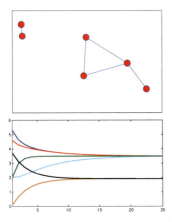 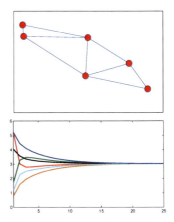

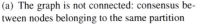

(a) The graph is not connected: consensus between nodes belonging to the same partition

(b) The graph is connected: all nodes reach consensus in a global common value

Figure 2.9: A necessary condition for all nodes to reach consensus in a global common value is that the graph is connected. If the graph changes at each iteration, the connectivity condition applies to the average of the instantaneous graphs.

node i refines its own state $x_i(n)$ by exchanging information only with those nodes belonging to the set $\Omega_i(n)$ at time n. This procedure continues until all the nodes agree asymptotically on a global common value, i.e., $\lim_{n\to\infty} \mathbf{x}[n] = \alpha \mathbf{1}$ (see Figure 2.8).

If we denote by $\mathbf{W}(n)$ the weight matrix used by the nodes to mix their values at time n, the process can be expressed as follows:

$$\mathbf{x}(n) = \mathbf{W}(n)\ldots\mathbf{W}(0)\mathbf{x}(0) = \mathbf{M}(n)\mathbf{x}(0) \quad (2.22)$$

This expression asymptotically reaches consensus if the product matrix $\mathbf{M}(n)$ converges to a matrix with equal rows, i.e., $\lim_{n\to\infty} \mathbf{M}(n) = \mathbf{1}\mathbf{m}^T$. It is known [17] that this occurs as long as every matrix $\mathbf{W}(n)$ is row stochastic, that is, $\mathbf{W}(n)\mathbf{1} = \mathbf{1}$, and the graph is connected on average (see Figure 2.9). In this case, we have that $\alpha = \mathbf{m}^T\mathbf{x}(0)$. These conditions for consensus can be easily fulfilled even in a distributed fashion by means of an appropriate construction of each matrix $\mathbf{W}(n)$ [20]. Depending on the number of received packets by each node at each iteration, each node gives weights to the incoming information and its own such that the weights add up to one. Furthermore, if every matrix $\mathbf{W}(n)$ is also column stochastic, that is, $\mathbf{1}^T\mathbf{W}(n) = \mathbf{1}^T$, we have that $\lim_{n\to\infty} \mathbf{M}(n) = \frac{1}{N}\mathbf{1}\mathbf{1}^T$, which ensures that the consensus value is the average of the initial values, hence $\alpha = \frac{1}{N}\sum_{i=1}^{N} x_i(0)$ (see Figure 2.10).

Nevertheless, in a real setting where the connectivity is random due to interferences and packet losses, each instantaneous underlying topology defined by $\mathbf{A}(n)$ is also random, and so is the corresponding weight matrix $\mathbf{W}(n)$. Therefore, this last

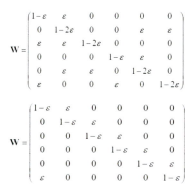
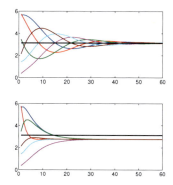

Figure 2.10: If the weight matrix **W** is row stochastic (up), all nodes converge to the same value and consensus is achieved. Additionally, if the weight matrix is also column stochastic (down), the consensus value is the average of the initial values of the nodes.

condition cannot be guaranteed, and $\mathbf{m} = (m_1 \ldots m_N)$ is a random vector, which is, in general, different from $\frac{1}{N}\mathbf{1}$. Close-form expressions for the first two moments of $\mathbf{m}$, namely its expectation $\boldsymbol{\mu}_m = (\mathbb{E}[m_1] \ldots \mathbb{E}[m_N])$, and covariance matrix $\mathbf{Q}_m$, have been derived in [18], as a function of the set of weight matrices $\mathbf{W}(n)$. Furthermore, by applying a conjecture in [5], we can also conjecture that each component of $\mathbf{m}$ follows a log-normal distribution. By considering all the above, we say that the iterative consensus process presents no deviation from the average if the following two conditions hold:

$$\boldsymbol{\mu}_m = \frac{1}{N}\mathbf{1}$$
$$\mathbf{Q}_m = 0 \qquad (2.23)$$

which imply that the average of the initial values is always obtained.

2.4.2 CONSENSUS-BASED DISTRIBUTED DETECTION

In order to perform the detection task in a complete distributed fashion, every node must be able to compute $T(\boldsymbol{\varphi})$ in (2.15) by just exchanging local information with their one hop neighbors available at each moment n. To accomplish that, we propose the use of the dynamic system presented in (2.22), and where the state of each node i is initialized with $x_i(0) = N\varphi_i s_i$. If all weight matrices are double stochastic the result of the consensus process is the average of initial values, hence all nodes compute the Neyman-Pearson detector. However, and due to the random link failures, all nodes asymptotically reach consensus in the following random value:

$$T_c(\boldsymbol{\varphi}, \mathbf{m}) = N \sum_{i=1}^{N} \varphi_i s_i m_i = N \boldsymbol{\varphi}^T \boldsymbol{\Delta}_m \mathbf{s} \tag{2.24}$$

where $\boldsymbol{\Delta}_m = \mathrm{diag}(\mathbf{m})$ is a random matrix, different, in general, from $\frac{1}{N}\mathbf{I}$. The properties of this consensus-based distributed detector are specified in the following lemma:

Lemma 1. *The detector $T_c(\boldsymbol{\varphi}, \mathbf{m})$ in (2.24) has a deflection coefficient given by the expression:*

$$d_c^2 = \frac{1}{\sigma^2} \frac{\left(\mathbf{s}^T \mathbb{E}[\boldsymbol{\Delta}_m]\mathbf{s}\right)^2}{\mathbf{s}^T \mathbb{E}\left[\boldsymbol{\Delta}_m^2\right]\mathbf{s}} \tag{2.25}$$

and is, in general, suboptimal. The optimality is achieved if and only if both conditions in (2.23) hold.

Proof: See [3].

In order to improve the performance of the distributed detector expressed in (2.25), we focus on increasing its deflection coefficient. To this end, in the next section we propose a variation of the detector, whose design considers not only the signal to be detected, but also the performance of the consensus process given by $\boldsymbol{\mu}_m$ and $\mathbf{C}_m$. In addition, we prove that the performance of the proposed detector outperforms the one of the Neyman-Pearson detector when both are implemented in a distributed fashion by means of the described iterative consensus. Finally, we show how to maximize this performance.

Adaptive distributed detector: The basis of the Neyman-Pearson detector is to confront the observation $\boldsymbol{\varphi}$ with the desired signal $\mathbf{s}$, such that the signal-to-noise ratio is maximized when $\mathbf{s}$ is present. However, if the detector is implemented in a distributed way by means of iterative consensus, the observation is not correlated with the original signal, but with a distorted version of it, given by $N\boldsymbol{\Delta}_m\mathbf{s}$. Therefore, our proposal consists in choosing another vector $\mathbf{r}$, rather than the original signal, to be confronted with the observations, such that it compensates the consensus error, and improves the final performance when implemented in a distributed fashion. Accordingly, each node i takes as initial value $x_i(0) = N\varphi_i r_i$, so that it asymptotically reaches consensus in the following proposed detector:

$$T_e(\boldsymbol{\varphi}, \mathbf{m}, \mathbf{r}) = N \boldsymbol{\varphi}^T \boldsymbol{\Delta}_m \mathbf{r} \tag{2.26}$$

for some vector $\mathbf{r}$. The moments of $T_e(\boldsymbol{\varphi}, \mathbf{m}, \mathbf{r})$ are given by:

$$\begin{aligned}
\mathbb{E}[T_e; \mathcal{H}_0] &= 0 \\
\mathbb{E}[T_e; \mathcal{H}_1] &= N \mathbf{s}^T \mathbb{E}[\boldsymbol{\Delta}_m]\mathbf{r} \\
\mathrm{var}\,(T_e; \mathcal{H}_0) &= N^2 \sigma^2 \mathbf{r}^T \mathbb{E}\left[\boldsymbol{\Delta}_m^2\right]\mathbf{r}
\end{aligned} \tag{2.27}$$

By considering the above, we express the performance of this detector by means of its deflection coefficient, as follows:

$$d_e^2(\mathbf{r}) = \frac{1}{\sigma^2} \frac{\left(\mathbf{s}^T \mathbb{E}[\Delta_m]\mathbf{r}\right)^2}{\mathbf{r}^T \mathbb{E}\left[\Delta_m^2\right]\mathbf{r}} \qquad (2.28)$$

which depends on the vector $\mathbf{r}$ that we use to confront the observations. If the consensus presents a deviation from the average, that is, condition in (2.23) is not satisfied, we have the following result:

Theorem 1. *There exists, at least, one N-dimensional vector $\mathbf{r}$ satisfying the energy constraint $\sum_{i=1}^{N} r_i^2 = \mathcal{E}$, such that, for $N >> \mathcal{E}$, the detector $T_e(\boldsymbol{\varphi}, \mathbf{m}, \mathbf{r})$ in (2.26) outperforms the detector $T_c(\boldsymbol{\varphi}, \mathbf{m})$ in (2.24), or, similarly:*

$$d_e^2(\mathbf{r}) > d_c^2.$$

Proof: See [3].

Given the log-normality of the entries of $\mathbf{m}$, the distribution of $T_c(\boldsymbol{\varphi}, \mathbf{m})$ and $T_e(\boldsymbol{\varphi}, \mathbf{m}, \mathbf{r})$ is the sum of N normal log-normal mixtures (NLNM) [21]. The optimal threshold for a given false alarm probability can be computed by numerical methods. However, it is well known that the log-normal distribution approximates a Gaussian as the ratio between the variance and the expectation tends to zero. In our case, it can be shown that m_i follows a normal distribution for small enough values of $\sigma_{m_i}^2$. On the other hand, the product of two Gaussian variables is non Gaussian. However, as the variance of one of the factors tends to zero, the product approximates a normal distribution. Again, in our case, it can be shown that for values of $\sigma_{m_i}^2$ small enough to make m_i Gaussian, the distribution of the product $N\varphi_i m_i$ is also Gaussian. Finally, if several Gaussian variables are jointly Gaussian, the sum of them is also Gaussian. Therefore, and considering the above, for values of $[\sigma_{m_1}^2, \ldots, \sigma_{m_i}^2]$ small enough the distribution of both $T_c(\boldsymbol{\varphi}, \mathbf{m})$ and $T_e(\boldsymbol{\varphi}, \mathbf{m}, \mathbf{r})$ can be approximated by a Gaussian, and the threshold for a given false alarm probability can be computed by (2.17) using the variance $\mathrm{var}(T_e; \mathcal{H}_0)$ in (2.27).

In order to compute the vector $\mathbf{r}^*$ that maximizes the performance expressed in (2.28), we state the following optimization problem:

$$\mathbf{r}^* = \min_{\mathbf{r}, \|\mathbf{r}\|_2^2 = \mathcal{E}} \frac{\mathbf{r}^T \mathbb{E}[\Delta_m^2]\mathbf{r}}{\mathbf{r}^T \mathbb{E}[\Delta_m]\mathbf{s}\mathbf{s}^T \mathbb{E}[\Delta_m]\mathbf{r}}$$

To solve this non-convex problem, we consider the Jagannathan's Theorem proposed in [11]. This theorem states that given the following general fractional problem:

$$\min\left\{\mu(x) = \frac{\Phi(x)}{\Psi(x)} : x \in X\right\}$$

where X is a nonempty compact subset of $\mathbb{R}^N$, and $\Psi(x) > 0$ for all $x \in X$, then x^* is an optimal solution for this problem if and only if it is also an optimal solution for:

$$\min\left\{\Phi(x) - \mu(x^*)\Psi(x) : x \in X\right\}$$

Based on that, the work in [8] presents the following iterative method for solving nonlinear fractional problems where the function $\Psi(x)$ is concave and $\Phi(x)$ is convex:

1. Take any feasible point x_1 from X, do $\mu_1 = \Phi(x_1)/\Psi(x_1)$ and set $k = 1$.
2. Solve the following convex subproblem:

$$f(\mu_k) = \min\left\{\Phi(x) - \mu_k\Psi(x) : x \in X\right\}$$

   and denote by x_{k+1} its solution.
3. If $f(\mu_k) = 0$, then stop and take x_{k+1} as the solution. Otherwise, set $\mu_{k+1} = \Phi(x_{k+1})/\Psi(x_{k+1})$, let $k = k+1$ and go to step 2.

In [16], the authors prove the convergence of this iterative algorithm for the case of general nonconvex fractional problem, provided that the optimization subproblem of step 2 can be solved. Since the matrix $\mathbb{E}[\boldsymbol{\Delta}_m]\mathbf{ss}^T\mathbb{E}[\boldsymbol{\Delta}_m]$ is positive semidefinite, in our case the condition $\Psi(x) > 0$ holds for all $x \in X$. Then, our optimization subproblem in step 2 is formulated as follows:

$$f(\mu_k) = \min_{\mathbf{r}, \|\mathbf{r}\|_2^2 = \mathcal{E}} \mathbf{r}^T \left\{\mathbb{E}[\boldsymbol{\Delta}_m^2] - \mu_k\mathbb{E}[\boldsymbol{\Delta}_m]\mathbf{ss}^T\mathbb{E}[\boldsymbol{\Delta}_m]\right\}\mathbf{r}$$

If we denote by $\mathbf{q}_1(\cdot)$ the normalized eigenvector associated to the smallest eigenvalue, the vector that minimizes the previous expression is the following:

$$\mathbf{r}_k = \mathcal{E}\mathbf{q}_1\left(\mathbb{E}[\boldsymbol{\Delta}_m^2] - \mu_k\mathbb{E}[\boldsymbol{\Delta}_m]\mathbf{ss}^T\mathbb{E}[\boldsymbol{\Delta}_m]\right)$$

so that the subproblem can be solved. The iterative algorithm adapted to our case is shown in Algorithm 12. The whole scheme of the implementation of this consensus-based distributed detector is shown in Figure 2.11.

2.4.3 CONSENSUS-BASED DISTRIBUTED STATE ESTIMATION

The Kalman filter introduced in Section 2.3.3 can be straightforwardly implemented in a centralized way, where all the necessary information is available at a central entity that computes (2.18) at each time step. Nevertheless, in the distributed scenario, which appeared in the monitoring of the biofilm evolution by a WSN, each node i has only access to its own observation $\varphi_i(t)$. The objective here is the design, following a complete distributed approach, of the Kalman filter described in Section 2.3.3, for a WSN under unreliable links. Accordingly, each node must be able to compute the expression in (2.18) based only on its own observation and by exchanging information just with one-hop neighbors. We assume that the network is much faster than

Algorithm 12

Require: $\mathbb{E}[\mathbf{m}], \mathbb{E}[\mathbf{mm}^T], \mathbf{s}, \delta$
  Set $\mu_1 = \lambda_{\min}(\mathbb{E}[\Delta_m^2])$ as initial value
  while $\left| \mathbf{r}_k^T \left\{ \mathbb{E}[\Delta_m^2] - \mu_k \mathbb{E}[\Delta_m] \mathbf{ss}^T \mathbb{E}[\Delta_m] \right\} \mathbf{r}_k \right| > \delta$ **do**
    Set $\mathbf{r}_k = \mathcal{E}\mathbf{q}_1(\mathbb{E}[\Delta_m^2] - \mu_k \mathbb{E}[\Delta_m]\mathbf{ss}^T \mathbb{E}[\Delta_m])$,
    $\mathbf{q}_1$ normalized eigenvector associated to the smallest eigenvalue
    Set $\mu_{k+1} = \dfrac{\mathbf{r}_k^T \mathbb{E}[\Delta_m^2] \mathbf{r}_k}{\mathbb{E}[\Delta_m]\mathbf{ss}^T \mathbb{E}[\Delta_m]}$
    Set $k = k+1$
  end while

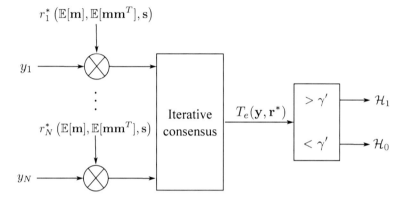

Figure 2.11: Implementation of the proposed distributed consensus-based detector for a known signal **s**. Previously to start the iterative consensus process, each node i confronts its observation y_i with the corresponding component of a vector $\mathbf{r}^*$, computed by considering both the error of the consensus represented by the two first moments of the random vector **m**, and the target signal **s** [3].

the evolution of the system, that is, between k and $k+1$, the network can iterate such that a consensus process like the one described in (2.22) asymptotically converges. Then, if each node i takes $Ng_i(k)\varphi_i(k)$ as its initial value, where $g_i(k)$ is the i-entry of $\mathbf{g}(k)$, and the process in (2.22) is applied, all nodes asymptotically converge to the value $\sum_{i=1}^{N} m_i N g_i(k)\varphi_i(k)$. If both conditions in (2.23) hold, thus $m_i = 1/N$ for all i, the consensus value reached by all nodes is $\sum_{i=1}^{N} g_i(k)\varphi_i(k) = \mathbf{g}^T(k)\boldsymbol{\varphi}(k)$. In this case, the expression in (2.18) can be computed by all nodes in a distributed way, and each node can run a local version of the filter identical to the optimal one.

Unfortunately, since the network is subject to random link failures, then $m_i \neq 1/N$ in general, and expression (2.18) becomes:

$$\hat{X}_F^+(k) = \hat{X}_F^-(k) + \mathbf{g}^T(k)\Delta_m \boldsymbol{\varphi}(k) - \mathbf{g}^T(k)\mathbf{h}\hat{X}_F^-(k) \qquad (2.29)$$

where $\mathbf{\Delta}_m = \mathrm{diag}(m_1 \ldots m_N)N$. If we evaluate the *a posteriori* estimation error of this filter, we have the following [4] :

$$
\begin{aligned}
\varepsilon^+(k) &= X_F(k) - \hat{X}_F^+(k) \\
&= \left[1 - \mathbf{g}^T(k)\mathbf{h}\right] a\varepsilon^+(k-1) - \mathbf{g}^T(k)\left[\mathbf{\Delta}_m - \mathbf{I}\right]\mathbf{h} a X_F(k-1) \\
&\quad + \left[1 - \mathbf{g}^T(k)\mathbf{\Delta}_m\mathbf{h}\right] w(k-1) - \mathbf{g}^T(k)\mathbf{\Delta}_m \mathcal{V}(k)
\end{aligned}
$$

By taking expectations at both sides, we obtain the estimation error mean:

$$
\begin{aligned}
\mathbb{E}\left[\varepsilon^+(k)\right] &= \left[1 - \mathbf{g}^T(k)\mathbf{h}\right] a\mathbb{E}\left[\varepsilon^+(k-1)\right] - \\
&\quad \mathbf{g}^T(k)\left(\mathbb{E}\left[\mathbf{\Delta}_m\right] - \mathbf{I}\right)\mathbf{h} a\mathbb{E}\left[X_F(k-1)\right]
\end{aligned}
$$

where we have considered that $\mathbf{\Delta}_m$ is independent of X_F and $w(k)$ and $\mathcal{V}(k)$ are zero-mean processes. Therefore, the expectation of the error at each step depends not only on the mean at the previous step, but also on a scaled version of the expectation of the state. Given that, in general, none of the two conditions in (2.23) is met, then $\mathbb{E}[\mathbf{\Delta}_m] \neq \mathbf{I}$. It entails that, unless $\mathbb{E}[X_F(0)] = 0$, this second term never vanishes, hence the error mean is always different from zero. Consequently, this filter is a biased estimator. In addition, the general gain computed in (2.21) is not optimal for this filter, since it has been obtained by considering expression (2.18). Then, the error variance is not minimized either. In the sequel, we explain how this distributed design can be reformulated in order to approximate the performance of the centralized approach. First, we force the filter to be unbiased. Then, and by considering the statistical properties of the consensus, we compute the optimal gain that minimizes the error variance.

Adaptive consensus-based distributed filter: Given the previously mentioned scenario, we propose the following alternative for the state update expression:

$$
\hat{X}_F^+(k) = \hat{X}_F^-(k) + \mathbf{g}^T(k)\mathbf{\Delta}_m\boldsymbol{\varphi}(k) - \mathbf{g}^T(k)\mathbf{\Delta}_m\mathbf{h}\hat{X}_F^-(k) \tag{2.30}
$$

that is, an additional consensus process is applied for the nodes to compute in a distributed way the term $\mathbf{g}^T(k)\mathbf{h}\hat{\boldsymbol{\varphi}}^-(k)$. Again, the evaluation of the *a posteriori* estimation error leads us to the following recursive expression:

$$
\begin{aligned}
\varepsilon^+(k) &= \left[1 - \mathbf{g}^T(k)\mathbf{\Delta}_m\mathbf{h}\right] a\varepsilon^+(k-1) \\
&\quad + \left[1 - \mathbf{g}^T(k)\mathbf{\Delta}_m\mathbf{h}\right] w(k-1) - \mathbf{g}^T(k)\mathbf{\Delta}_m\mathcal{V}(k)
\end{aligned} \tag{2.31}
$$

which depends only on the error at the previous step, and on the process and observation noises. By taking expectations at both sides, we obtain the estimation error mean:

$$
\mathbb{E}\left[\varepsilon^+(k)\right] = \left[1 - \mathbf{g}^T(k)\mathbb{E}\left[\mathbf{\Delta}_m\right]\mathbf{h}\right] a\mathbb{E}\left[\varepsilon^+(k-1)\right]
$$

Therefore, if we make $\hat{X}_F^+(0) = \mathbb{E}[X_F(0)]$, then $\mathbb{E}[\varepsilon^+(k)] = 0$, for all k. Consequently, as long as the initial estimate of X_F is equal to the expected value of $X_F(0)$, the expected value of $\hat{X}_F^+(k)$ is equal to $X_F(k)$, and the proposed estimator is unbiased regardless of the value of the gain matrix $\mathbf{g}(k)$.

Distributed Detection and State Estimation of Biofilm in RO Membranes

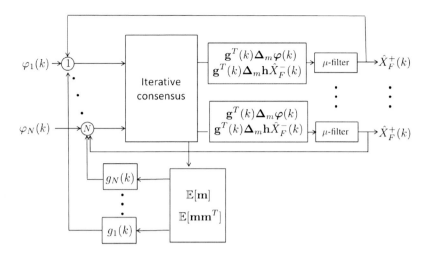

Figure 2.12: Block diagram of the practical implementation of the proposed adaptive filter for the tracking of the evolution of the biomass concentration. Given the observations φ_i and the previous estimation $\hat{X}_F^+(k-1)$, and by considering the statistical properties of random consensus, nodes cooperate on a consensus basis such that each one can compute locally expression (2.30), and get a new estimation $\hat{X}_F^+(k)$ of the biomass concentration, by means of a μ-filter [4].

In order to compute the optimal value of the filter gain $\mathbf{g}(k)$, we aim at minimizing the trace of the *a posteriori* error variance $p^+(k) = \mathbb{E}\left[\left(\varepsilon^+(k)\right)^2\right]$. By using (2.31), $p^+(k)$ can be expressed as:

$$p^+(k) = \mathbb{E}\left[\left(1 - \mathbf{g}^T(k)\mathbf{\Delta}_m\mathbf{h}\right)^2 p^-(k)\right] + \mathbb{E}\left[\left(\mathbf{g}^T(k)\mathbf{\Delta}_m \mathcal{V}(k)\right)^2\right] \quad (2.32)$$

where we have applied the whiteness of both $w(k)$ and $\mathcal{V}(k)$, their independence of $\varepsilon^+(k)$, and the projection of the variance in (2.20). If we derivate the previous expression with respect to the gain $\mathbf{g}(k)$, we have that:

$$\frac{\partial p^+(k)}{\partial \mathbf{g}(k)} = 2\mathbf{g}^T(k)\mathbb{E}\left[\mathbf{\Delta}_m\mathbf{h}\mathbf{h}^T\mathbf{\Delta}_m p^-(k) + \sigma_v^2 \mathbf{\Delta}_m^2\right] - 2p^-(k)\mathbf{h}^T \mathbb{E}\left[\blacksquare_m\right]$$

If we define the matrix $\mathbf{C}_m = N^2 \mathbb{E}\left[\mathbf{mm}^T\right]$, equal to zero, and solve for $\mathbf{g}(k)$, we obtain the following expression for the optimal filter gain:

$$\mathbf{g}(k) = p^-(k)\mathbb{E}[\mathbf{\Delta}_m]\left(\mathbf{C}_m \circ \mathbf{h}\mathbf{h}^T p^-(k) + \sigma_v^2 \mathbb{E}[\mathbf{\Delta}_m^2]\right)^{-1} \quad (2.33)$$

where $\circ$ stands for the Hadamard product, namely $(\mathbf{A} \circ \mathbf{B})_{i,j} = (\mathbf{A})_{i,j}(\mathbf{B})_{i,j}$. This expression gives the optimal gain of the filter as a function of the first- and second-order statistics of the consensus process. It is worth pointing out that if the consensus process always computes the average, that is, both conditions in (2.23) hold, then the expression in (2.33) becomes the same optimal gain as that for the centralized case expressed in (2.21).

To obtain the evolution of the error variance, we insert in (2.32) the optimal value of the gain obtained in (2.33), and rearrange terms to get the following:

$$p^+(k) = \left(1 - \mathbf{g}^T(k)\mathbb{E}[\mathbf{\Delta}_m]\mathbf{h}\right) p^-(k) \tag{2.34}$$

Again, the expression in (2.34) becomes that for the centralized case if both conditions in (2.23) hold.

Therefore, our consensus-based Kalman filter is completely defined by the update state expression in (2.30), the optimal filter gain in (2.33), and the error variance evolution in (2.34). Provided that the nodes know the statistical moments of the consensus process, they are able to compute expressions (2.33) and (2.34), and by exchanging information just with local neighbors on a consensus basis, to update their local states based on (2.30). Furthermore, the gain of the filter is adapted to the consensus features in order to minimize the error variance at each step. The whole scheme of the implementation of the proposed adaptive distributed filter is shown in Figure 2.12.

2.5 NUMERICAL RESULTS

Throughout this section we present numerical results from MATLAB® simulations that confirm the validity of our approach. The setup includes $N = 40$ nodes randomly deployed in a square area of $L = 50$ meters per side, and transmission power and attenuation are determined such that the average degree of the network is equal to 15. We assume that the power of the signal is $\mathcal{E} = 1$. Without loss of generality, the consensus process is based on the Laplacian matrix, such that $\mathbf{W}(n) = \mathbf{I} - \varepsilon\mathbf{L}(n)$, with a value of the step size $\varepsilon = 1/N$ to ensure convergence [20]. We also assume that the initial value of the signal is $X_F(0) = 0$, and the process noise has a variance $\sigma_v^2 = 10^{-1}$. The observation at each node i is attenuated by a factor $h_i < 1$ and the variance of the observation noise is $\sigma_w^2 = 10$.

We study first the detection process. Figure 2.13 shows, as a function of the consensus features, the performance of the different detectors: centralized detector, consensus-based distributed detector, and adaptive distributed detector. We have also included a detector, similar to the one based on consensus, but where each node i performs a trivial precoding, consisting in dividing its initial observation y_i by μ_{m_i} to compensate the average deviation. In Figure 2.13(a) the bias of vector $\mathbf{m}$, defined by the distance between $\boldsymbol{\mu}_m$ and vector $\frac{1}{N}\mathbf{1}$, is increased, while the variances $(\sigma_{m_1}^2, \ldots, \sigma_{m_N}^2)$ are kept approximately constant, in the order of $5 * 10^{-4}$. It can also be seen that the adaptive detector clearly outperforms the consensus-based detector as the bias increases. Logically enough, the performance of the precoded detector follows the one of the adaptive detector, but starts dropping for large biases. In Figure 2.13(b), the variance of the consensus process increases while the bias of $\mathbf{m}$ remains constant. Although both the adaptive and the precoded detectors compensate

Distributed Detection and State Estimation of Biofilm in RO Membranes 65

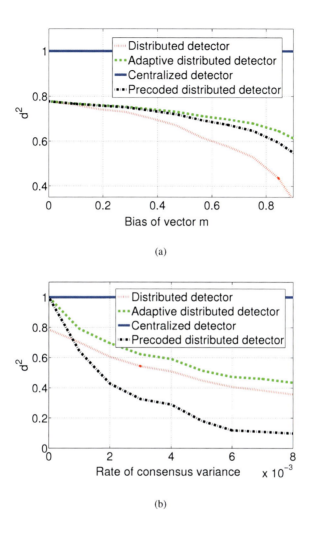

Figure 2.13: Performance of the different detectors, as a function of the properties of the consensus process, for a network composed of $N = 40$ nodes. In (a), the deflection coefficient is shown as a function of the distance between the consensus expectation $\boldsymbol{\mu}_m$ and vector $\frac{1}{N}\mathbf{1}$, for a rate of consensus variance equal to 0.0005. In (b), the deflection coefficient is depicted as a function of the variance of the consensus process, for a distance between the consensus expectation $\boldsymbol{\mu}_m$ and vector $\frac{1}{N}\mathbf{1}$ of 0.2 [3].

the bias and attain the optimal performance for a variance equal to zero, as this variance increases the performance of the precoded detector falls, while the gap between performances of the adaptive and the consensus-based detectors remains constant.

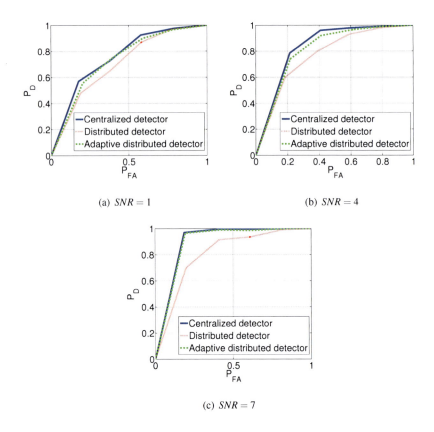

Figure 2.14: ROC curves of the different detectors. Each curve shows the probability of false alarm as a function of the probability of detection, for a given SNR. Results obtained for $N = 40$, a rate of consensus variance rate equal to 0.004, and a bias of **m** equal to 0.2 [3].

Figure 2.14 compares the receiver operating characteristic (ROC) curves of the centralized detector with those of the consensus-based and adaptive detectors, for different values of $SNR = \frac{\mathcal{E}}{\sigma^2}$. To this end we have assumed a Gaussian distribution of both $T_c(\boldsymbol{\varphi}, \mathbf{m})$ and $T_e(\boldsymbol{\varphi}, \mathbf{m}, \mathbf{r}^*)$, and computed the different thresholds accordingly. It can be seen that despite the Gaussian approximation, the thresholds give a reasonable performance for both distributed detectors.

As with respect to the state estimation process, Figure 2.15 shows the result of applying, to this state estimation problem, three different Kalman filters: the optimal filter based on a centralized setting, the general consensus-based distributed filter expressed in (2.29), and the distributed adaptive filter proposed in this work. More precisely, the tracking capabilities can be appreciated in Figure 2.15(a), where the

Distributed Detection and State Estimation of Biofilm in RO Membranes 67

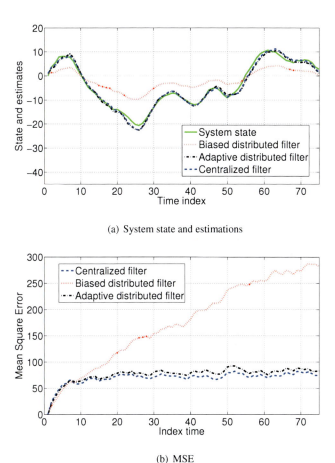

Figure 2.15: State estimation of a scalar parameter via three different versions of the Kalman filter. In (a), both the parameter evolution and the three estimators are shown for a single realization. It can be seen how the tracking capability of the proposed adaptive filter approaches the optimal one. The square error averaged over $R_2 = 200$ realizations is depicted in (b), for the three estimators. Its biasedness and non-adaptability to the randomness of the consensus process makes the general distributed filter diverge [4].

evolution of the system, as well as the different estimates, are depicted for a single realization of the experiment. Although the centralized solution shows the best performance, the consensus-based distributed filter proposed in the present work exhibits a very similar behavior. It implies that, due to its unbiasedness and its adaptive nature, this filter compensates the error introduced by the random consensus and approximates the optimal estimator. On the other hand, the biased and nonadaptive distributed filter is barely able to track properly the evolution of the parameter. This pattern is more easily appreciated in Figure 2.15(b), which shows the mean square error of the three estimators for $R_2 = 200$ realizations of the experiment, and where it can be seen how the biased filter diverges.

2.6 CONCLUDING REMARKS AND FUTURE DIRECTIONS

The biofilm formation implies a serious threat to the correct operation of reverse osmosis membrane based systems. The detection of the biofilm in an early stage and its posterior monitoring are crucial so that the corresponding control actions can be performed before an irreversible state is reached.

In this chapter, we have explained how the formation and the evolution of the biofilm can be properly monitored by a wireless sensor network. This scheme, which is generally part of a complete control loop, is addressed in a complete distributed fashion by means of an iterative consensus algorithm. Although a generic distributed approach implies several advantages in terms of scalability, robustness, and flexibility, it is also highly dependent on the failures of the communication. To this end, we propose a redesign of both the detection and the estate estimation processes such that the effects of the random communications on the consensus process are considered. We show how our approach outperforms a generic distributed implementation without these considerations, thus being more suitable for a realistic scenario where links are subject to random failures.

Further extensions to this work may involve the adoption of more elaborated models of the biofilm evolution, beyond the linear and Gaussian assumptions, which may lead to envision more complex detectors and filters, together with their distributed design under realistic constraints of random link failures. In addition, the computation of the statistical features of the consensus could be implemented on a machine learning basis aligned with the processes to be monitored. Along with those improvements, and in order to cover the whole control loop, the action of the biocide should also be modelled and incorporated to the state equation of the system. Finally, complete distributed control where each local controller has access only to the observations of a set of neighbors, instead of the whole set of observations, is still an open problem for which no tractable algorithm is known. The adoption of suboptimal solutions specially conceived for the specific case of the biofilm evolution would provide a useful framework for the future implementation of similar solutions.

ACKNOWLEDGMENTS

The work in this chapter was supported by the PETROMAKS "Smart-Rig" grant 244205/E30, the SFI "Offshore Mechatronics" grant 237896/O30, and the IKT-PLUSS "INDURB" grant 270730/O70, all from the Research Council of Norway.

2.7 REFERENCES

1. Hydrobionets: Autonomous control of large-scale water treatment plants based on self-organized wireless biomem sensor and actuator networks. http://www.hydrobionets.eu.
2. D Alonso-Roman, C Asensio-Marco, E Celada-Funes, and B Beferull-Lozano. Consensus based distributed estimation of biomass concentration in reverse osmosis membranes. In *Proceedings of the 1st ACM International Workshop on Cyber-Physical Systems for Smart Water Networks*, page 1. ACM, 2015.
3. D Alonso-Roman, B Beferull-Lozano, et al. Adaptive consensus-based distributed detection in wsn with unreliable links. In *2016 IEEE International Conference on Acoustics, Speech and Signal Processing (ICASSP)*, pages 4438–4442. IEEE, 2016.
4. B Beferull-Lozano et al. Adaptive consensus-based distributed kalman filter for wsns with random link failures. In *Distributed Computing in Sensor Systems (DCOSS), 2016 International Conference on*, pages 187–192. IEEE, 2016.
5. R Bellman. Limit theorems for non-commutative operations. i. *Duke Math. J.*, 21(3):491–500, 09 1954.
6. D P Bertsekas and J N Tsitsiklis. *Parallel and distributed computation: numerical methods*, volume 23. Prentice Hall Englewood Cliffs, NJ, 1989.
7. S P Boyd and C H Barratt. *Linear controller design: Limits of performance*. Prentice Hall Englewood Cliffs, NJ, 1991.
8. W Dinkelbach. On nonlinear fractional programming. *Management Science*, 13(7):492–498, 1967.
9. H Eberl et al. *Mathematical modelling of biofilms*, volume 18. IWA Publishing, 2006.
10. J P Hespanha, P Naghshtabrizi, and Y Xu. A survey of recent results in networked control systems. *Proceedings of the IEEE*, 95(1):138–162, 2007.
11. R Jagannathan. On some properties of programming problems in parametric form pertaining to fractional programming. *Management Science*, 12(7):609–615, 1966.
12. S M Kay. *Fundamentals of statistical signal processing. Detection theory*. Prentice Hall, 1998.
13. B Picinbono. On deflection as a performance criterion in detection. *IEEE Transactions on Aerospace and Electronic Systems*, 31(3):1072–1081, 1995.
14. A I Radu, J S Vrouwenvelder, M C M Van Loosdrecht, and C Picioreanu. Modelling the effect of biofilm formation on reverse osmosis performance: Flux, feed channel pressure drop and solute passage. *Journal of Membrane Science*, 365(1):1–15, 2010.
15. B E Rittmann and P L McCarty. *Environmental biotechnology: Principles and applications*. McGraw-Hill, 2001.
16. R G Ródenas, M Luz López, and D Verastegui. Extensions of Dinkelbach's algorithm for solving non-linear fractional programming problems. *Top*, 7(1):33–70, 1999.
17. A Tahbaz-Salehi and A Jadbabaie. A necessary and sufficient condition for consensus over random networks. *Automatic Control, IEEE Transactions on*, 53(3):791–795, April 2008.
18. A Tahbaz-Salehi and A Jadbabaie. Consensus over ergodic stationary graph processes. *IEEE Transactions on Automatic Control*, 55(1):225–230, 2010.
19. G Welch and G Bishop. An introduction to the Kalman filter, 1995.
20. L Xiao and S Boyd. Fast linear iterations for distributed averaging. *Systems & Control Letters*, 53(1):65–78, 2004.
21. M Yang. Normal log-normal mixture, leptokurtosis and skewness. *Applied Economics Letters*, 15(9):737–742, 2008.

3 Communication Optimization and Edge Analytics for Smart Water Grids

SOKRATIS KARTAKIS, JULIE A. MCCANN

> **CHAPTER HIGHLIGHTS**
>
> In this chapter, the authors introduce communication optimization and edge processing techniques for smart water grids, focused on three directions:
>
> - Define the limitations of the state-of-the-art low power wide area communication technologies by providing experimental results based on various deployments including a real-world smart water grid.
>
> - Examine the performance and impact of data reduction techniques in the communication within a smart water grid, exploiting lossless/lossy compression and compressive sensing.
>
> - Reveal the importance of the edge processing and offline big data analytics incorporation in communication minimization.

3.1 INTRODUCTION AND MOTIVATION

Over the last decade, there has been a trend where water utility companies aim to make water distribution networks more intelligent in order to improve their quality of service, reduce water waste, and minimize maintenance costs, etc., by incorporating IoT technologies. Current state of the art solutions use expensive power hungry deployments to monitor and transmit water network states periodically in order to detect anomalous behaviours such as water leakage and bursts (e.g., in Figure 3.1). However, more than 97% of water network assets are away from power and are often in geographically remote underpopulated areas: facts that make current approaches unsuitable for next generation dynamic adaptive water networks.

S Kartakis ✉

Imperial College London & Intel Labs Europe, UK. E-mail: sokratis.kartakis@intel.com

J A McCann

Imperial College London, UK

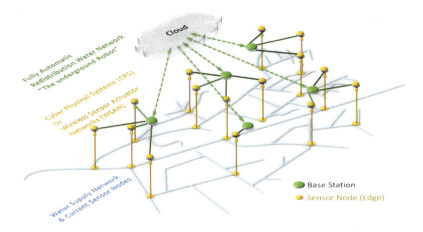

Figure 3.1: Smart water grids and communication: sensors, actuators, based stations, and cloud.

Water security is currently a hot topic. Water demands are not being met in regions of the world, both developed and underdeveloped, where climate change and economic water scarcity are two issues that have the largest impact. The former sees areas of the planet less able to generate enough water for its people, whereas in economic scarcity the government is unable to build or maintain a water distribution network to continuously meet demands. Drought prone areas such as California in the USA have had severe water restrictions in place for some time. Wet countries, such as the UK, have been experiencing what has been termed "wettest droughts" over the past few years. Notwithstanding the 7.5bn investment in UK water distribution networks, 3.3bn litres of water were lost per day in 2010 [22]. Following this trend, 46bn litres of drinking water were lost globally in 2015 [32].

The use of sensing systems to identify water leakage has been around for some time [51], but uptake has not been prolific. ICT to support WDN typically consists of remote or online battery-powered telemetry units (data loggers) that record water data such as flow and pressure, over numbers of minutes, then aggregate this data and send to a server periodically – typically via mobile phone networks. Contemporary approaches use wireless sensor network (WSN) [4,5,43,51,54] technologies to monitor the status of the water network and detect leakage or water bursts closer to real time. The main drawbacks of these approaches are: (a) the analysis of the data takes place offline, in base-stations or servers meaning that optimal real-time decision-making for control would be unrealistic and (b) the sensor nodes require a lot of energy, which places upper bounds on the amounts of data that can be sensed and relayed for analysis. The latter issue particularly impacts what is important to water companies, leakage localization. Given the cost of digging up roads to fix leaks, timely and accurate determination of the location of a leak is now more important than identifying all leakages. However, to carry out localization high-fidelity sensing and analytics are required to triangulate leakage accurately; current systems are not quite there yet.

In parallel, civil engineers advocate that next generation water networks will not be passive water deliverers, but active highly-distributed control systems that route the water intelligently to match demand and route round failures, etc. [1]. Such a dynamical system will heavily rely on sensing and actuation and will effect a dual control system of a water distribution network coupled with a smart distributed sensor/actuator network. This will require that both networks interact in a complex cyber-physical way and make use of state-of-the-art, low-powered high-precision sensing and processing technologies. State-of-the-art high-precision sensors can help build such a system, but it is too costly, and not very useful, to send all this higher precision data back to offline servers through a long-range communication for processing.

In this chapter, we tackle the high communication demands found in the harsh underground environment of smart water grids, focusing on three directions: hardware communication infrastructures based on long-range low-power technologies, data reduction techniques involving lossless compression and compressive sensing, and edge processing incorporation coupling with big data offline algorithms.

An overview of these directions follows:

Low Power Wide Area (LPWA): Low power wide area (LPWA) communication technologies have recently emerged as a viable alternative to cellular and mesh networks for providing cost effective Internet of Things (IoT) connectivity in cities. With several manufacturers and consortia now announcing national rollouts using different LPWA technologies, IoT solution and connectivity providers are now faced with the dilemma of what technology may be applicable in different environments or scenarios. Unfortunately, there is little empirical evidence that provides a good understanding of the performance trade-offs of these solutions applied to different application requirements and deployment contexts [35]. In this chapter, we provide a good understanding of the performance trade-offs of these solutions applied to urban [23] and smart water grid environments.

Lossless Compression and Compressive Sensing: To alleviate the problem of the demanding communication, efficient data compression enables high-rate sampling, while reducing significantly the required storage and bandwidth resources without sacrificing meaningful information. Lossless compression enables data reduction while ensuring the complete reconstruction of the water data streams [26]. However, data can be reduced further by applying compressive sensing [28] which introduces a low level reconstruction error. In this chapter, we provide not only results from both the compression approach over water data but also algorithms which combine the accuracy of standard lossless compression with the efficiency of a compressive sensing framework.

Edge Processing: In current smart water networks, physical states such as water pressure or pipe vibration are sampled at a reasonably high frequency (e.g., more than 128 samples per second) and stored in underground battery-driven sensor nodes. Currently, such raw data transmission is via cellular networks which is very costly.

The alternative is to transmit over long-range (several kilometres) wireless communications; however, high transmission power is required that leads to fast battery depletion. Current state of the art solutions [5] are unable to support underground battery driven sensor nodes and therefore the typical choice remains the expensive over ground sensor nodes directly connected to power and transmit data periodically every 15min using 3G. Furthermore, these solutions use offline computationally intensive algorithms to detect anomalies and require large amounts of historical data, leading to the inability of the algorithm to be distributed and deployable on sensor nodes. For example, [50] describes a wavelet-based algorithm which requires data continuously from the sensor nodes. To overcome these drawbacks, we present an efficient stream-based edge data processing and decision-making algorithm [26] which enables lightweight event-based communication and at the same time covers the needs of the server side burst localization algorithm in [29].

3.2 LOW POWER WIDE AREA COMMUNICATION TECHNOLOGIES

IoT deployments in smart city applications, such as smart water grids, provide one of the most appropriate scenarios for LPWA technologies (see Figure 3.2), due to the long-range communication requirements. However, they also impose diverse requirements on LPWA technologies in terms of application demands and deployment contexts. To illustrate the different scenarios, let us briefly look at three simple deployment environments of smart water grids as Figure 3.2 presents.

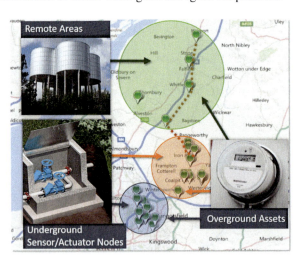

Figure 3.2: Smart water grid communication deployment scenarios.

Typical water quality monitoring to reservoirs or tanks requires a small number of samples to be taken at regular intervals sparsely distributed throughout the day. In this case, the sensor node deployments are in rural environments and the main commu-

Table 3.1

Major notations used in this chapter

Symbols	Meanings
$\mathbf{x}$	discrete time signal (vector)
$\widehat{\mathbf{x}}$	reconstruction of original signal (vector)
$\mathbf{\Psi}$	transform basis matrix
$\mathbf{w}$	coefficient vector
$\widehat{\mathbf{w}}$	estimated sparse coefficient vector
$\mathbf{y}$	vector of compressive measurements
$\mathbf{\Phi}$	measurement matrix
SR	sample rate
ρ	vector of SR
ε	error
$RMSRE$	mean square relative error
t_j	time that represents the j-th communication period
K_p	threshold gain for the selection of compression technique
B_{max}	maximum battery level
$B_i(t_j)$	battery level at the period t_j
$CR_i^{ls}(t_j)$	lossless compression rate of sensor node i at the period t_j
$E_i^{tr}(t_j)$	energy consumption of sensor node i to transmit the uncompressed data at the period t_j
$E_i^{in}(t_j)$	energy consumption of sensor node i due to other operations at the period t_j
G	graph representation of a water network
V_J	vertex set of pipe junctions
V_S	vertex set of deployed sensor locations
E	edge set of pipe sections connecting two vertices
A	carries the length of each pipe section
L	number of hops
$\mathbf{S}^{(L)}$	average distance matrix
u, v	graph vertices
t_u, t_v	time of anomaly occurrence in u and v graph vertices
$\hat{X}$	hyperbolic curves

nication obstruction can be caused by vegetation or people. Therefore, the wireless communication can be characterized either as line of sight or semi-line of sight (LoS or Semi-LoS, respectively). On the other hand, the monitoring of delay-tolerant overground water assets (i.e., smart meters) within a built environment brings additional challenges to developments where the penetration of radio through built spaces can greatly impact the design choice of communication. This communication is characterized as non-line of sight (NLoS). In contrast to the previous scenarios, pipe monitoring of critical infrastructures (e.g., leak detection in a water network) has quite different requirements. Data transmission is event-based in nature, and only occurs if critical events or warning signs are detected. The penetration challenge here is exacerbated due to underground deployments of nodes, which places them in a very harsh radio environment. As the effects of leaks or contamination can be severe, the delivery time of such events is critical and communication delays are not tolerated. To achieve long-range LoS, Semi-LoS, N-LoS and underground communications, LPWA technologies are emerging with new physical layer devices capable of leveraging trade-offs between range and data-rates [21].

In order to increase the communication range, a wireless device essentially needs to increase either the transmission power of the transmitter or sensitivity of the receiver. This in turn increases the link budget of the communication system consisting of transmission (Tx) and receive (Rx), which technically implies enhancing the signal-to-noise ratio (SNR)[1] at the receiver. Spread-Spectrum and Ultra-Narrow-Band (UNB) are two candidate technologies offering LPWA communication, and based on these technologies there are some specially tailored products being offered by various vendors. For example, LoRa by Semtech [45] uses Chirp Spread-Spectrum modulation, and NWave [38], Sigfox [49], etc., are offering modules based on Ultra-Narrow-Band technology.

3.2.1 THEORY BEHIND LPWA TECHNOLOGIES

The background theory of spread-spectrum and UNB follows, while the conventional narrow-band systems are presented as well. The comparison between LPWA and conventional technologies is necessary in order to understand the limitations and trade-offs of the contemporary technologies.

Spread-spectrum technologies spread the energy of the signal over a wide band, which means that signal will appear as background noise to other devices in the same part of the spectrum (see Figure 3.3). Therefore, isolating that signal makes it less prone to interference and also adds the advantage of making the communication more secure. That is, the signal is harder to jam, detect, intercept, or demodulate due to the pseudo-random nature of the code sequence used in modulation. One example, Semtech LoRa is a product that uses a proprietary spread-spectrum modulation scheme.

Alternative approaches exploit **Ultra Narrow Band** (UNB) communication technology which uses a narrow channel width to attain higher receiver sensitivity, which in turn increases the range achievable at the cost of data rates. Because the channels are ultra-narrow, they are more volatile and prone to interference. Furthermore, there are more channels in the spectrum. Therefore, to maximize spectrum utilization and link reliability such systems typically have complex back-end infrastructures that

---

[1] Signal-to-noise ratio is defined as the ratio of the power of a signal (meaningful information) and the power of background noise (unwanted signal).

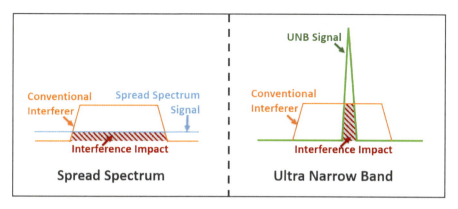

Figure 3.3: Spread spectrum and ultra narrow band spectrum utilization.

manage the MAC in terms of scheduling, transmission power levels, frequencies, modulation, etc. Examples of such emerging technologies are SIGFOX [49] and Weightless protocol [55], the latter being initially developed for the unused licensed TV band. Based on Weightless, nWave implemented Weightless-N and Platanus has started the development of Weightless-P. The above ultra-narrowband products provide only uplink communication (mainly because the receiver requires much power so low powered terminal nodes are optimized for send). However, SIGFOX and nWave have promised downlink support soon.

Theoretically, both technologies can achieve similar ranges for given data rates, however filters used to realize UNB are only now practical and the products are still maturing compared with the products that perform spread-spectrum modulation. Furthermore, this makes their comparison invalid unless an application is specified. From a long-term scalability perspective, Ultra-Narrow-Band seems to be more promising. However, many current early stage IoT deployments have opted for LoRa because of their ability to withstand interference and their security features. Yet narrowband approaches have the same scope expected to scale better by supporting many more edge nodes exploiting complex spectrum orchestration in the back end infrastructure.

Spread Spectrum and UNB products are designed to provide long-range communications with low power consumption, but data rates suffer. For applications that require higher data rates one would turn to conventional narrow-band products such as XBee868. XBee868 also has the advantage of providing bi-directional communications and mesh capabilities. In this chapter, we summarize the experimental results of [23] related to the performance of LPWA technologies in urban environments. Furthermore, we compare those with new evaluation results which were acquired from real experiments into the smart water grid of Welsh Water.

3.2.2 EVALUATION PLATFORM

In this evaluation, we focus on two contemporary technologies which developed specifically for long-range communication (i.e., LoRa and nWave), and the well-established Xbee868 module. The selection of these technologies was made due to the current maturity of the products in the market. To evaluate these technologies in both [23] and Welsh Water network experiments, BentoBox, a testbed node that allows hybrid communication experiments incorporating LoRa, XBee868, and nWave, is exploited. This platform enables the fair comparison of the collocated modules.

Figure 3.4: BentoBox: Hybrid LPWA Communication Box (Fig. 2 in [23]).

Figure 3.4 illustrates BentoBox. Specifically, BentoBox is based on Intel Edison development board which provides a Linux-based OS environment (i.e., Ubilinux) and a custom made hardware shield which incorporate numerous hardware interfaces including I2C, digital and analogue I/O, etc. In addition to other features, the hardware shield supports real-time energy monitoring of the board and for each communication module separately. Furthermore, the experimental data is logged into an external micro SD card. Having a hardware platform which incorporates multiple communication modules, the hosting case constitutes a critical aspect of the infrastructure. In order to reduce the signal interference from external RF environment and maximize the performance of the communication modules, the core of BentoBox is shielded into a aluminum box. The antenna socket of each communication module is connected to a metallic SMA socket on the surface of the box, while a metallic USB socket is installed as well to allow the power supply and connection with Intel Edison board. To maintain the same fair condition for all modules, a single 4.5 dBi 868-900 SMAM-RP antenna is used during the experiments.

At this point, the analysis of the main characteristics of LoRa, NWave, and Xbee868 is necessary. LoRa determines its transmission mode based on three config-

Table 3.2

LoRa modes and hardware configuration

LoRa Mode	Bandwidth	Coding Rate	Spreading Factor	Sensitivity (dB)
LM1 (max range)	125	4/5	12	-134
LM2	250	4/5	12	-131
LM3	125	4/5	10	-129
LM4	500	4/5	12	-128
LM5	250	4/5	10	-126
LM6	500	4/5	11	-125.5
LM7	250	4/5	9	-123
LM8	500	4/5	9	-120
LM9	500	4/5	8	-117
LM10 (min range)	500	4/5	7	-114

urable parameters: bandwidth (BW), coding rate[2] (CR) and spreading factor[3] (SF). We use 10 different modes (see Table 3.2) as provided by the Libelium API [19] where mode-1 corresponds to the largest distance with lowest data rate (0.2 kbps), and mode-10 corresponds to a higher data rate (6.45 kbps) and shortest distance; the remaining 8 intermediary modes represent a trade-off between link distance and data rate. Gradual variation of LoRa from mode-10 to mode-1 effectively results in enhancement in sensitivity from -114 dBm to -134 dBm (see Table 3.3), which increases the link budget for the same transmission power. On the other hand, the other modules, XBee868 and nWave, provide less configuration freedom by allowing only the selection of radio frequency and power level.

To conduct each experiment, a pair of BentoBoxes (receiver and transmitter) are used. Each BentoBox is connected through USB to a laptop to provide real-time monitoring of the communication and experiment progress. Having the reliable sensor nodes and real-time monitoring feature available, the next step is to set up a fair installation that minimizes the impact of tester human body and ground. Therefore, the tester is located at least two meters away from the BentoBox node which is installed on the top of a tripod, at least 1.2 meters above ground. In the scenarios of underground deployments, the BentoBox is suspended by a rope into chambers with the antenna pointing downward[4]. For our experiments, a number of BentoBoxes were

---

[2]Coding rate is an error detection/correction mechanism in wireless communication. Specifically, the transmitter generates n bits of data for every m bits of useful information (where n and m are redundant) and then the code rate is m/n.

[3]If a chip is the shortest modulated signal and a transmitted symbol is spread over multiple chips then the Spreading Factor (SF) indicates the number of chips used to transmit a symbol. Intuitively, higher SF leads to wider spread of the signal over the spectrum achieving longer ranges but lower data rates.

[4]The position of the antenna was selected after a number of experiments that allowed us to define the best ranges for our application.

Table 3.3
Technical parameters LPWA modules

Frequency: 868 MHz Power Supply: 3.3V		LoRa (Spread Spectrum)	Xbee Pro (Narrow Band)	nWave (Ultra-Narrow Band)
Data Rate (kbps) per datasheet		0.3-50 flexible	24	0.1
Power Consumption	Transmit	18 mA - 125 mA	85 mA - 500 mA	22 mA - 54 mA
	Receive	11 mA	65 mA	N/A
	Sleep	1 uA	55 uA	1.5 uA
Rx Sensitivity		-137 dBm	-112 dBm	N/A
Max Link Budget		157 dBm	137 dBm	N/A
Payload in our Evaluation		10, 100, 180, 250 bytes	10, 50, 100 bytes	10, 20 bytes

Reprinted from *Proceedings of the 2016 ACM Wireless Network Testbeds, Experimental Evaluation, and Characterization (WiNTECH) Workshop.* Kartakis, Sokratis, et al. "Demystifying low-power wide-area communications for city IoT applications," Copyright 2016, with permission from WiNTECH.

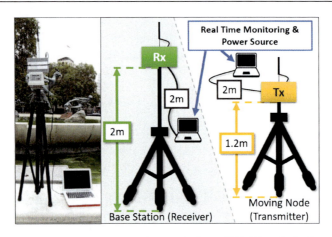

Figure 3.5: Base station and moving node from [23] (Fig. 3) deployments.

installed in and around the Imperial College area in different buildings and roof tops, in Hyde Park, and in manholes, while experiments in real water industry environments were conducted in Welsh Water network outside Cardiff.

3.2.3 EXPERIMENTAL METHODOLOGY AND ENVIRONMENTS[5]

---

[5]Partially reprinted from *Proceedings of the 2016 ACM Wireless Network Testbeds, Experimental Evaluation, and Characterization (WiNTECH) Workshop.* Kartakis, Sokratis, et al. "Demystifying low-power wide-area communications for city IoT applications," Copyright 2016, with permission from WiNTECH.

Communication Optimization and Edge Analytics for Smart Water Grids 81

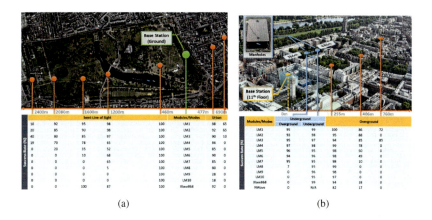

Figure 3.6: Experimental urban deployments and results as presented in [23], Fig. 4: (a) G2G Evaluations in Semi-LoS and Urban scenario; (b) RT2U and Overground Evaluations in Urban scenario.

Five main sensor node deployments (four deployments from [23] and a new deployment in Welsh Water network, see Figure 3.7) were used to conduct the experiments: ground to ground (G2G), ground to roof top (G2RT), underground to roof top (U2RT), underground to underground (U2U), and underground to ground (U2G). For each deployment, two BentoBoxes are used; one as transmitter and one as receiver. To define the communication range of each technology, the one node is located at a fixed point while the second is moved away. In addition, all the modules were set at maximum transmission power so that their coverage limitations were identified. In each new position, the transmitter sends 100 packets for each XBee868 and LoRa experiment and 20 packets for nWave. nWave was run fewer times due to significant delays (i.e. more than 10 minutes per packet) in getting experimental results from the cloud server. In addition to packet number per experiment, the transmission payload size varied (see Table 3.3). LoRa has 10 modes as described earlier, each of which was tested per each experiment. All experiments for each technology ran over 868MHz, and were interleaved automatically using the BentoBox software in a round-robin fashion to minimize the difference in medium properties for each technology at a given moment.

Deployment 1 - G2G: emulates the scenario where both receiver and transmitter are located close to ground level and deployed in a semi-LoS configuration (see Figure 3.6(a)). The receiver was located at the highest altitude of Hyde Park and 2m above ground (Figure 3.5 – left). The transmitter was placed on a tripod at 1.20m height (Figure 3.5 – right). Due to the location of the receiver at the edge of the park, two different environments are explored: open green space and built environment. To conduct the experiments for multiple locations, we increased the distance of the transmitter from the receiver. Specifically, in the open green space area, we increased the distance, in 400m increments, until none of the LPWA technologies were able to communicate successfully. In the built environment, the increment step is to 100m. The experiment is performed only for LoRa and XBee868. We could not consider nWave, for this experiment due to the requirements of internet connection to a cloud-based backend and constant 220V power source.

Deployment 2 - G2RT: considers a ground to rooftop communication case, where the receiver is placed on top of the Electrical Engineering Department building (i.e., floor eleven). On the other side, the transmitter node is placed on the ground at a height of 1.20m. As depicted in Figure 3.6(b), the distance between the transmitter and receiver is varied in 100m steps until no reception was possible. Due to the existence of an internet connection, we include nWave in this experiment.

Deployment 3 - U2RT: represents the communication scenario where a transmitter is placed underground and the receiver node is located at the roof top. Figure 3.6(b) shows underground (manhole) deployment. The BentoBox transmitter is placed in a manhole (i.e. 1m depth) with the antenna pointing downwards[6]. The manhole is covered with cast iron and all the available LPWA technologies are evaluated.

Deployment 4 - U2U: indicates the underground to underground communication where both transmitter and receiver are placed in two manholes at a distance of 80m (see Figure 3.6(b)). Both of the manholes are located in a built environment. In this scenario, the utilization of nWave base station is impossible in such a harsh underground environment.

Deployment 5 - U2G: extends the previous deployments to a real water network environment. This deployment emulates the underground to ground communication case where the transmitter is placed into one of the chambers of Welsh Water network and the receiver is installed overground on a tripod 2m above ground. Figure 3.7 illustrates the interior of the chamber which was almost full of water. BentoBox was suspended from the metallic lead with a rope by pointing the antenna downwards without contacting the water. Furthermore, as depicted in Figure 3.7, we vary the distance between the transmitter and receiver in 20 to 60m steps until no reception was possible. In this deployment only Xbee868 and LoRa are evaluated.

3.2.4 EVALUATION RESULTS[7]

The multidimensional approach of our study results in an empirical characterization of the performance limitations and trade-offs of the different LPWA technologies,

---

[6]The antenna direction was decided after optimal performance observation through exhaustive experimentation.

[7]Partially reprinted from *Proceedings of the 2016 ACM Wireless Network Testbeds, Experimental Evaluation, and Characterization (WiNTECH) Workshop.* Kartakis, Sokratis, et al. "Demystifying low-power wide-area communications for city IoT applications". Copyright 2016, with permission from WiNTECH.

Communication Optimization and Edge Analytics for Smart Water Grids 83

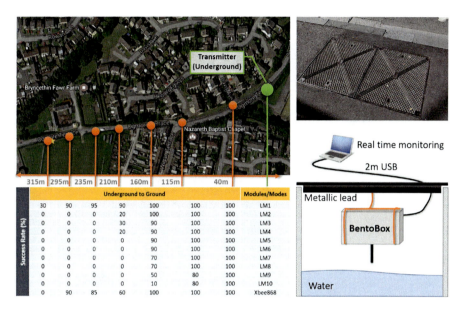

Figure 3.7: Underground to ground experimental environment in Welsh Water network.

and allows us to define the appropriate communication setup for smart water grid applications and deployment contexts. To achieve that, during the aforementioned experiments, BentoBoxes report the following metrics: data rate, success rate, transmission energy and power, and communication range. This section presents the experimental results[8] which are in line with communication theory, intuition, and initial controlled lab tests, while providing useful insights. In addition to the deployment environments, Figure 3.6 and Figure 3.7 illustrate the success rate and the maximum range of each communication module and configuration setup (i.e. LoRa modes) as [23] analyzes.

To summarize the results of [23] and provide an easier interpretation of the experimental results, Figure 3.8 and Figure 3.9 aggregate key information for the different communication modules under various environments in terms of range, success rates (reliability), power and energy consumption, and data rates. To ensure the comparison validity, the presented results refer to the experiment with 10 byte payload, a size which all modules can support. Based on these figures three main lessons are inferred:

Insight 1: From Figure 3.9, we infer that the data rate achieved by the XBee868 module is equivalent to that of LoRa LM5. Figure 3.8 illustrates that for semi-LoS communication, XBee868 has a much longer range (around 2km) than LoRa LM5

---

[8]Note that the results presented here are specific to our particular experimental setup; other environments and hardware support for these communications technologies may produce different figures.

84 Smart Water Grids: A Cyber-Physical Systems Approach

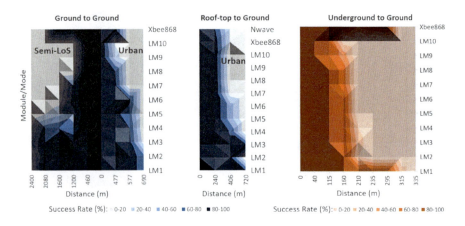

Figure 3.8: Comparison of success rate and range between [23] (Fig. 5) and Welsh Water network experiments.

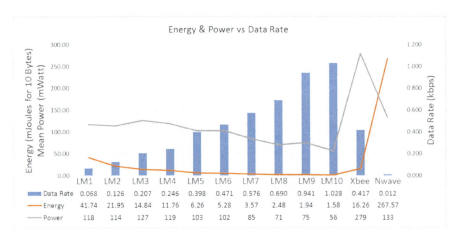

Figure 3.9: Experimentally measured data rate, power and energy consumption (Fig. 6b in [23]).

(around 1.3km). This implies that the narrow-band XBee868 can provide higher data rates for the same link distances. However, the main drawback of XBee868 is the power consumption which is 300mW while LoRa utilizes only 20-40mW power.

Insight 2: Figure 3.9 presents that XBee868 consumes significantly higher power (279mW) compared to the other modules. However, due to the fast transmission period, the total energy consumption is at the same level as LoRa LM3. Due to the fact that energy consumption is the main indicator of battery life, the above implies that battery-operated nodes which use XBee or LoRa LM3 will operate for the same duration.

Insight 3: Based on Figure 3.9, the nWave module consumes similar average

Communication Optimization and Edge Analytics for Smart Water Grids **85**

power as the LoRa module in transmission mode. However, the nWave transmission process lasts 34 times longer than XBee868 and similarly, 5 times and 85 times than LoRa LM1 and LM10, respectively, for the same payload. As such, nWave module consumes 94% more energy in transmission modes compared to the power hungry XBee868 and 84% to 99.5% against LoRa.

Due to the promising results of U2RT and U2U deployments, we extend the work in [23] and conduct experiments placing BentoBox into real water network chambers in Welsh Water network. The difference between U2RT and this deployment was the thickness of cast lead (i.e., five times thicker) and the existence of water in the chamber. These two factors reduce the communication performance by almost 25% for LoRa considering both range and success rate, achieving 315m in LM1 (see Figure 3.7). On the other hand, Xbee868 performed worse by achieving 295m with high reliability. At this point, we have to notice that between the points in 160m and 210m, the success rate of the modules and especially LM5 to LM10 drop significantly. The reason for this drop is the difference in altitude. The point at 160m has the highest altitude. Overall, the results in Figure 3.7 reveal that reliable communication based on LPWA modules under the context of a smart water network can be achieved only if a base station exists in every 160m. However, this limitation can increase the deployment cost due to the high number of required nodes. On the other hand, by utilizing LPWA modules, utility companies can reduce cost by avoiding overground deployments and by changing all the metallic leads with other materials (i.e. plastic) within a city to accommodate other wireless communication techniques with less signal penetration capabilities. An appropriate solution to incorporate LPWA modules in a smart water grid, especially within an urban environment, is the creation of hybrid communication smart meters for each household which exploit the theory of opportunistic networking and relay the information from underground to a base station or cloud.

The last aspect of our evaluation is related to the impact of payload size on communication. Figure 3.10 illustrates the average success rate of the 100 independent experiments for each LM1 to LM10, and Xbee868 per location per payload size as described in Table 3.3 (i.e. for LoRa $\{10, 100, 180, 250\}$ and Xbee868 $\{10, 50, 100\}$ bytes)[9]. In the cases of LM1 (long-range configuration), LM2, LM3, and Xbee868, the payload size has greater impact than LM4 to LM10 for distances greater than 100m. Specifically, Xbee868 has a 70% higher success rate in the longest range when the payload size is 10 bytes compared to 50 bytes. Similarly and in the same distance, LM1 performs 60% more reliably with a 10 byte payload size than 100 bytes. Even in the LM10 case, in which the communication has similar performance for each payload size, the smallest payload can achieve around a 30% better success rate. This fact results in the following:

Insight 4: The communication modules perform more reliably (i.e. up to 70%) with smaller packet/payload sizes (i.e. up to 10 bytes). For this reason, the split of information into relatively small chunks is necessary.

---

[9]The evaluation required more than 35,200 repetitions and 10 hours of continuous experimentation (excluding the setup time) to ensure similar environmental and weather conditions.

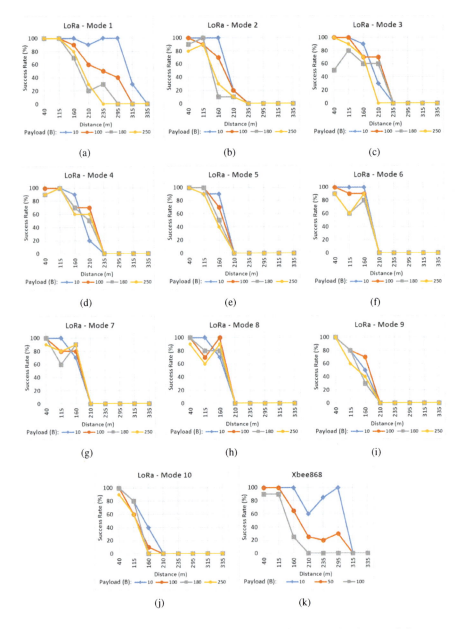

Figure 3.10: Success rate per payload size and communication module.

Another observation that can be derived from Figure 3.10 is that in LM3 and LM4 the success rate is lower for 10 byte payload size than the others at 210m. The reason for this drop is the condition of the environment during the experiments. Specifically, during that time the traffic and the number of people around the area was increased dramatically due to the finish of a nearby school. From this fact, we can derive that data recovery and estimation mechanisms are necessary for any kind of city scale wireless deployments.

The evaluation results of this section reveal that LPWA technologies can enable the communication in city scale deployment, such as smart water grids, by providing lower energy and less hardware infrastructure compared to other conventional communication solutions, i.e., WiFi or 4G, which are inappropriate for deployments in harsh environments and low power hardware infrastructures. However, the main drawback of the LPWA technologies is the reduction of the communication data rates. In the case of relatively large amounts of transmitted data (e.g. more than 1MB), the slow data rates may lead to high energy consumption due to the long transmission time. Therefore, the incorporation of data reduction techniques (i.e. compression) and edge processing with LPWA technologies is vital.

3.3 LOSSLESS COMPRESSION AND COMPRESSIVE SENSING

Smart water grids ideally consist of low power, inexpensive sensor nodes, which are responsible for retrieving data from one or more sensors, make computations, store data in local memory, and send them to the network. The nodes of a smart water grid can be installed in a variety of environments (e.g. with high pressure, temperature, radiation, and electromagnetic noise). However, node installation in harsh environments like underground or underwater [53], makes access impossible. The majority of sensor nodes are battery-driven and consequently the per-node and the total reduction (especially in distributed architectures) of power consumption is vital to its lifetime. This section, provides a variety of data reduction techniques that minimize the energy consumption over wireless sensor networks and can be applied to smart water grids.

According to [7], the wireless transmission of a single bit can require over 1000 times more energy than a single 32-bit computation. Therefore, many researchers have approached data reduction from many different points of view, by using: (a) probabilistic/ machine learning techniques (data fusion, aggregation), and (b) data compression (lossless and lossy).

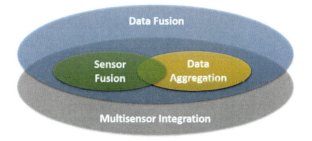

Figure 3.11: Data fusion terms.

Data fusion (Figure 3.11) is the process of gathering and transforming data into an accurate, consistent, and useful representation by increasing the significance of such a mass of data [36]. In this process the system overcomes the sensor failures, technological limitations, and spatial and temporal coverage problems, while in WSN the main objectives are the improvement of accuracy and energy saving. Consistent with [36] and [3] the data fusion methods can be categorized by their objective in: inference, estimation, feature maps, abstraction sensors, aggregation, and compression.

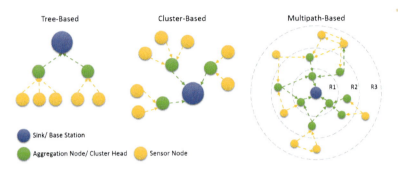

Figure 3.12: Aggregation routing approaches.

Data aggregation techniques are considered the main data fusion method, which aims to solve the problem of data reduction in WSNs. These techniques consist of routing protocols (tree-based (or classical), cluster-based, multipath-based and hybrid-based; Figure 3.12), the aggregation function (Temporal coherency-aware in-Network Aggregation (TiNA) [46], Data Aggregation and Dilution by Modulus Addressing (DADMA) [15], Adaptive Application-Independent Data Aggregation (AIDA) [20], Synopsis Diffusion Framework [37], and the Quartile Digest [48]), and data representation (this refers mostly to data compression). Routing protocols are responsible for selecting the next hop based on data content; that is they are data centric (e.g., position of the most suitable aggregation point, data type, priority of information), and to provide timing strategies, while the aggregation functions fuse the data. The efficiency of these algorithms depends on the correlations among the data generated by different information sources. The correlation algorithms can be classified in spatial (the pattern of sensor readings is the same or similar), temporal (the pattern of future sensor readings is equal to or similar to that of previous sensor readings in a sensor), and semantic (the correlation of the data reported by multiple sensors due to the semantics of the query).

3.3.1 LOSSLESS AND LOSSY COMPRESSION

Until this point, we have overviewed data fusion techniques, data aggregation methods (including routing approaches and data aggregation functions), and how to enhance them by using correlation algorithms. In addition, data reduction can be achieved by using appropriate data representations. More specifically, the meaning of data representation is how to organize the resulting data from the data fusion process in order to minimize the data size. Compression techniques are a solution to the representation problem. There are two different compression techniques: (a) lossless, and (b) lossy.

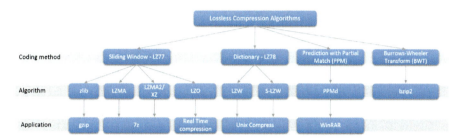

Figure 3.13: Lossless compression methods, algorithms, and applications.

Lossless compression occurs when compressed data can be perfectly reconstructed on decompression, for example the compression of a text or binary file. Figure 3.13 presents the four fundamental coding methods (Sliding Window-LZ77 [56], Dictionary-LZ78 [57], Prediction with Partial Match-PPM [17], and Burrows-Wheeler Transform (BWT) [10]), and the standard algorithms and applications which are based on these methods.

The selection of the appropriate lossless data reduction technique depends on the hardware limitations (i.e., memory size or cache availability, CPU computational power), power consumption per part of the node, the node topologies (centric or distributed), the raw data types and content (i.e., bytes, text, web data), and the correlations between raw data. Thus, the analysis, simulation, and evaluation are necessary to understand if the above techniques are applicable because the optimal compression techniques in the powerful computers are not necessarily feasible for the smart water grid and generally in restricted wireless sensor networks.

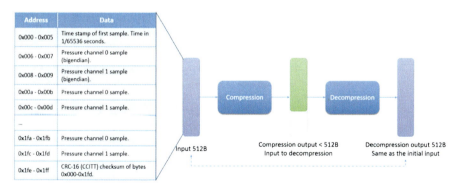

Figure 3.14: Lossless compression algorithms evaluation process.

By using Atmel Studio 6 simulator, we wrote a program to simulate the process described in Figure 3.14. In this process, the system structures the data in 512 bytes including: (a) the timestamp of the first measurement, (b) data from two input channels, and (c) CRC. This structure was used as input to compression algorithms which compressed the data to less than 512 bytes, then the data was decompressed and the final output was compared with the initial input. To evaluate the memory usage of each compression algorithm, water pressure data from the large-scale testbed [1] was used.

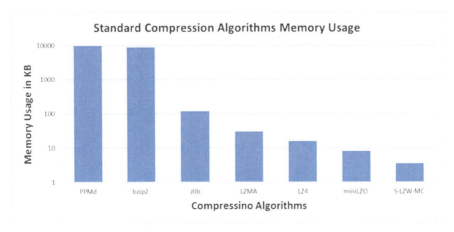

Figure 3.15: Standard compression algorithms memory usage.

Based on Figure 3.15, the quantitative results of memory consumption per algorithm are the following: PPMd consumes more memory than the other algorithms (10MB), bzip2 [11] requires around 9MB of memory and data segments of 100KB, zlib requires 120 KB, for LZMA [2] the lowest initial configuration values requires 6.5KB and the minimum dictionary size is 4096 B. In total, during the execution, the minimum memory usage can be 30KB (the decompression process 1/10 of the compression memory usage). LZ4 requires at least 16 kB of memory, miniLZO [39] requires memory only during the compression (dictionary creation) and the minimum memory requirement is 8192B, and the minimum required memory for S-LZW-MC [41] is 3.25KB. This algorithm is an embedded version of LZW, therefore further evaluation is not required.

Current ultra low power MCUs, that can be utilized in battery-driven sensor nodes, have 64Kbytes memory [6]. Therefore, only the lightweight compression algorithms are appropriate for smart water grids. Based on the above memory requirements, only MiniLZO [30] (coding method is sliding window-LZ77) and S-LZW-MC [42] (coding method is dictionary-LZ78), which require 8.192KB and 3.250KB of memory correspondingly (Figure 3.15), are applicable. We evaluated the compression rate of each algorithm using three different real pressure datasets of the smart water network [1]. Each dataset consists of 5.5 million data pairs (i.e., timestamp, pressure value). Similarly to the memory usage evaluation (see Figure 3.14),

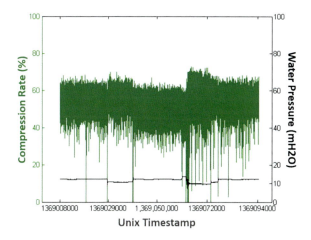

Figure 3.16: Raw water pressure data and compression rates (Fig. 1 in [25]).

the input stream is converted into 512-byte packets, with the following structure: (a) timestamp (8-byte double data type), (b) 62 measurements (8-byte double * 62 = 469 bytes), and (c) CRC (8-byte double data type). Both MiniLZO and S-LZW-MC performed similarly (i.e., average 54% and 55% accordingly) by achieving almost the same compression rates for the datasets. Figure 3.16 illustrates the raw pressure data (black line) and compression rates (green line) of a representative dataset.

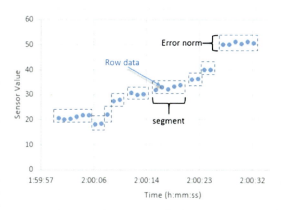

Figure 3.17: Lossy compression and raw data segmentation.

In addition to lossless compression, lossy compression constitutes another data reduction technique. During the lossy compression process, some loss or degradation of the data can be tolerated, for instance in the image, sound, and video compression. Lossy compression often exploits spatial-temporal correlations to compress the data within a specific error norm (approximation error). Furthermore, dimensionality data reduction techniques like standard orthogonal transformation methods (e.g., Fourier, wavelet transform) can be included with lossy compression. The main goal of lossy compression algorithms is to cluster raw data into the longest segments, based on an error norm (Figure 3.17). Instead of returning the raw data values, the algorithms (like Swing, Polynomial Regression, and MidRange) return the start and the end time of data segments, and a representative value (or other attributes like the slope of regression line) of this segment. Compressive sensing constitutes another state-of-the-art lossy compression technique which will be analyzed in a later section.

Lossless compression techniques provide the most accurate sensor values compared to lossy compression techniques. However, according to [44], lossless compression techniques cause more energy and node resource consumption than lossy approaches for the following reasons: (a) lossless compression and decompression are usually more computationally intensive than lossy, (b) indexing cannot be applied in lossless compression, and (c) some lossless compression techniques (like gzip) achieve lower compression rates than lossy techniques (the specific evaluation result was 50%).

Overall, there is no perfect answer to which data reduction technique is the optimal or appropriate solution for different scenarios in smart water grids. The reason is that the data reduction techniques are strongly coupled with the scenario or application. For example, in centralized network approaches the most suitable routing and data aggregation algorithms are different from the distributed network approaches because in the centralized architecture the system has more information about the topology and correlation relationships between the sensor nodes. In addition, in applications with low precision tolerance lossy compression could be inappropriate in spite of the data reduction efficiency. For all these reasons before the selection of a specific technique, the physical/ hardware limitations, requirements, and architecture should be declared a priori.

3.3.2 COMPRESSIVE SENSING

For decades, the sampling process has been largely dominated by the classical Nyquist-Shannon theories. However, several studies have shown that many natural signals are amenable to highly sparse representations in appropriate transform domains (e.g. wavelets and sinusoids) [18, 33]. Compressive sensing (CS) provides a powerful framework for simultaneous sensing and compression [12], enabling a significant reduction in the sampling and computation costs on a sensor node with limited memory and power resources.

Figure 3.18: Compressive sensing process (Fig. 2 in [25]).

According to the theory of CS (see Figure 3.18), a signal having a sparse representation in a suitable transform domain (e.g. FFT, wavelet, cosine) can be reconstructed from a small set of incoherent random projections. Specifically, consider a finite length, discrete time signal $\mathbf{x} \in \mathbb{R}^N$ and a matrix $\mathbf{\Psi} \in \mathbb{R}^{N \times L}$ whose columns correspond to a *transform basis*, in general overcomplete (i.e. $N \leq L$). Alternatively, in the literature of CS (i.e. [47]), researchers refer to $\mathbf{\Psi}$ as *dictionary* and define it as a set of elementary waveforms used to decompose the signal.

Definition 1. *Sparsity of a signal is defined as the number of zero-valued (or smaller than a certain minor threshold) elements divided by the total number of elements in the signal under some representation.*

The signal $\mathbf{x}$ has a sparse representation over a known $\mathbf{\Psi}$, if a sparse coefficient vector $\mathbf{w} \in \mathbb{R}^L$ exists such that $\mathbf{x} = \mathbf{\Psi}^T \mathbf{w}$, with $\|\mathbf{w}\|_{l_0} = K << L$, where $\|\cdot\|_{l_0}$ indicates the non-zero elements in $\mathbf{w}$ and $\mathbf{\Psi}^T$ denotes the inverse transform. Based on CS, an unknown vector $\mathbf{x}$ is recovered from an underdetermined system of linear equations

$$\mathbf{y} = \mathbf{\Phi}\mathbf{x} = \mathbf{\Phi}\mathbf{\Psi}^T\mathbf{w}, \tag{3.1}$$

where $\mathbf{y} \in \mathbb{R}^M$ is the vector of compressive measurements and $\mathbf{\Phi} \in \mathbb{R}^{M \times N}$ with $K < M < N$ is the sensing or measurement matrix. The sample rate SR can be derived from $\mathbf{\Phi}$ by calculating the following trace:

$$SR = Tr(\mathbf{\Phi}^T\mathbf{\Phi}) \tag{3.2}$$

Due to the fact that $M < L$, the system (3.1) is not invertible and has an infinite number of solutions from which the sparsest are the most important. However, direct minimization of $\|\mathbf{w}\|_{l_0}$ is an NP-hard problem and therefore all the l_p-norms for $p \geq 1$ are special cases of interest for convexity. Particularly, l_1-norm is the most popular having the tendency to sparsify the solution. Based on [16], a sparse signal $\mathbf{w}$ can be recovered from only $M = \mathcal{O}(Klog(\frac{N}{K}))$ non-adaptive linear projections onto a second measurement basis by solving the following relaxation

$$\min_{\mathbf{w} \in \mathbb{R}^L} \|\mathbf{w}\|_{l_1} \; subject \; to \; \|\mathbf{y} - \mathbf{\Phi}\mathbf{\Psi}^T\mathbf{w}\|_{l_2} \leq \varepsilon, \tag{3.3}$$

where $\|\mathbf{w}\|_{l_1} = \sum_{i=0}^{L} |\mathbf{w}_i|$ and $\varepsilon > 0$. Solving the (3.3) optimization problem, the estimated sparse coefficient vector $\widehat{\mathbf{w}}$ can be used to reconstruct the original signal by simply taking the inverse transformation

$$\widehat{\mathbf{x}} = \mathbf{\Psi}^T \widehat{\mathbf{w}}. \tag{3.4}$$

In the literature, two main approaches have been used to solve the (3.3) optimization problem. In the first approach, the so-called Basis Pursuit (BP) techniques [13, 14, 16] transform the problem to Linear Program (LP) [9]. In the second approach, the Matching Pursuit (MP) greedy methods [34, 52] use an iterative way to solve the optimization problem. In this study, the NESTA algorithm is used, which was shown to achieve a very good trade-off between reconstruction accuracy and computation time [7]. We emphasize that the scope of this section is to illustrate the efficiency of the CS framework in achieving highly compact, yet very accurate, representations of real high-resolution sensor data recorded in water distribution networks. As such, an exhaustive comparison with the various reconstruction algorithms for finding the optimal solution is left as a separate study.

In the subsequent experimental evaluation, the discrete wavelet transform (DWT) [33] will be applied to the raw sensor data due to its computational efficiency and sparse representation accuracy. However, we note that the CS-based approach is generic and can be applied with alternative sparsifying transformations (e.g. cosine, FFT), other than the DWT.

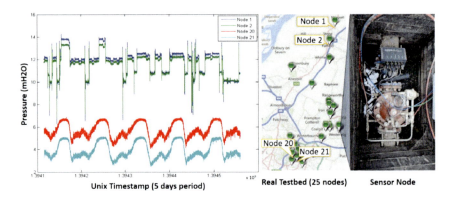

Figure 3.19: Large-scale testbed [1] and pressure data from 4 sensor nodes.

The performance of the CS-based approach is evaluated and compared with well-established lossless compression techniques, as has been presented earlier in this section, for compressing data streams recorded by a set of sensors deployed in the smart water distribution network of [1]. More specifically, the available dataset consists of high sample-rate pressure data (64 samples per second) from 25 sensor nodes for a 170-day period. For the sake of brevity, this section presents the evaluation results for four pressure data streams from an equal number of sensor nodes as shown in Figure 3.19.

In the previous test case, data were compressed using various lossless compression techniques. The applicable compression method for the current hardware infrastructure was MiniLZO [30], which uses sliding window as its coding method (LZ77). We repeated the experiments of the previous section by dividing each pressure data stream into non-overlapping windows of length N=1024 (or, equivalently, 1024/64 = 16 sec) and applied MiniLZO algorithm for each data chunk achieving 55% average compression rate.

Despite the fact that lossless compression allows the original data to be perfectly reconstructed from the compressed data, this is typically achieved at lower compression rates. To overcome these limitations, the acquired data are compressed via

Communication Optimization and Edge Analytics for Smart Water Grids

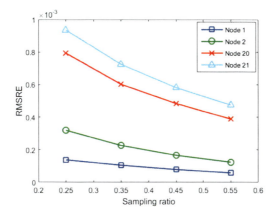

Figure 3.20: Average RMSRE per CS sampling rate for the four nodes.

highly reduced sets of random measurements (ref. equation (3.1)) and reconstructed by solving the constrained optimization problem expressed by equations (3.3) and (3.4). To this end, each pressure data stream is divided again into non-overlapping windows of length N=1024. Then, a multi-scale decomposition in 10 levels is applied to each window using the DWT with the "db4" wavelet function. Subsequently, the resulting wavelet coefficients are projected onto the rows of a measurement matrix Φ with i.i.d. standard Gaussian entries (mean 0, and standard deviation 1). The efficiency of the proposed CS-based scheme is tested for a varying number of measurements, that is rows of Φ, $M = \rho \cdot N$, where $\rho \in \{0.25, 0.35, 0.45, 0.55\}$. This is equivalent to compressing the data in each window at compression rates $\{75\%, 65\%, 55\%, 45\%\}$, respectively.

The reconstruction accuracy is measured in terms of the root mean squared relative error (RMSRE). Specifically, if $\mathbf{x}$ and $\widehat{\mathbf{x}}$ are the original and reconstructed data, respectively, their RMSRE is defined as

$$RMSRE(\mathbf{x}, \widehat{\mathbf{x}}) = \sqrt{\frac{1}{N} \sum_{j=1}^{N} \left(\frac{x_j - \widehat{x}_j}{x_j}\right)^2}. \quad (3.5)$$

Figure 3.20 shows the RMSRE, averaged over all individual windows, as a function of the CS sampling rate for each one of the four nodes. As can be seen, the reconstruction accuracy is already high even for high compression rates (75%), while it improves (that is, the RMSRE decreases) as the compression rate decreases (or, equivalently, the CS sampling rate increases).

Finally, to illustrate the approximation accuracy of the original pressure data, Figure 3.21 shows segments of the original and reconstructed data streams for sensor node 2. As expected, the approximation accuracy improves as the CS sampling rate increases (see Figure 3.21(b)); however, a highly accurate reconstruction is achieved even for low sampling rates (see Figure 3.21(a) - approximately ± 0.0025 mH2O er-

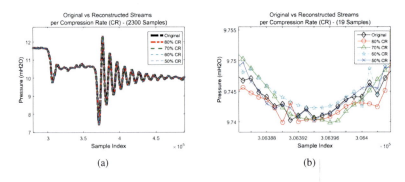

Figure 3.21: Original and reconstructed segments for sensor node 2 using NESTA algorithm (Fig. 4 in [25]): (a) 2300 samples (36 sec), (b) 19 samples (296 msec).

ror) with a result of significantly higher compression rates than the lossless approach which was 55% and consequently resulted in proportional battery life extension.

3.3.3 EDGE ADAPTIVE COMPRESSION

Up to this point, we have proposed two solutions to reduce data volume: (a) lossless compression [27] and (b) lossy compression by using the powerful framework of compressive sensing [28], both of which minimize the communication by covering high information level needs to the server-side. As has been described, each compression technique has trade-offs. Specifically, in lossless compression, the initial stream can be reconstructed completely without losing information, while compressive sensing introduces a reconstruction error. On the other hand, compressive sensing can significantly reduce the amount of data, and consequently the communication. Lossless compression places an upper bound on the compression performance.

This section summarizes the benefits of the lossless compression and compressive sensing combination in the communication of smart water grids as presented in our previous work. Specifically, [25] proposes a system that balances the trade-offs of the two types of compression methods and presents an optimal algorithm which selects the best compression mode in a dynamic and distributed fashion by minimizing the reconstruction error, given the current sensor node battery state. The main idea behind this algorithm is the following:

In every communication period t_j, the sensor node selects to transmit losslessly compressed data if and only if the required energy to transmit this data is lower than the current battery level $B_i(t_j)$ weighted by a gain $K_p \in (0, 1]$ [10], or in other words if the following inequality holds:

$$K_p \cdot B_i(t_{j-1}) > (1 - CR_i^{ls}(t_j))E_i^{tr}(t_j) + E_i^{in}(t_j) \qquad (3.6)$$

---

[10] K_p indicates a battery level threshold in which the system selects compressive sensing as the compression technique. Otherwise, the systems would select compressive sensing only when the battery is close to depletion level.

where $CR_i^{ls}(t_j)$ represents the achieved lossless compression rate, $E_i^{tr}(t_j)$ is the energy consumption to transmit the uncompressed information, and $E_i^{in}(t_j)$ is the energy consumption of sensor node due to other operations over the last period, i.e., sensing and idle.

If inequality (3.6) is violated then the sensor node is unable to transmit losslessly compressed data due to lack of energy and has to select the optimal compression rate $CR_i(t_j)$ for the compressive sensing approach by solving a multi-objective optimization problem with two confronted goals. The objective functions of the system are related to the reconstruction error (i.e. RMSRE) and the new battery level $B_i(t_j)$ and can be described as follows:

$$\min_{CR_i^{ls}(t_j)\leq C_i(t_j)\leq CR_{max}} \quad f_1 = RMSRE(CR_i(t_j)) \tag{3.7}$$

$$\max \quad f_2 = B_i(t_j) \tag{3.8}$$

$$\text{subject to} \quad \forall i \in \mathcal{S}, \forall j \in \mathbb{N}^+ t, RMSRE_i(CR_i(t_j))$$
$$\leq RMSRE_i(CR_i'(t_j)) \tag{3.9}$$

$$\forall i \in \mathcal{S}, \forall j \in \mathbb{N}^+ t, 0 \leq B_i(t_j) \leq B_{max} \tag{3.10}$$

$$B_i(t_j) = |B_i(t_{j-1}) - CR_i(t_j)E_i^{tr}(t_j)$$
$$-E_i^{in}(t_j)|_+ \tag{3.11}$$

The $RMSRE(CR_i(t_j))$ function maps the produced RMSRE for the specific compression rate $CR_i(t_j)$. The constraint (3.9) maintains the monotonically non-decreasing ratio between the CR and the error[11] where $CR_i(t_j) \leq CR_i'(t_j)$. In order to perform the compression and transmission, the battery must not run out of the capacity at any point in time. The battery level at the specific time instant must not exceed the maximal allowed capacity. As a consequence of those system needs, the inequality constraint (3.10) is formulated. Equality constraint (3.11) represents the battery dynamics for each communication period. The above algorithm is executed locally in every node for every communication period. In [25], we solve the optimization problem by using multi-objective genetic algorithm.

Table 3.4, aggregates the savings of the algorithm in terms of battery life and communication while representing the introduced error. Based on the evaluation results with real water pressure data, this algorithm can reduce the communication by around 66% and extend the battery life by 46% compared to the traditional periodic communication approaches without compression capabilities. Furthermore, the adaptive compression algorithm performs 37% and almost 30% better in terms of battery life and communication, respectively, compared to a smart water grid in which the data are being compressed by using lossless compression techniques. Lastly, the adaptive algorithm introduces $2.80 \cdot 10^{-4}$ RMSRE, which is considered significantly low for the smart water applications.

To sum up, until this section, we described the applicability of LPWA communication technologies and the necessity of data reduction techniques to overcome

---

[11] As the compression rate increases the error increases as well (see Figure 3.20).

Table 3.4

Savings compared to periodic communication (a) without compression and (b) with only lossless compression

Compared to	Optimal Compression Selection Savings		
	Battery life	Communication	Error (RMSRE)
Without Compression	46%	66.37%	$-2.80 \cdot 10^{-4}$
Only Lossless Compression	37%	29.15%	

Reprinted from *Proceedings of the 2016 Cyber-physical Systems for Smart Water Networks (CySWater) Workshop*, Kartakis, Sokratis, et al. "Energy-based adaptive compression in water network control systems," Copyright 2016, with permission from CySWater.

their slow data rate performance. In addition to the data reduction techniques, edge processing can improve the reduction of information that has to be transmitted wirelessly over smart water grids. Under this context, the following section presents an edge anomaly detection based on compression rates and its incorporation into an offline graph-based burst localization algorithm. Having the ability to detect anomalies to the edge allows the system to minimize the communication and consequently the energy consumption significantly.

3.4 EDGE PROCESSING

Due to the low data rates, LPWA technologies can be enabled in smart water grids only when they are combined with data reduction techniques, i.e., compression. To reduce the information further in-node or edge processing is necessary. In this section, we discuss the impact of edge processing in the communication within a smart water grid by presenting our edge anomaly detection algorithm [27] and its combination with our offline burst localization algorithm [29].

3.4.1 ADAPTIVE THRESHOLDING-BASED REAL-TIME ANOMALY DETECTION EXPLOITING COMPRESSION RATES

In the previous section, we introduce data reduction techniques to reduce the energy cost related to the communication without sacrificing the precision of the data. In the same section, we evaluated a number of lossless compression algorithms. During this evaluation, using real data, we observed a correlation between compression rate and data value fluctuation, and from this derived a scheme that enables the identification of transients or failures in the smart water grids. This means that instead of compressing raw sensor data and sending it to servers to be decompressed and then analyzed for anomalies, we can use the compression rate to detect anomalies and outliers directly on the sensor node (i.e., the edge). This is faster, more lightweight

Communication Optimization and Edge Analytics for Smart Water Grids 99

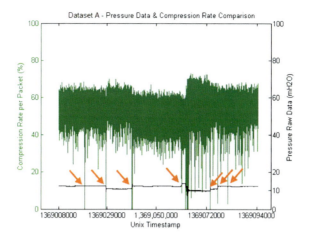

Figure 3.22: Compression rate and raw water pressure data correlation.

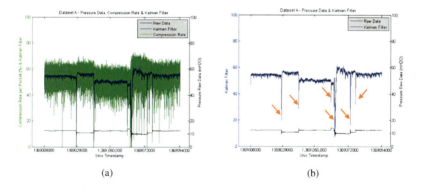

Figure 3.23: Removing noise from compression rate stream.

and provides early indications of an issue, which can be fed directly into the control function without having to communicate via servers saving time and energy.

By following the same methodology of the previous section, we apply MiniLZO [30] (coding method is sliding window-LZ77) in different real water pressure datasets (e.g. Figure 3.22) of the smart water network [1]. The results of the compression rate per packet can be seen in the chart in Figure 3.22 (green line), while the original water pressure data is also overlaid (black line). From Figure 3.22, we observe that data anomalies can be indicated (orange arrows) from the drops of the compression rates. We verified offline that the anomalies indicated on our graphs were true due to changes in actuators (i.e., valves) from water technicians. At these points, due to high pressure fluctuation, the compression algorithm is unable to compress the data so the compression rate falls to 0%. In Figure 3.22, the drop in compression rate isolates the areas of raw data where the fluctuation pattern is changeable.

Based on the above observation, we derived an adaptive thresholding-based anomaly detection algorithm in [27] to automatically detect significant changes in compression rate and therefore identify the timestamps of anomalies. In this algorithm, the first step is to remove the noise of the compression rates. Therefore, we apply a one-dimension Kalman filter [31, 40] to the compression rates indicated in Figure 3.23b with a blue line.

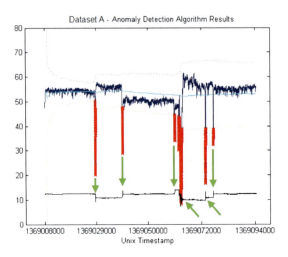

Figure 3.24: Anomaly detection results over water pressure data (Fig. 2b in [29]).

After noise removal, the anomalies can be detected accurately because according to Figure 3.23b (which presents the x value of Kalman filter state and raw data) the anomalies are presented as great drops (orange arrows). The drops are being detected by using the average and the standard deviation of the compression rates. Figure 3.24 illustrates the produced thresholds (upper and lower bound). Every measurement that lies outside these thresholds is considered as an anomaly.

The accuracy of the proposed algorithm depends on the configuration of the different parameters. [27] presents a mechanism based on supervised learning to configure the algorithm optimally by minimizing the fault-positives and true-negatives. Based on the evaluation results in [27], we show that our edge anomaly detection algorithm reduces computation by 98%; data volumes and, consequently, the communication by 55%; while requiring only 11KB of memory at runtime (including the compression algorithm). Furthermore, due to the fact that the compression rate is a normalized value within $[0, 1]$, the algorithm can be applied to any kind of data. We verified this assumption by applying our algorithm to water pressure data in this section while pipe vibration is used in the following section.

3.4.2 INCORPORATING EDGE ANALYTICS AND OFFLINE GRAPH-BASED BURST LOCALIZATION ALGORITHM

Having the ability to detect anomalies into the edge, the next step is to find ways to exploit this feature by considering the requirements of offline big data algorithms. Under this context, in [29], we identified the requirements of our offline burst localization algorithm and, combined with our edge processing, we minimized the communication. The overview of this approach follows.

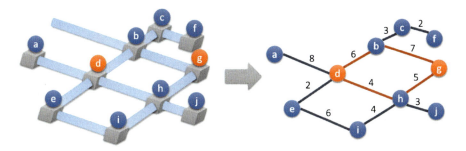

Figure 3.25: Modelling a water network (left) as a weighted graph (right)

Our offline algorithm considers a smart water network as an attributed graph $G = (V_J \cup V_S, E, A)$, where V_J is a vertex set of pipe junctions, V_S is a vertex set of deployed sensor locations, E denotes an edge set of pipe sections connecting two vertices, and A carries the length of each pipe section. Given the graph representation G of the water network, [29] defines the "average distance" matrix $\mathbf{S}^{(L)}$ within L hops ($L = 2, 3, \cdots$) for each sensor node which intuitively captures the weighted average distance within L hops between vertices u and v.

Based on realistic values for a water network, an example of the "average length" matrix $\mathbf{S}^{(5)}$ can be obtained as follows:

$$\mathbf{S}^{(5)} = \begin{array}{c} a \\ b \\ c \\ d \\ e \\ f \\ g \\ h \\ i \\ j \end{array} \begin{bmatrix} \begin{array}{cccccccccc} a & b & c & d & e & f & g & h & i & j \\ 0 & 17.902 & 20.609 & 14.996 & 14.162 & 19.000 & 22.707 & 15.729 & 19.902 & 18.729 \\ 17.902 & 0 & 9.667 & 12.999 & 12.214 & 8.512 & 13.998 & 14.512 & 18.424 & 17.512 \\ 20.609 & 9.667 & 0 & 12.609 & 14.973 & 6.159 & 13.512 & 17.306 & 17.666 & 17.000 \\ 14.996 & 12.999 & 12.609 & 0 & 9.395 & 14.609 & 14.707 & 10.792 & 11.902 & 10.729 \\ 14.162 & 12.214 & 14.973 & 9.395 & 0 & 13.000 & 17.465 & 11.434 & 11.271 & 14.434 \\ 19.000 & 8.512 & 6.159 & 14.609 & 13.000 & 0 & 15.512 & 16.000 & 19.666 & 19.000 \\ 22.707 & 13.998 & 13.512 & 14.707 & 17.465 & 15.512 & 0 & 11.853 & 12.902 & 11.772 \\ 15.729 & 14.512 & 17.306 & 10.792 & 11.434 & 16.000 & 11.853 & 0 & 10.273 & 8.951 \\ 19.902 & 18.424 & 17.666 & 11.902 & 11.271 & 19.666 & 12.902 & 10.273 & 0 & 10.382 \\ 18.729 & 17.512 & 17.000 & 10.729 & 14.434 & 19.000 & 11.772 & 8.951 & 10.382 & 0 \end{array} \end{bmatrix}$$

Overall, $\mathbf{S}^{(L)}$ can capture multiple paths of different lengths between every two sensor locations by fully exploiting water network topology information. Thus, if the "average length" $\mathbf{S}^{(L)}$ is used to quantify the wave distance propagated from a burst location to a sensor location, water loss events can be positioned more accurately.

Let's assume that a burst has occurred at location x. Then, the anomaly is observed by two sensor nodes, u and v, at different times t_u and t_v, respectively. Based on this anomaly arrival time difference, [29] proves that the burst can be localized accurately by forming for multiple sensor nodes u and v the following "hyperbolic curves" and finding their intersection:

$$\hat{X} = \arg(\text{top-}k)\min\{|\bar{v}(t_u - t_v) - ([\mathbf{S}^{(L)}]_{u,x} - [\mathbf{S}^{(L)}]_{v,x})|\} \quad (3.12)$$

where k is often set to 3–5 in practice. The above graph-based algorithm requires only the anomaly arrival time per node. Due to the ability to detect the anomalies on the edge, we exploit this fact and force the sensor nodes to transmit only their anomaly arrival time. Figure 3.26 illustrates the new end-to-end system architecture.

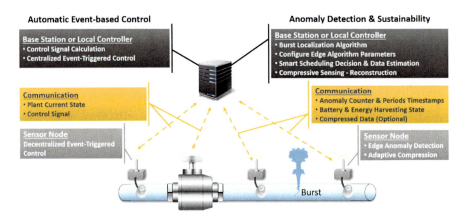

Figure 3.26: System architecture and future research directions for the water grids and communication.

Table 3.5
Edge Processing Evaluation Results

	Average Compression Rate	Observed Anomalies	Real Anomalies	Anomaly Detection Accuracy	Communication Savings (Transmit Compressed Data)	Communication Savings (Transmit only Timestamps)
Node A	28.01%	18	20	90%	79.26%	99.88%
Node B	30.06%	11	12	91%	87.69%	99.93%
Node C	39.25%	8	8	100%	92.22%	99.95%

Reprinted from *Proceedings of the 2016 Internet-of-Things Design and Implementation (IoTDI) Conference*, Kartakis, Sokratis, et al. "Adaptive edge analytics for distributed networked control of water systems," Copyright 2016, with permission from IoTDI.

The setup of Welsh Water test rig [29] was used to evaluate our end-to-end system by exploiting pipe vibration data. Table 3.5 presents the evaluation results from three sensor nodes which reveal more than 99% communication reduction and 0.5m

Communication Optimization and Edge Analytics for Smart Water Grids **103**

burst localization accuracy. Furthermore, in the case of the raw data transmission, the communication savings are more than 79%.

Based on the evaluation results, the incorporation of edge processing and offline algorithms is crucial to the communication within smart water grids. Additionally, the 99% reduction of communication enables the LPWA communication technologies to be applicable to the harsh and underground environments of smart water grids by maximizing the battery life of the sensor nodes.

3.5 CONCLUDING REMARKS AND FUTURE DIRECTIONS

In this chapter, we explored the different aspects of wireless communication in the harsh environments of smart water grids. In order to define the limitations of the appropriate wireless communication technologies for smart water grids, we explore the contemporary LPWA communication modules, i.e., LoRa, Nwave, and Xbee868, by conducting experiments on the Welsh Water network under real harsh environmental conditions. The experiments reveal that the performance of LPWA communication modules drops by 50% compared to overground deployments in urban environments. Having identified slow data rates as the main drawback of LPWA, we introduce a variety of data reduction techniques focusing on lossless compression and compressive sensing. Lossless compression reduces the water pressure data by 55% while compressive sensing compresses the data more than 75%, introducing an error (i.e., RMSRE) that is less than 10^{-3}. In both cases the communication is reduced proportionally. To fully exploit the benefits of both lossless compression and compressive sensing, we presented a distributed algorithm which allows each sensor node in a smart water grid to decide the appropriate compression technique achieving up to 46% battery life extension and 66% communication reduction. Edge processing constitutes another approach to reduce further the necessary information that has to be exchanged through the network. Based on our previous work, the edge anomaly detection algorithm based on compression rates and its combination with our offline burst localization algorithm, the evaluation results show that the communication can be reduced more than 99%. In the future, under the context of communication for smart water grids, as Figure 3.26 illustrates, we will focus on smart communication scheduling algorithms and decentralized aperiodic communication techniques based on control theory [24].

ACKNOWLEDGMENTS

The authors would like to thank Welsh Water, which provided the access to the chambers and test rig for our BentoBox deployments. Furthermore, we thank Yad Tahir, Michael Breza, and Zhijin Qin, who participated actively in the LPWA communication experiments over the Welsh Water network.

3.6 REFERENCES

1. NEC and Imperial College London Smart Water Project. `https://wp.doc.ic.ac.uk/aese/project/smart-water/`, 2016. [Online; accessed 10-February-2016].

2. 7zip. LZMA SDK (Software Development Kit). `http://www.7-zip.org/sdk.html`, 2015. [Online; accessed 20-September-2016].

3. A Abdelgawad and M Bayoumi. *Resource-Aware data fusion algorithms for wireless sensor networks*, volume 118. Springer Science & Business Media, 2012.

4. B Aghaei. Using wireless sensor network in water, electricity and gas industry. In *Proc. IEEE ICECT*, volume 2, pages 14–17, 2011.

5. M Allen, A Preis, M Iqbal, and A J Whittle. Water distribution system monitoring and decision support using a wireless sensor network. In *Proc. IEEE SNPD*, pages 641–646, 2013.

6. Atmel. Atmel AVR 8-bit and 32-bit Microcontrollers. `http://www.atmel.com/products/microcontrollers/avr/default.aspx`, 2015. [Online; accessed 9-February-2016].

7. K C Barr and K Asanović. Energy-aware lossless data compression. *ACM Trans. Comput. Syst.*, 24(3):250–291, 2006.

8. S Becker, J Bobin, and E J Candès. Nesta: A fast and accurate first-order method for sparse recovery. *SIAM Journal on Imaging Sciences*, 4(1):1–39, 2011.

9. S Boyd and L Vandenberghe. *Convex optimization*. Cambridge University Press, 2004.

10. M Burrows and D J Wheeler. A block-sorting lossless data compression algorithm. *Tech. Report in CiteSeerX*, vol. 124, 1994.

11. bzip2. bzip2 and libbzip2. `http://www.bzip.org`, 2010. [Online; accessed 20-September-2016].

12. F J Candès, J Romberg, and T Tao. Robust uncertainty principles: Exact signal reconstruction from highly incomplete frequency information. *IEEE Trans. Inf. Theory*, 52(2):489–509, 2006.

13. E J Candès, J Romberg, and T Tao. Robust uncertainty principles: Exact signal reconstruction from highly incomplete frequency information. *IEEE Trans. Inf. Theory*, 52(2):489–509, 2006.

14. E J Candes and T Tao. Near-optimal signal recovery from random projections: Universal encoding strategies? *IEEE Trans. Inf. Theory*, 52(12):5406–5425, 2006.

15. E Cayirci. Data aggregation and dilution by modulus addressing in wireless sensor networks. *IEEE Comm. Letters*, 7(8):355–357, 2003.

16. S S Chen, D L Donoho, and M A Saunders. Atomic decomposition by basis pursuit. *SIAM Review*, 43(1):129–159, 2001.

17. J Cleary and I Witten. Data compression using adaptive coding and partial string matching. *IEEE Trans. Comm.*, 32(4):396–402, 1984.

18. I Daubechies. Ten lectures on wavelets. *SIAM*, 1992.

19. C Hacks. Lora experiments, 2016. `www.cooking-hacks.com/documentation/tutorials/extreme-range-lora-sx1272-/module-shield-arduino-raspberry-pi-intel-galileo`.

20. T He, B M Blum, J A Stankovic, and T Abdelzaher. Aida: Adaptive application-independent data aggregation in wireless sensor networks. *ACM Trans. Embed. Comput. S.*, 3(2):426–457, 2004.

21. Indigoo. Mobile and Wireless LPWAN. `https://www.indigoo.com/dox/itdp/12_MobileWireless/LPWAN.pdf`, 2017. [Online; accessed 10-June-

2017].

22. A Johnson and J Burton. Water torture: 3,300,000,000 litres are lost every single day through leakage. *The Independent*, 2010.

23. S Kartakis, B D Choudhary, A D Gluhak, L Lambrinos, and J A McCann. Demystifying low-power wide-area communications for city iot applications. In *Proc. ACM WiNTECH*, pages 2–8, 2016.

24. S Kartakis, A Fu, M Mazo Jr., and J A McCann. Evaluation of decentralized event-triggered control strategies for cyber-physical systems. *arXiv preprint arXiv:1611.04366*, 2016.

25. S Kartakis, M Milojevic Jevric, G Tzagkarakis, and J A McCann. Energy-based adaptive compression in water network control systems. In *IEEE CySWater*, pages 43–48, 2016.

26. S Kartakis and J A McCann. Real-time edge analytics for cyber physical systems using compression rates. In *IEEE ICAC*, pages 153–159, 2014.

27. S Kartakis and J A McCann. Real-time edge analytics for cyber physical systems using compression rates. In *ICAC'14, USENIX*, pages 153–159, 2014.

28. S Kartakis, G Tzagkarakis, and J A McCann. Adaptive compressive sensing in smart water networks. In *2nd International Electronic Conference on Sensors and Applications*, 2015.

29. S Kartakis, W Yu, R Akhavan, and J A McCann. Adaptive edge analytics for distributed networked control of water systems. In *Proc. IEEE IoTDI*, pages 72–82, 2016.

30. J Kraus and V Bubla. Optimal methods for data storage in performance measuring and monitoring devices. In *Proceedings of Electronic Power Engineering Conference*, 2008.

31. Interactive Matter Lab. Filtering Sensor Data with a Kalman Filter, 2009. `http://interactive-matter.eu/blog/2009/12/18/ filtering-sensor-data-with-a-kalman-filter/`. [Online; accessed 20-March-2014].

32. S LaBrecque. Water loss: seven things you need to know about an invisible global problem. *The Guardian*, 2015.

33. S Mallat. *A wavelet tour of signal processing: The sparse way.* Academic Press, 2009.

34. S G Mallat and Z Zhang. Matching pursuits with time-frequency dictionaries. *IEEE Trans. Signal Process.*, 41(12):3397–3415, 1993.

35. B Moyer. Low power, wide area: A survey of longer-range IoT wireless protocols. *Electronic Engineering Journal*, September 2015.

36. E F Nakamura, A AF Loureiro, and A C Frery. Information fusion for wireless sensor networks: Methods, models, and classifications. *ACM CSUR*, 39(3):9, 2007.

37. S Nath, P B Gibbons, S Seshan, and R Anderson. Synopsis diffusion for robust aggregation in sensor networks. In *Proc. ACM SenSys*, pages 250–262, 2004.

38. NWave. Nwave radio modules. `http://www.nwave.io/`, 2017. [Online; accessed 26-January-2017].

39. Oberhumer. LZO Version 2.09. `http://www.oberhumer.com/opensource/ lzo/`, 2015. [Online; accessed 20-September-2016].

40. R Olfati-Saber. Distributed Kalman filter with embedded consensus filters. In *Proc. IEEE CDC*, pages 8179–8184, 2005.

41. C Sadler. LZW. `https://sites.google.com/site/cmsadler/`, 2015. [Online; accessed 20-September-2016].

42. C M Sadler and M Martonosi. Data compression algorithms for energy-constrained devices in delay tolerant networks. In *Proc. ACM SenSys*, pages 265–278, 2006.

43. A Santos and M Younis. A sensor network for non-intrusive and efficient leak detection

in long pipelines. In *Proc. IEEE WD*, pages 1–6, 2011.

44. S Sathe, T G Papaioannou, H Jeung, and K Aberer. A survey of model-based sensor data acquisition and management. In *Managing and Mining Sensor Data*, pages 9–50. Springer, 2013.

45. Semtech. LoRa: The ultimate long range wireless solution. `http://www.semtech.com/wireless-rf/lora.html`, 2017. [Online; accessed 26-January-2017].

46. A Sharaf, J Beaver, A Labrinidis, and K Chrysanthis. Balancing energy efficiency and quality of aggregate data in sensor networks. *VLDB*, 13(4):384–403, 2004.

47. S K Sharma, E Lagunas, S Chatzinotas, and B Ottersten. Application of compressive sensing in cognitive radio communications: A survey. *IEEE Communications Surveys & Tutorials*, 18(3):1838–1860, 2016.

48. N Shrivastava, C Buragohain, D Agrawal, and S Suri. Medians and beyond: new aggregation techniques for sensor networks. In *Proc. ACM SenSys*, pages 239–249, 2004.

49. Sigfox. Sigfox radio modules. `https://www.sigfox.com/`, 2017. [Online; accessed 26-January-2017].

50. S Srirangarajan, M Allen, A Preis, M Iqbal, H B Lim, and A J Whittle. Wavelet-based burst event detection and localization in water distribution systems. In *Journal of Signal Processing Systems*, pages 1–16, 2013.

51. I Stoianov, L Nachman, S Madden, T Tokmouline, and M Csail. Pipenet: A wireless sensor network for pipeline monitoring. In *Proc. IPSN*, pages 264–273, 2007.

52. J A Tropp. Greed is good: Algorithmic results for sparse approximation. *IEEE Trans. Inf. Theory*, 50(10):2231–2242, 2004.

53. I Vasilescu, K Kotay, D Rus, M Dunbabin, and P Corke. Data collection, storage, and retrieval with an underwater sensor network. In *Proc. ACM SenSys*, pages 154–165, 2005.

54. Z Wang, X Hao, and D Wei. Remote water quality monitoring system based on WSN and GPRS. *Instrument Technique and Sensor*, 1:018, 2010.

55. Weightless. What is Weightless? `http://www.weightless.org/about/what-is-weightless`, 2012. [Online; accessed 12-September-2016].

56. J Ziv and A Lempel. A universal algorithm for sequential data compression. *IEEE Trans. Inf. Theory*, 23(3):337–343, 1977.

57. J Ziv and A Lempel. Compression of individual sequences via variable-rate coding. *IEEE Trans. Inf. Theory*, 24(5):530–536, 1978.

4 Matrix and Tensor Signal Modelling in Cyber Physical Systems

GRIGORIOS TSAGKATAKIS, KONSTANTINA FOTIADOU, MICHALIS GIANNOPOULOS, ANASTASIA AIDINI, ATHANASIA PANOUSOPOULOU, PANAGIOTIS TSAKALIDES

CHAPTER HIGHLIGHTS

- The consideration of the sparsity of a signal representation and how it can be learned through dictionary learning is discussed, while its utilization in the recovery of compressed measurements is provided.
- The cutting edge signal paradigm of low rank modelling is described and considered for properly exploiting the inherent correlations among space and time and for removing noise.
- Extension of state-of-the-art vector- and matrix-based approaches to high-order structured measurements using tensors is reported and potential in tasks like missing measurements recovery is demonstrated.

4.1 INTRODUCTION AND MOTIVATION

Signal acquisition, processing, and understanding play a fundamental role in cyber-physical systems (CPS), where networks of sensors actively monitor the natural environment, infrastructure, even user behaviors [36]. To properly analyze these observations and extract actionable information, measurements must first be encoded

---

G Tsagkatakis ✉

Institute of Computer Science, Foundation for Research and Technology-Hellas, GR.

E-mail: greg@ics.forth.gr

K Fotiadou • M Giannopoulos • A Aidini • P Tsakalides

Institute of Computer Science, Foundation for Research and Technology-Hellas,

and Department of Computer Science, University of Crete, GR.

A Panousopoulou

Foundation for Research and Technology-Hellas, GR.

in appropriate structures. For example, measurements from a sensor at a specific location acquiring observations over time lead to a 1D measurement vector. Measurements from multiple sensors at different locations and time instances can be encoded in a 2D sensor/time matrix. When each sensor is equipped with multiple sensing modalities, measurements are naturally encoded in a 3D sensor/modality/time structure. Exploitation of the structural characteristics of the observations can support numerous tasks, from anomaly detection to long-term dynamics modelling.

In many cases, however, lightweight and battery-limited sensors acquire observations with a significant amount of noise due to challenges in the environment or complexity of the sensing process [47]. Furthermore, even when noisy measurements are acquired, packets containing these measurements often fail to reach the end host due to low bandwidth wireless channels and dysynchonization among sensors [35]. Moreover, for cases where successful communication is possible, power management schemes may restrict the sensing frequency or the sleep/awake periods might dictate the intentional abstaining from sensing. All these factors lead to scenarios involving significant numbers of missing measurements. Fortunately, correlations among observations due to the spatio-temporal characteristics of the underlying processes can significantly help in combating the challenges of sensing.

In this chapter, we present key ideas from high dimensional signal processing and analysis, involving the processing of observations ranging from single time series (1D) to multi-sensor, multi-modal observation sequences (3D). To achieve this goal, core concepts including sparsity of representation and low rank characterization are presented. Based on these mathematical concepts, algorithmic frameworks that allow high-quality recovery from compressed, noisy, and missing measurements are presented, focusing on important methods from matrix and tensor factorizations. To demonstrate the capabilities of these frameworks, recovery of real measurements from a soil moisture sensing network under challenging conditions is explored.

The rest of this chapter is organized as follows: Section 4.2 provides an overview of the state-of-the-art framework of sparse representations, including efficient optimization and dictionary learning. Section 4.3 considers matrices and examines their analysis via low rank modelling. Section 4.4 extends the ideas of low rank models to high-order tensor structures. Section 4.5 presents some results on the application of these methods in sensors data, while some remarks are provided in Section 4.6. In terms of notation, we denote scalar values with the lower case letter (x), vectors with boldface ($\mathbf{x}$), matrices with boldface capital ($\mathbf{X}$), and tensors with calligraphic symbols ($\mathcal{X}$).

4.2 SPARSE SIGNAL CODING

A core aspiration in signal modelling is to express a measurement vector $\mathbf{y} \in \mathbb{R}^n$ in terms of elements contained in a representation matrix $\mathbf{D} \in \mathbb{R}^{n \times k}$, through a vector of coefficients $\mathbf{x} \in \mathbb{R}^k$ such that $\mathbf{y} = \mathbf{D}\mathbf{x}$. Depending of the application at hand, the representation matrix $\mathbf{D}$ can be populated using predefined functions such as wavelets or may be directly learned from the data including previously acquired observations. Depending on the dimensionality of $\mathbf{D}$, the aforementioned system of linear equa-

Matrix and Tensor Signal Modelling in Cyber Physical Systems 109

tions can have: (i) infinite solutions when $n > k$, (ii) exactly one solution when $n = k$, (iii) no solution when $n < k$.

In real-life signal processing tasks, the two extreme cases where $n < k$ or $n > k$ are most frequently encountered. In the first case, a much smaller number of observations compared to the unknown parameters is available leading to an under-determined system. The challenges associated with providing high-quality estimations from limited sets of observations is known as *compressed sensing*, in which case $\mathbf{D}$ (typically written as $\boldsymbol{\Phi}$) is called the sensing matrix [8, 17]. The other case corresponds to the situation where a large number of measurements or models is available leading to ill-posed inverse problems. In this case, prior knowledge regarding the signal must be imposed on the representation of $\mathbf{x}$ with respect to elements from matrix $\mathbf{D}$, the so-called *dictionary* [20].

To address these scenarios, a regularization term is introduced encoding prior knowledge, leading to a regularized regression problem given by

$$\min_{\mathbf{x}} \|\mathbf{y} - \mathbf{D}\mathbf{x}\|_2^2 \quad \text{subject to } \mathcal{R}(\mathbf{x}) \tag{4.1}$$

The main issue now becomes the proper selection of the regularization function $\mathcal{R}(\mathbf{x})$. Cases of particular interest propose the introduction of convex constraints in the form of l_p-norms, $\|\mathbf{x}\|_p = \sum_i \|\mathbf{x}_i\|^p$ A typical choice is the squared euclidean norm $\|\mathbf{x}\|_2^2$, leading to the well-known ridge regression problem, for which a number of algorithmic frameworks exists [32]. However, when the linear system in (4.1) is under-defined, corresponding to the case where the number of observations is less than the number of parameters which need to be estimated, or over-defined where an infinite number of solutions exist, one has to impose other types of regularization to the solution. One case of particular importance is the sparsity regularization where one seeks the solution $\mathbf{x}$ containing the smallest number of non-zero elements.

Depending on the characteristics of the representation process, sparsity can effectively address a number of issues. According to the compressed sensing theory, a signal can be perfectly reconstructed from a dramatically smaller number of observations compared to its dimensionality, provided a sufficient number of measurements is acquired and the signal is sparse enough when expressed on an appropriate basis. Furthermore, for the case of overcomplete representation, sparsity offers robustness to noise and allows better generalization of the generative process.

Formally, to enforce sparsity, an l_0 norm can be introduced as the regularizer in equation (4.1), which is defined as $\|\mathbf{x}\|_0 = \sum_{i=1}^{k} \|x_k\|^0 = \#\{i : x_i \neq 0\}$. Consequently, minimizing the l_0-norm leads to the following optimization problem

$$\min_{\mathbf{x}} \|\mathbf{y} - \mathbf{D}\mathbf{x}\|_2^2 \quad \text{subject to } \|\mathbf{x}\|_0 \leq T_0 \tag{4.2}$$

where T_0 is the requested sparsity level. While the l_0 norm guarantees sparsity, the minimization in equation (4.2) is NP-hard, i.e., the complexity of the problem increases exponentially with the number of parameters, making this problem impractical for real-world applications. State-of-the-art techniques for solving the sparsity minimization problem can be broadly classified in two classes: the greedy and the relaxation methods.

Greedy algorithms seek an approximation of the signal vector $\mathbf{x}$ through a sequence of incremental updates, by suitably selecting elements from $\mathbf{D}$, called atoms. A representative example of this class is the orthogonal matching pursuit (OMP) [43] algorithm and its variants [18, 31]. OMP is an iterative algorithm which searches for the sparse representation of the signal $\mathbf{x}$ through a sequence of incremental single-atom approximations. Each iteration involves two steps: the atom selection and the residual update. These algorithms terminate when a halting criterion is satisfied, i.e., when the number of distinct atoms in the approximation drops below a pre-defined threshold. The main advantages of the OMP-family of techniques is their moderate computational requirements, the straightforward implementation, and theoretical quarantines regarding the recovery of signals over redudant dictionaries [42].

Relaxation techniques on the other hand, replace the non-convex l_0-norm with its convex surrogate l_1-norm and solve the problem through linear programming. The unconstrained version of the minimization problem in equation (4.2), with a fixed regularization parameter $\lambda > 0$, becomes

$$\min_{\mathbf{x}} \|\mathbf{y} - \mathbf{D}\mathbf{x}\|_2^2 + \lambda \|\mathbf{x}\|_1 \tag{4.3}$$

which is known as LASSO [41] and Basis pursuit [11]. In many cases, due to the existence of noise, a perfect solution to the linear system of equations $\mathbf{y} = \mathbf{D}\mathbf{x}$ cannot be achieved or may not even be desired. For this purpose, the sufficient similarity between $\mathbf{D}\mathbf{x}$ and $\mathbf{y}$ in the l_2-norm is required, in which case equation (4.3) can be alternatively formulated as an l_1 regularized regression called basis pursuit denoising given by

$$\min_{\mathbf{x}} \|\mathbf{x}\|_1 \quad \text{subject to} \quad \|\mathbf{D}\mathbf{x} - \mathbf{y}\|_2^2 \leq \varepsilon \tag{4.4}$$

where ε in threshold chosen according to the properties of noise.

4.2.1 DICTIONARY LEARNING

A crucial point to the sparse representation of a signal, especially in the over-determined case, is the proper selection of the dictionary matrix $\mathbf{D}$. In general, a dictionary can be populated using analytic functions or using training examples. In the former case, analytic dictionaries arise from a set of predefined mathematical formulations, such as wavelets and variants, e.g., curvelets, countourlets, etc. [30]. Analytic dictionaries contain sufficient elements to span the subspace where the signals of interest reside while they encode prior knowledge such as the presence of lines, curves, and other types of structures, while efficient methods exist for fast application.

In many cases, however, the complexity of the problem prevents the definition of specific functions for populating the dictionary. In such cases, available examples can be utilized for generating a dictionary directly from the observations, a process called dictionary learning [20]. The objective in dictionary learning is to utilize training examples encoded in matrix $\mathbf{Y} = \{\mathbf{y}_i\}_{i=1}^m$, containing m signal examples such that all

Matrix and Tensor Signal Modelling in Cyber Physical Systems **111**

elements admin a sparse representation when expressed in the dictionary $\mathbf{D}$

$$\min_{\mathbf{D},\mathbf{X}} \; \|\mathbf{Y} - \mathbf{D}\mathbf{X}\|_F^2 \quad \text{subject to } \|\mathbf{x}_i\|_0 \leq T_0, \quad 1 \leq i \leq M \tag{4.5}$$

where $\|\mathbf{D}\|_F = \sqrt{\Sigma_{ij} d_{ij}^2}$ is the Frobenius norm, while often a unit l_2-norm constraint on the dictionary elements is imposed such that $\|\mathbf{d}_j\| \leq 1, 1 \leq j \leq k$. Unfortunately, the optimization problem is not jointly convex for both variables $\mathbf{D}$ and $\mathbf{X}$. It is convex, however (or can be approximated), when we minimize with respect to one variable and keep the other variable fixed.

Given a set of signal measurements, $\mathbf{Y}$, the K-SVD algorithm for dictionary learning [3], searches for a dictionary matrix $\mathbf{D}$ that can represent efficiently each example in this set, using strict sparsity constraints. K-SVD follows an iterative approach which involves alternating between the sparse coding of the examples based on the estimated dictionary, and the dictionary update step, in order to find a better representation for the data. Given an initialized dictionary $\mathbf{D}^{(0)}$, the K-SVD algorithm performs a two-stage procedure at each iteration k, namely

Sparse Representation: $\quad \mathbf{X}^{(k+1)} = \arg\min_{\mathbf{X}} \|\mathbf{Y} - \mathbf{D}^{(k)}\mathbf{X}\|_F^2$

Dictionary Update: $\quad \mathbf{D}^{(k+1)} = \arg\min_{\mathbf{D}} \|\mathbf{Y} - \mathbf{D}\mathbf{X}^{(k+1)}\|_F^2$

The first stage of this procedure is a simple sparse coding problem which is solved using the OMP method. The dictionary update step is more challenging and is applied at one atom at a time by generating an error matrix using all atoms but the one under consideration, and updating the atoms according to the orthogonal components of an error matrix.

The discussion thus far has focused on the case where vector-structured observations need to be analyzed. Although dictionary learning operates for data matrices, the sparsity constraints are still imposed on individual vector elements. To address this issue, various extensions have been proposed including the multiple measurement vectors (MMV) [15] and structure sparsity models [23]. While sparsity has had a profound impact on signal processing, a significant portion of its power stems from the design or learning of a proper dictionary. An alternative approach involves analyzing all available observations and directly extracting useful information by exploiting the 2-dimensional matrix structure of the observations.

4.3 LOW RANK MODELLING OF MATRICES

Generalizing the single sensor case, encoding measurements over k sampling instances from n sensors at different locations can be naturally encoded in a 2D measurement matrix $\mathbf{M} \in \mathbb{R}^{n \times k}$. A key property that characterizes such a matrix, and every matrix for that matter, is the *matrix rank*, which corresponds to the number of linearly independent columns or rows (whichever is the smaller of the two). Beyond the strict mathematical formulation, the rank of a matrix acts as a proxy to the underlying redundancies and correlations that exist in the data. Matrices with rank close to their (smaller) dimension, indicate that the observations have little correlation while low rank matrices indicate significant amounts of redundancies, which can be exploited. Figure 4.1 showcases a full rank and low rank approximation of a measurements matrix.

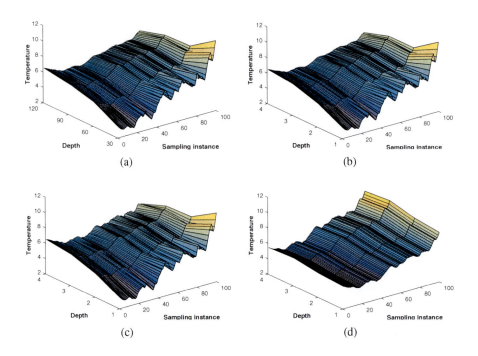

Figure 4.1: Original (a) and rank three (b), two (c) and one (d) approximations of a measurements matrix.

To demonstrate the effects of rank on a typical measurement matrix, Figure 4.1 showcases a real set of temperature observations from a single location at multiple depths over a period of 100 sampling instances (a) and three low rank approximations (c,b,d). Notice that in this case, the rank of the row measurements is four, equal to the smallest dimension. We observe that reducing the rank, even by half, preserves key features in the true signal, e.g., relatively abrupt increase at lower depths and latter sampling instances. In many cases, rank one approximations could also be of importance since they capture the most prominent trends, e.g., increasing temperatures at latter sampling instances, allowing large-scale analysis.

Rank is a dimensionless number which although reliable offers little intuition on the actual characteristics of the correlation, e.g., magnitudes and scales. To obtain a more informed view, the singular value decomposition (SVD) method has been extensively employed [38]. Formally, SVD factorizes a matrix $\mathbf{M} \in \mathbb{R}^{n \times m}$ where

Matrix and Tensor Signal Modelling in Cyber Physical Systems

$n \leq m$ as the product of three matrices $\mathbf{Y} = \mathbf{U}\mathbf{S}\mathbf{V}^T$ where $\mathbf{U} \in \mathbb{R}^{n \times n}$ and $\mathbf{V} \in \mathbb{R}^{m \times m}$ are matrices containing orthogonal vectors, while $\mathbf{S} \in \mathbb{R}^{n \times m}$ is a diagonal matrix. The vectors in $\mathbf{U}$ and $\mathbf{V}$ are called left and right singular vectors, respectively, and they contain elements which span the row and column spaces of the matrix respectively. The diagonal elements in $\mathbf{S}$, typically ordered based on their magnitude, contain the singular values which represent the contribution of each right-left singular vector pair. The singular values of a matrix can provide valuable information regarding the entries of the matrix, while the number of non-zero singular values is also the rank of the matrix.

A rank k matrix is characterized as a low rank matrix when the effective number of degrees of freedom is significantly smaller compared to its dimensionality, i.e., $k \ll \min(n, m)$. Low rank matrices arise when the underlying phenomena that generate the entries are correlated, effectively reducing the dimensionality of the row and column space. From a cyber-physical point-of-view, when matrices are populated by measurements from independent yet neighboring sensors, or acquired at different time instances, correlations due to proximity in space and time lead to low rank matrices. At the same time, noise of the measurement process, sensor faults, and clock desynchronization lead to high rank matrices. Formally, the problem of low rank modelling of a matrix $\mathcal{X}$ corresponds to the following optimization problem

$$\min_{\mathbf{X}} \ \text{rank}(\mathbf{X}) \quad \text{subject to} \ \|\mathbf{Y} - \mathbf{X}\|_F \leq \varepsilon \tag{4.6}$$

The solution to this problem can be identified by exploiting the Eckart–Young theorem which states that a low rank approximation $\hat{\mathbf{X}}$ of the matrix $\mathbf{X}$ is given by $\hat{\mathbf{X}} = \mathbf{U}\mathcal{S}(\mathbf{S})\mathbf{V}'$, where $\mathbf{U}, \mathbf{S}, \mathbf{V}$ are obtained through the SVD decomposition of $\mathbf{X}$, while $\mathcal{S}(\cdot)$ is a thresholding operator. The thresholding operator is responsible for preserving a small number of singular values leading to low rank reconstructions.

4.3.1 MATRIX RECOVERY

The problem of recovering a matrix from an incomplete sampling of its entries is known as *matrix completion* (MC) [7]. There are many applications of practical significance that need to complete a partially known matrix, as we saw above, and usually there is a very small number of available samples. The question is whether it is possible to recover the missing elements in a matrix from a small number of known entries. According to matrix completion, such a recovery is possible provided the matrix is characterized by a small rank (compared to its dimensions) and enough randomly selected entries of the matrix are acquired.

Formally, we consider a matrix $\mathbf{M} \in \mathbb{R}^{n_1 \times n_2}$ of measurements where for the case of a cyber-physical sensor network, n_1 corresponds to the number of sensors and n_2 to the number of sampling instances ($n_1 \leq n_2$). The object of matrix completion is to fill in the missing entries of the matrix from a significantly reduced subset q of its entries, i.e., $q << (n_1 \times n_2)$, such that the resulting matrix $\mathbf{X}$ (which is a reliable approximation of $\mathbf{M}$) has the lowest possible rank. This problem arises in machine learning scenarios where we are given partially observed examples of a process with a low-rank covariance matrix and would like to estimate the missing data. In many of these settings, some prior probability distribution (such as a Bernoulli model or uniform distribution on subsets) is assumed to generate the set of available entries.

Specifically, accurate approximation $\mathbf{X}$ of the matrix $\mathbf{M}$ from a small number of entries can be obtained via the solution of the minimization problem

$$\min_{\mathbf{X}} \ \text{rank}(\mathbf{X}) \quad \text{subject to} \ \mathcal{P}_\Omega(\mathbf{X}) = \mathcal{P}_\Omega(\mathbf{M}) \tag{4.7}$$

where $\mathcal{P}_\Omega$ is a random sampling operator which retains only a small number of entries from the matrix $\mathbf{M}$, i.e.,

$$\mathcal{P}_\Omega(\mathbf{M}) = \begin{cases} m_{ij} & \text{if } i, j \in \Omega \\ 0 & \text{otherwise} \end{cases} \tag{4.8}$$

and Ω is the sampling set. In general, to solve the matrix completion problem, the sampling operator $\mathcal{P}$ must satisfy the modified restricted isometry property, which is the case when uniform random sparse sampling is employed in both rows and columns of matrix $\mathbf{M}$ [7].

Although by solving the above optimization problem, we will recover an approximation of the matrix, rank minimization is an NP-hard problem and the calculation time is of the order $O(n^n)$, where $n = \min(n_1, n_2)$. Since the rank function is simply the number of nonzero singular values, it can be replaced by the nuclear norm, which is defined as $\|\mathbf{X}\|_* = \sum_{i=1}^n \sigma_i$ where σ_i is the ith singular value of $\mathbf{X}$. The nuclear norm heuristic was introduced by Fazel and is alternatively known by several other names including the Schatten 1-norm, the Ky Fan r-norm, and the trace class norm. Since the singular values are all positive, the nuclear norm is also equal to the l_1 norm of the vector of singular values. So, the optimization problem arises and is given by

$$\min_{\mathbf{X}} \ \|\mathbf{X}\|_* \quad \text{subject to} \ \mathcal{P}_\Omega(\mathbf{X}) = \mathcal{P}_\Omega(\mathbf{M}) \tag{4.9}$$

The nuclear norm is a convex function which makes the above problem easy to solve, while under mild conditions, the nuclear norm minimization can estimate the same matrix as the rank minimization with high probability provided $q \geq C \cdot \max(n_1, n_2)^{\frac{6}{5}} \cdot r \cdot \log(\max(n_1, n_2))$ randomly selected entries of the rank r matrix are acquired [7]. To solve the nuclear norm minimization problem, various approaches have been proposed including the singular value thresholding [5] and the augmented Lagrange multiplier method [27].

In addition to the case of recovering missing measurements, low rank modelling has also been considered for removing the noise that may have affected the signals. The robust principal components analysis (RPCA) [6] framework assumes a generative process where the observed measurement matrix is the summation of a low rank matrix and a sparse noise matrix. Similar to the case of MC and CS, the rank is replaced with the nuclear norm, while the l_0 norm is replaced with the l_1, leading to the minimization problem

$$\min_{\mathbf{L},\mathbf{S}} \ \|\mathbf{L}\|_* + \lambda \|\mathbf{E}\|_1 \quad \text{subject to} \ \mathbf{Y} = \mathbf{L} + \mathbf{E} \tag{4.10}$$

where λ is a parameter weighting the contribution of the low rank and the sparse components.

4.4 FROM MATRICES TO TENSORS

Although matrix-based modelling as presented in the aforementioned section provides a concrete framework for dealing with various scientific tasks (i.e., classification, feature extraction, blind source separation), quite frequently we have to deal with more involved structures which arise from modern sensor modalities and consist of more than two dimensions.

As an example, one can consider a network of sensing instruments where each instrument acquires multiple observations, e.g., temperature, humidity, pressure, over time. To encode such observations a 3-way tensor is required, one dimension per instrument, one per monitored physical parameter, and one for time. From a linear algebra point of view, tensors consist of a generalization of scalars, vectors, and matrices such that a scalar is a zero-order tensor, a vector is a first-order tensor, a matrix is a second-order tensor, and a third-order tensor is a higher-order tensor [40]. Such data can emanate from diverse scientific as well as industrial fields, such as smart water networks, hyper-spectral imaging, wireless networks, social networks, telecommunications, to name but a few .

Formally, a tensor $\mathcal{M} \in \mathbb{R}^{n_1 \times n_2 \times \ldots \times n_N}$ of order N is a multi-index numerical array with elements $m_{n_1, n_2, \ldots n_N}$. Just as one can form submatrices given a specific matrix, a subtensor is obtained from a given tensor by fixing a subset of its indices. An important example involves the reordering of the elements of a tensor into a matrix in order to support matrix-based processing. Such a transformation is called "matricization" or "unfolding" and is not unique since different ways of stacking the horizontal, lateral, and frontal slices of a tensor in either column-wise or row-wise arrays exists [14]. Formally, the mode-n unfolding of a tensor stacks its mode-n fibers as columns to a matrix $\mathbf{M}_{(n)}$. Figure 4.2 illustrates a tensor and its matrix unfoldings.

4.4.1 TENSOR DECOMPOSITIONS

A key process which can be applied in tensors involves the three types of tensor products [24]. The outer product of two tensors $\mathcal{A} \in \mathbb{R}^{n_1 \times n_2, \ldots, \times n_k}$ and $\mathcal{B} \in \mathbb{R}^{n_1 \times n_2, \ldots, \times n_l}$ corresponds to the element-wise multiplication of the tensors and is defined as $\mathcal{M} = \mathcal{A} \circ \mathcal{B} \in \mathbb{R}^{n_1 \times n_2, \ldots, n_k \times n_1 \times n_2, \ldots, \times n_l}$. The inner product, also known as the Kronecker product, is defined between two matrices $\mathbf{A} \in \mathbb{R}^{I \times J}$ and $\mathbf{B} \in \mathbb{R}^{K \times L}$ as

$$\mathcal{M} = \mathbf{A} \otimes \mathbf{B} \in \mathbb{R}^{IK \times JL} = [\mathbf{a}_1 \otimes \mathbf{b}_1 \mathbf{a}_1 \otimes \mathbf{b}_2 \mathbf{a}_1 \otimes \mathbf{b}_3 \cdots \mathbf{a}_J \otimes \mathbf{b}_{K-1} \mathbf{a}_J \otimes \mathbf{b}_K]$$

What should be highlighted at this point is the fact that generally the Kronecker product of vectors yields a vector, as opposed to the outer product which ends up as a tensor. Last, the Khatri-Rao product, also known as the "matching columnwise" product [24], given two matrices with the same number of columns, i.e. $\mathbf{A} \in \mathbb{R}^{I \times J}$ and $\mathbf{B} \in \mathbb{R}^{K \times J}$, their Khatri-Rao product is defined as follows:

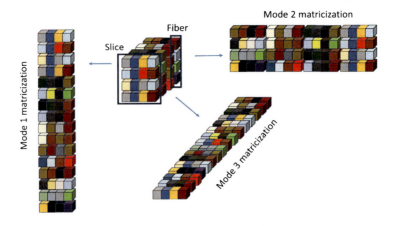

Figure 4.2: The three matrix unfoldings of a 3D tensor, and the associated fibers and slices.

$$\mathcal{M} = \mathbf{A} \odot \mathbf{B} \in \mathbb{R}^{IK \times J} = \begin{bmatrix} \mathbf{a}_1 \otimes \mathbf{b}_1 & \mathbf{a}_2 \otimes \mathbf{b}_2 & \mathbf{a}_3 \otimes \mathbf{b}_3 ... & \mathbf{a}_J \otimes \mathbf{b}_J \end{bmatrix}$$

A corollary to the above product definitions is that if **a** and **b** are vectors, then Kronecker and Khatri-Rao products are identical. Note that although for exposition purposes, we explore the two-way tensor, the corresponding properties and operation can generalize to an arbitrary number of dimensions [13, 40].

Tensors have been gaining increased attention since the 1960s mainly due to their various decompositions which capture the underlying latent structure in diverse scientific problems. Two of well-known decomposition, namely CANonical DECOMPosition / PARAllel FACtors (CANDECOMP/PARAFAC) decomposition [9,22] and the Tucker decomposition [46], were inspired from human sciences, such as psychometrics and linguistics.

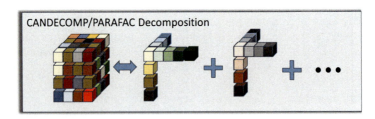

Figure 4.3: CANDECOMP/PARAFAC decomposition expresses the observed tensor as the summation of rank-1 tensors.

Given a third-order tensor $\mathcal{M} \in \mathbb{R}^{n_1 \times n_2 \times n_3}$, applying the PARAFAC/CANDECOMP decomposition allows one to express the tensor as

$$\mathcal{M} = \sum_{r=1}^{J} \lambda_r \mathbf{a}_r \circ \mathbf{b}_r \circ \mathbf{c}_r \qquad (4.11)$$

where $\mathbf{a}_r \in \mathbb{R}^{n_1}$, $\mathbf{b}_r \in \mathbb{R}^{n_2}$, $\mathbf{c}_r \in \mathbb{R}^{n_3}$ are the vector components and λ_r a normalization parameter (can be absorbed). In PARAFAC/CANDECOMP, vectors $\mathbf{a}_r, \mathbf{b}_r, \mathbf{c}_r$ are treated as column vectors of factor (hence the naming Parallel Factors-PARAFAC) matrices $\mathbf{A} = \{\mathbf{a}_1, \mathbf{a}_2, ... \mathbf{a}_J\}$, $\mathbf{B} = \{\mathbf{b}_1, \mathbf{b}_2, ... \mathbf{b}_J\}$, $\mathbf{C} = \{\mathbf{c}_1, \mathbf{c}_2, ... \mathbf{c}_J\}$, which combined together via outer product produce the original tensor $\mathcal{M}$. The number of terms J in the summation in equation (4.11) can be defined as the rank of the tensor and plays a pivotal role in tensor algebra. A visual illustration of the decomposition is shown in Figure 4.3.

To formally define the rank of a tensor, one must employ rank-1 tensors, a specific form which has a key-position in multi-way analysis. Formally, an N-way tensor $\mathcal{M} \in \mathbb{R}^{n_1 \times n_2 \times ... \times n_N}$ is called a rank-1 tensor if it can be written as the outer product of N vectors: $\mathcal{M} = \mathbf{a}^{(1)} \circ \mathbf{a}^{(2)} \circ ... \circ \mathbf{a}^{(N)}$. Alternatively, each element of $\mathcal{M}$ consists of the product of the corresponding vector elements $t_{d_1, d_2, ..., d_N} = a_{d_1}^{(1)}, a_{d_2}^{(2)}, ..., a_{d_N}^{(N)}$ for all $1 \leq d_n \leq D_n$. Then, the rank of the tensor is defined as the smallest number of rank-1 tensors whose sum generate $\mathcal{M}$ [25], in which case the decomposition in equation (4.11) is called canonical polyadic (CP) decomposition. In addition to the intuitive interpretation of the CP decomposition, such decompositions are unique under mild conditions for third [25] and higher [39] order tensors. A major algorithmic framework for CP involves the alternating least squares (ALS) method [13] which iteratively estimates one factor while keeping the other fixed. Other scenarios have also recently been explored, including extension to distributed non-negative factorization via the ADMM methods [26].

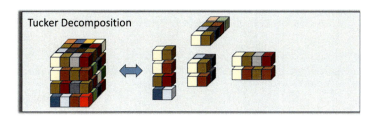

Figure 4.4: Tucker decomposition expressed the tensor as the product of a core tensor and mode-specific factors.

The Tucker decomposition is named for Ledyard R. Tucker [46] and may be considered as the extension of the classical SVD method to higher dimensions [16]. Tucker decomposition decomposes a tensor into a set of matrices and one core tensor. Assuming a third-order tensor $\mathcal{M} \in \mathbb{R}^{D_1 \times D_2 \times D_3}$, the Tucker decomposition expresses the tensor as

$$\mathcal{M} = \sum_{j=1}^{n_1} \sum_{r=1}^{n_2} \sum_{p=1}^{n_3} \mathcal{G}_{jrp} \mathbf{a}_j \circ \mathbf{b}_r \circ \mathbf{c}_p$$

where $\mathcal{G} \in \mathbb{R}^{J \times R \times P}$ is the core tensor, $\mathbf{A} = \{\mathbf{a}_1, \mathbf{a}_2, ..., \mathbf{a}_J\}$, $\mathbf{B} = \{\mathbf{b}_1, \mathbf{b}_2, ..., \mathbf{b}_R\}$, and $\mathbf{C} = \{\mathbf{c}_1, \mathbf{c}_2, ..., \mathbf{c}_P\}$ are the factor matrices, which can be interpreted as the principal components along each mode. A visual illustration of the decomposition is presented in Figure 4.4. The CP and Tucker decompositions can be utilized in a number of problems including low rank tensor approximations, multivariate regression, and component separation [13].

4.4.2 TENSOR COMPLETION

Similar to the case of matrix completion, tensor completion (TC) addresses the problem of recovering missing measurements encoded in tensor structures. One major line of research involves the unfolding of a high-order tensor into a lower-order matrix and its recovery through low rank MC [28]. To process the data using a matrix method, a matrix unfolding process needs to be considered. Figure 4.5 presents the two different unfoldings of the tensor as well as the associated singular values. One can easily notice that different unfoldings lead to dramatically different ratios between high magnitude and low magnitude singular values. As a rule of thumb, one should consider the matrix with the largest dimensions, since such matrices are typically characterized by limited degrees of freedom compared to their dimension. Although matrix recovery is a viable option, such an approach encodes the low-rankness with respect to a single mode and fails to capture the interdimensional variances, leading to low reconstruction quality. To address this issue, a formal approach to low tensor-rank models needs to be considered.

Equation (4.7) for the matrix case, i.e., the two-order tensor, is extended to higher-order tensors by solving an optimization problem which tries to estimate the lowest-rank tensor $\mathcal{X}$ which agrees with the observed data given by

$$\min_{\mathcal{X}} \ \|\mathcal{X}\|_* \ \text{ subject to } \ \mathcal{P}_\Omega(\mathcal{X}) = \mathcal{P}_\Omega(\mathcal{M}) \tag{4.12}$$

Similar to the matrix case, Ω is the set of index set (i_1, i_2, i_3) corresponding to the observed entries and the linear map $\mathcal{P}$ is the random subsampling operator

$$\mathcal{P}_\Omega(\mathcal{M}) = \begin{cases} m_{i_1, i_2, ... i_k} & \text{if } (i_1, i_2, ..., i_k) \in \Omega \\ 0 & \text{otherwise} \end{cases} \tag{4.13}$$

However, the optimization regime now is tougher than before, as the tensor nuclear norm is not defined as the tightest convex relaxation of the tensor rank, as was the case with matrices. The tensor nuclear norm can be defined as

$$\|\mathcal{X}\|_* = \sum_{i=1}^n \alpha_i \|\mathcal{X}_{(i)}\|_* \tag{4.14}$$

where α_i's are constants satisfying $\alpha_i \geq 0$ and $\sum_{i=1}^n \alpha_i = 1$ [28]. Thus, the nuclear norm for a general tensor case can be defined as the convex combination of the

Matrix and Tensor Signal Modelling in Cyber Physical Systems 119

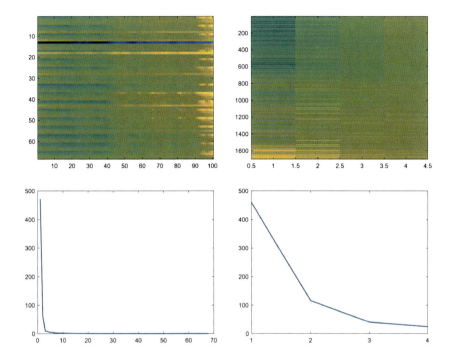

Figure 4.5: Two representative matrix unfoldings of a 3D tensor (top row) and associated distributions of singular values (bottom row).

nuclear norms of all matrices unfolded along each of its modes. Under this definition, equation (4.14) can be written as

$$\min_{\mathcal{X}} \sum_{i=1}^{n} \alpha_i \|\mathcal{X}_{(i)}\|_* \quad \text{subject to} \quad \mathcal{P}_\Omega(\mathcal{X}) = \mathcal{P}_\Omega(\mathcal{M}) \qquad (4.15)$$

Similar to the matrix completion case, the reconstruction of the sub-populated tensor $\mathcal{M} \in \mathbb{R}^{n_1 \times n_2 \times n_3}$ relies on the availability of $k << n_1 \times n_2 \times n_3$ observations.

An efficient approach for the recovery of low-rank tensor based on parallel matrix factorization [50] was recently proposed which offers lower computational complexity and high-quality recovery of tensors via a factorization of different matrix unfoldings. Specifically, the tensor $\mathcal{M}$ is unfolded in all modes to a set of matrix factors $\mathbf{X}_n$ $\mathbf{Y}_n$, such that $\mathbf{M}_n \approx \mathbf{X}_n \mathbf{Y}_n$, where $n = 1, 2, 3$ indicates the corresponding mode. To integrate the different mode factorizations, a shared variable $\mathcal{Z}$ is introduced, such that recovery of tensor $\mathcal{M}$ corresponds to

$$\min_{\mathbf{X},\mathbf{Y},\mathbf{Z}} \sum_{n=1}^{3} \frac{\alpha_n}{2} \|\mathbf{X}_n \mathbf{Y}_n - \mathbf{Z}_n\|_F^2 \quad \text{subject to} \quad \mathcal{P}_\Omega(\mathcal{Z}) = \mathcal{P}_\Omega(\mathcal{M}) \qquad (4.16)$$

where $\mathbf{X} = \{\mathbf{X}_1, \mathbf{X}_2, \mathbf{X}_3\}$, $\mathbf{Y} = \{\mathbf{Y}_1, \mathbf{Y}_2, \mathbf{Y}_3\}$ and $\mathbf{Z} = \{\mathbf{Z}_1, \mathbf{Z}_2, \mathbf{Z}_3\}$ correspond to the unfolding of the three-way tensors we consider in this work. The parameters α_n are weights which satisfy $\sum_n \alpha_n = 1$ and are introduced in order to properly weight the contribution of each unfolding. A limitation of this approach is that all three ranks of the tensor must be defined beforehand, an assumption which may limit the scope of the method. To address this issue, a rank adjustment scheme is introduced in [50]. Specifically, this scheme dynamically updates the ranks' estimates at progressing iteration, when slow progress in the updates of the singular values of their corresponding factor matrices is detected. Thus, equation (4.16) is solved by cyclically updating $\mathbf{X}_n, \mathbf{Y}_n$, and $\mathbf{Z}_n$. The procedure is performed for all modes in parallel, thereby improving both the speed of execution and its effectiveness, since it exploits the low-rank characteristics of tensor modes at the same time.

4.5 APPLICATIONS IN CPS SYSTEMS

4.5.1 STATE-OF-THE-ART IN CPS MONITORING

In the majority of cases, the enabling technology for CPS systems is based on wireless sensor and actuation networks [2]. Smart wireless sensor networks exist for sea water quality monitoring [1], marine environments [49], and water distribution networks [34] among others. From a signal processing and learning perspective, methods include sparse coding and dictionary learning schemes for recovery of WSN measurements [45], efficient activation of sensors in water distribution networks [19], data gathering on wireless sensor networks [12], online weather forecasting [48], and sensor localization [33]. Low rank modelling is useful for detection of faults and intrusions [29], monitoring large traffic networks [4], event detection in urban environments via analysis of crowd mobility and social activity data [10], as well as in the prediction of future observations [44]. Furthermore, tensor modelling of observations has also been considered for the recovery of environmental observations [21] and measurements from wearable sensors [37] among others.

4.5.2 RECOVERY FROM COMPRESSED MEASUREMENTS

To illustrate the potential of sparse coding, we consider a WSN deployed for a CPS monitoring task where nodes both acquire and relay measurements from other nodes to a sink. We focus on the case of sparse signal recovery and explore the potential of compressed sensing (CS) in data gathering as a framework which may encompass a number of multi-hop network protocols. We assume a network of $m = 100$ nodes where only k observes a non-zero value (1), while the rest do not (0). This situation involves sensors that detect and communicate events or anomalies like fire, flooding, etc. In the typical non-CS setting, each sensor will have to transmit its measurements to nodes in its path to the sink, which will forward its measurements to the sink. The total number of packets is thus equal to the number of sensors times the number of nodes in each path. In CS, when a node transmits its measurements to the sink through nodes, instead of simply forwarding the packet, each intermediate node injects its own measurement before transiting it to the next node in the path.

The injection involves the multiplication of the sensors' binary measurements with a random number and the addition to the value encoded in the received packet. For every packet, this information is encoded in a matrix $\boldsymbol{\Phi}$ of size $n \times m$ such that for any packet/node there exists a list of maximum m nodes which will be involved in the transmission. The recovery of the sparse network-wide measurements vectors of size n can then be found by solving the minimization problem/task $\mathbf{m}^* = \min \|\mathbf{y} - \boldsymbol{\Phi}\mathbf{m}\|_2^2$ subject to $\|\mathbf{m}\|_1 \leq k$. Once original measurements are reconstructed, the error will be measured by the number of different measurements between the original measurements and the reconstructed ones over the total number of measurements.

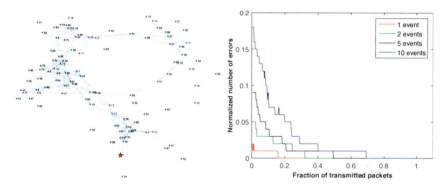

Figure 4.6: Network connectivity graph (left) and normalized number of detection errors with respect to number of packets (right).

Figure 4.6 presents the network's connectivity graph on the left and the detection capabilities of the CS architecture as a fraction of the transmitted packets over the total number of packets required by a non-CS architecture on the right. We consider four sparsity scenarios, where 1, 2, 5, and 10 events are generated. Experimental results suggest that for highly sparse signals, e.g., one event, as low as 20% of the number of packets that non-CS approach requires, are sufficient for accurate detection. Following the theoretical prescription, the more events generated, the less sparse the signal is and more packets have to be transmitted for reliable detection. Note that in this case, the natural sparsity of the signal is considered. The same philosophy can be invoked in more complex signals by leveraging the representational power of a proper sparsifying dictionary.

4.5.3 RECOVERY FROM NOISY MEASUREMENTS

The objective of the rest of this section is to demonstrate how theoretical models considered thus far can be exploited in the context of a real application dealing with environmental monitoring data. We explore two critical cases of significant impact, namely removing noise of observations and recovering measurements that were either lost or intentionally not transmitted due to energy management. The data were acquired by an automated insitu soil sensor network, and include hourly and daily measurements of volumetric water content, soil temperature, and bulk electrical conductivity, collected at 42 monitoring locations and 5 depths across Cook Agronomy Farm.[1] We consider temperate measurements at 4 depths (30m, 60m, 90m, and 120m) which were collected during the first 100 days of 2016 from a subset of 17 sensors where measurements were available for that period, leading to a $17 \times 4 \times 100$ tensor.

In this scenario, we investigate low rank matrix analysis for detecting and removing sparse errors in a low rank matrix via the Robust PCA (RPCA) method [6]. To generate the noisy data, 10% and 20% of the observations were contaminated with noise of standard deviation 1 and 2, respectively. The measurements are encoded into a matrix of size 68×100, which corresponds to the unfolding of the 3D $17 \times 4 \times 100$ tensor. Figure 4.7 presents two cases of noisy observation matrices (top row), as well as the respective sparse noisy (middle row) and low rank (bottom row) components. The results demonstrate that the RPCA method is able to accurately separate the clean low rank signal even in the presence of a significant amount of noise.

4.5.4 RECOVERY FROM MISSING MEASUREMENTS

In addition to the separation of true signal and noise as was demonstrated in the previous subsection, in many cases the problem involves the recovery of missing observations. In this subsection, we consider the recovery of high-order observations from a subset of measurements where a given sampling rate, i.e., 20%, quantifies the number of observed measurements over the total number of measurements.

The observations are encoded in a $17 \times 4 \times 100$ tensor, while the recovery involved the reconstruction of the three-way tensors [50] or the recovery of one of the three matrix unfoldings [27], and we report the average reconstruction quality in terms of normalized mean square error over ten independent realizations. In Figure 4.8, we present the input (a), and the reconstructions from 25% (b), and 50% (c) observations through tensor completion. We observe that when half the measurements are missing a very realistic estimation can be produced, while even when one quarter of the measurements is not available, the most significant features are preserved. Figure 4.8d presents the plot of the reconstruction quality in terms of NMSE as a function of the sampling rate (%) achieved by TC and MC applied on the three unfoldings of the tensor. One can observe that increasing the sampling rate leads to more accurate reconstruction in both modelling tensor and matrix scenarios. Furthermore, an observation related to the performance achieved by different methods is that results clearly demonstrate the superiority of the tensor approach, which enjoys a nearly monotonic decrease in terms of error, in contrast to the matrix cases where in two out of three unfolding scenarios, the reconstruction error reaches a plateau. This phenomenon

---

[1] Gasch, Caley; Brown, David (2017). Data from: A field-scale sensor network data set for monitoring and modelling the spatial and temporal variation of soil moisture in a dryland agricultural field. *Ag Data Commons.* http://dx.doi.org/10.15482/USDA.ADC/1349683

Matrix and Tensor Signal Modelling in Cyber Physical Systems 123

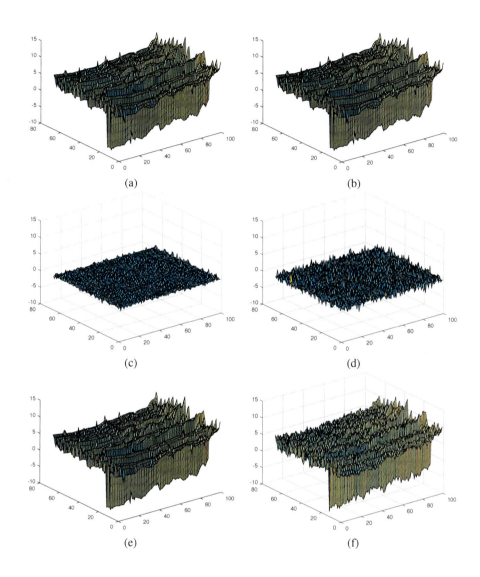

Figure 4.7: Noisy observations with moderate (a) and high noise (b). Recovery of corresponding sparse noise (c,d) and low rank (e,f) components.

can be attributed to the inability of specific unfoldings to generate sufficiently low rank matrices, leading to poor reconstruction.

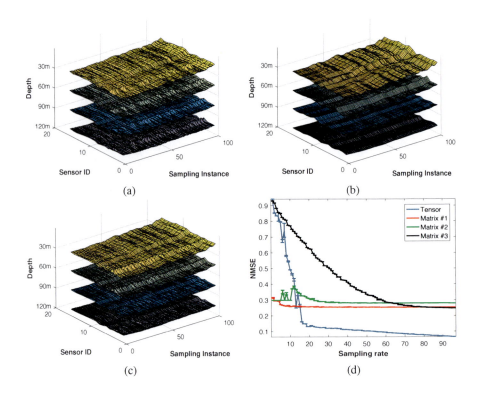

Figure 4.8: Timeseries of observations from multiple sensors at different depths (a) are recovered from half (b) and one quarter (c) of the respective measurements. Performance of tensor and matrix completion over different sampling rates (d).

4.6 CONCLUDING REMARKS AND FUTURE DIRECTIONS

A core process of CPS is the sensing and gathering of observations in order to obtain the most informative view of the physical environment. Wireless sensor networks and the Internet-of-Things are enabling technologies for CPS which have been drawing significant attention from both the academic and the industrial communities. However, the complexity of sensing, the processing limitations, and communications' breakdowns often lead to cases where observations are missing or heavily affected by noise. For this scenario, modern signal processing and learning approaches play a pivotal role in recovering and understanding CPS data from challenging conditions.

Taking a broader look at the scientific progress in the field, one can identify the significant potential in the realm of multidimensional structured observations modelling. Tensor modelling and processing, and their application in tasks such as recovery, denoising, and exploratory analysis are the tip of the spear due both to their mathematical justification and excellent practical performance. At the same time, processing algorithms such as tensor decompositions are becoming more stable and efficient for practical use. A reasonable prediction is that a tight integration with other CPS components like control and actuation systems is expected to provide significant gains in terms of efficiency, cost, robustness, and value.

4.7 ACKNOWLEDGMENTS

This work was funded by the DEDALE project, contract no. 665044, within the H2020 Framework Program of the European Commission.

4.8 REFERENCES

1. F Adamo, F Attivissimo, C G C Carducci, and A M L Lanzolla. A smart sensor network for sea water quality monitoring. *IEEE Sensors Journal*, 15(5):2514–2522, 2015.
2. K Sarpong A-Manu, C Tapparello, W Heinzelman, F Apietu Katsriku, and J-D Abdulai. Water quality monitoring using wireless sensor networks: Current trends and future research directions. *ACM Transactions on Sensor Networks (TOSN)*, 13(1):4, 2017.
3. M Aharon, M Elad, and A Bruckstein. K-svd: An algorithm for designing overcomplete dictionaries for sparse representation. *IEEE Transactions on signal processing*, 54(11):4311–4322, 2006.
4. M T Asif, N Mitrovic, J Dauwels, and P Jaillet. Matrix and tensor based methods for missing data estimation in large traffic networks. *IEEE Transactions on Intelligent Transportation Systems*, 17(7):1816–1825, 2016.
5. J-F Cai, E J Candès, and Z Shen. A singular value thresholding algorithm for matrix completion. *SIAM Journal on Optimization*, 20:1956–1982, 2010.
6. E J Candès, X Li, Y Ma, and J Wright. Robust principal component analysis? *Journal of the ACM (JACM)*, 58(3):11, 2011.
7. E J Candès and B Recht. Exact matrix completion via convex optimization. *Foundations of Computational Mathematics*, 9(6):717, 2009.
8. E J Candès, J Romberg, and T Tao. Robust uncertainty principles: Exact signal reconstruction from highly incomplete frequency information. *IEEE Transactions on Information Theory*, 52(2):489–509, 2006.
9. J D Carroll and J-J Chang. Analysis of individual differences in multidimensional scaling via an n-way generalization of eckart-young decomposition. *Psychometrika*, 35(3):283–319, 1970.
10. L Chen, J Jakubowicz, D Yang, D Zhang, and G Pan. Fine-grained urban event detection and characterization based on tensor cofactorization. *IEEE Transactions on Human-Machine Systems*, 47(3):380–391, 2017.
11. S Shaobing Chen, D L Donoho, and M A Saunders. Atomic decomposition by basis pursuit. *SIAM Review*, 43(1):129–159, 2001.
12. J Cheng, Q Ye, H Jiang, D Wang, and C Wang. Stcdg: An efficient data gathering algorithm based on matrix completion for wireless sensor networks. *IEEE Transactions on Wireless Communications*, 12(2):850–861, 2013.
13. A Cichocki, D Mandic, L De Lathauwer, G Zhou, Q Zhao, C Caiafa, and H APhan. Tensor decompositions for signal processing applications: From two-way to multiway component analysis. *IEEE Signal Processing Magazine*, 32(2):145–163, 2015.
14. A Cichocki, R Zdunek, A H Phan, and S Amari. *Nonnegative matrix and tensor factorizations: Applications to exploratory multi-way data analysis and blind source separation*. John Wiley & Sons, 2009.

15. S F Cotter, B D Rao, K Engan, and K Kreutz-Delgado. Sparse solutions to linear inverse problems with multiple measurement vectors. *IEEE Transactions on Signal Processing*, 53(7):2477–2488, 2005.

16. L De Lathauwer, B De Moor, and J Vandewalle. A multilinear singular value decomposition. *SIAM Journal on Matrix Analysis and Applications*, 21(4):1253–1278, 2000.

17. D L Donoho. Compressed sensing. *IEEE Transactions on Information Theory*, 52(4):1289–1306, 2006.

18. D L Donoho, Y Tsaig, I Drori, and J-L Starck. Sparse solution of underdetermined systems of linear equations by stagewise orthogonal matching pursuit. *IEEE Transactions on Information Theory*, 58(2):1094–1121, 2012.

19. R Du, L Gkatzikis, C Fischione, and M Xiao. Energy efficient sensor activation for water distribution networks based on compressive sensing. *IEEE Journal on Selected Areas in Communications*, 33(12):2997–3010, 2015.

20. M Elad. Prologue. In *Sparse and Redundant Representations*, pages 3–15. Springer, 2010.

21. M Giannopoulos, S Savvaki, G Tsagkatakis, and P Tsakalides. Application of tensor and matrix completion on environmental sensing data. In *European Symposium on Artificial Neural Networks, Computational Intelligence and Machine Learning*, 2017.

22. R A Harshman. Foundations of the parafac procedure: Models and conditions for an" explanatory" multi-modal factor analysis. 1970.

23. R Jenatton, J-Y Audibert, and F Bach. Structured variable selection with sparsity-inducing norms. *Journal of Machine Learning Research*, 12(Oct):2777–2824, 2011.

24. T G Kolda and B W Bader. Tensor decompositions and applications. *SIAM Review*, 51(3):455–500, 2009.

25. J B Kruskal. Three-way arrays: rank and uniqueness of trilinear decompositions, with application to arithmetic complexity and statistics. *Linear Algebra and Its Applications*, 18(2):95–138, 1977.

26. A P Liavas and N D Sidiropoulos. Parallel algorithms for constrained tensor factorization via alternating direction method of multipliers. *IEEE Transactions on Signal Processing*, 63(20):5450–5463, 2015.

27. Z Lin, M Chen, and Y Ma. The augmented lagrange multiplier method for exact recovery of corrupted low-rank matrices. *arXiv preprint arXiv:1009.5055*, 2010.

28. J Liu, P Musialski, P Wonka, and J Ye. Tensor completion for estimating missing values in visual data. *IEEE Transactions on Pattern Analysis and Machine Intelligence*, 35(1):208–220, 2013.

29. A Y Lokhov, N Lemons, T C McAndrew, A Hagberg, and S Backhaus. Detection of cyber-physical faults and intrusions from physical correlations. In *International Conference on Data Mining Workshops (ICDMW)*, pages 303–310. IEEE, 2016.

30. S Mallat. *A wavelet tour of signal processing: the sparse way*. Academic Press, 2008.

31. D Needell and J A Tropp. Cosamp: Iterative signal recovery from incomplete and inaccurate samples. *Applied and Computational Harmonic Analysis*, 26(3):301–321, 2009.

32. A Neumaier. Solving ill-conditioned and singular linear systems: A tutorial on regularization. *SIAM Review*, 40(3):636–666, 1998.

33. S Nikitaki, G Tsagkatakis, and P Tsakalides. Efficient multi-channel signal strength based localization via matrix completion and bayesian sparse learning. *IEEE Transactions on Mobile Computing*, 14(11):2244–2256, 2015.

34. A Ostfeld, J G Uber, E Salomons, J W Berry, W E Hart, C A Phillips, J-P Watson, G Dorini, P Jonkergouw, Z Kapelan, et al. The battle of the water sensor networks (bwsn):

A design challenge for engineers and algorithms. *Journal of Water Resources Planning and Management*, 134(6):556–568, 2008.

35. A Panousopoulou, M Azkune, and P Tsakalides. Feature selection for performance characterization in multi-hop wireless sensor networks. *Ad Hoc Networks*, 49:70–89, 2016.

36. D B Rawat, J JPC Rodrigues, and I Stojmenovic. *Cyber-physical systems: From theory to practice*. CRC Press, 2015.

37. S Savvaki, G Tsagkatakis, A Panousopoulou, and P Tsakalides. Matrix and tensor completion on a human activity recognition framework. *IEEE Journal of Biomedical and Health Informatics*, 2017.

38. T S Shores. *Applied linear algebra and matrix analysis*. Springer Science & Business Media, 2007.

39. N D Sidiropoulos and R Bro. On the uniqueness of multilinear decomposition of n-way arrays. *Journal of Chemometrics*, 14(3):229–239, 2000.

40. N D Sidiropoulos, L De Lathauwer, X Fu, K Huang, E E Papalexakis, and C Faloutsos. Tensor decomposition for signal processing and machine learning. *IEEE Transactions on Signal Processing*, 65(13):3551–3582, 2017.

41. R Tibshirani. Regression shrinkage and selection via the lasso. *Journal of the Royal Statistical Society. Series B (Methodological)*, pages 267–288, 1996.

42. J A Tropp. Greed is good: Algorithmic results for sparse approximation. *IEEE Transactions on Information Theory*, 50(10):2231–2242, 2004.

43. J A Tropp and C Gilbert. Signal recovery from random measurements via orthogonal matching pursuit. *IEEE Transactions on Information Theory*, 53(12):4655–4666, 2007.

44. G Tsagkatakis, B Beferull-Lozano, and P Tsakalides. Singular spectrum-based matrix completion for time series recovery and prediction. *EURASIP Journal on Advances in Signal Processing*, 2016(1):66, 2016.

45. G Tsagkatakis and P Tsakalides. Dictionary based reconstruction and classification of randomly sampled sensor network data. In *Sensor Array and Multichannel Signal Processing Workshop (SAM), 2012 IEEE 7th*, pages 117–120. IEEE, 2012.

46. L R Tucker. Implications of factor analysis of three-way matrices for measurement of change. *Problems in Measuring Change*, 122137, 1963.

47. G Tzagkarakis, G Tsagkatakis, D Alonso, E Celada, C Asensio, A Panousopoulou, P Tsakalides, and B Beferull-Lozano. Signal and data processing techniques for industrial cyber-physical systems. *Cyber physical systems: From theory to practice, CRC Press, USA*, 2015.

48. K Xie, L Wang, X Wang, J Wen, and G Xie. Learning from the past: Intelligent online weather monitoring based on matrix completion. In *International Conference on Distributed Computing Systems (ICDCS)*, pages 176–185. IEEE, 2014.

49. G Xu, W Shen, and X Wang. Applications of wireless sensor networks in marine environment monitoring: A survey. *Sensors*, 14(9):16932–16954, 2014.

50. Y Xu, R Hao, W Yin, and Z Su. Parallel matrix factorization for low-rank tensor completion. *arXiv preprint arXiv:1312.1254*, 2013.

5 Sensing-Based Leak Quantification Techniques in Water Distribution Systems

ARY MAZHARUDDIN SHIDDIQI, RACHEL CARDELL-OLIVER, AMITAVA DATTA

CHAPTER HIGHLIGHTS

- Methods for discovering the size and location of leaks in water distribution systems are reviewed.
- Machine learning algorithms classify leaks based on observed flows and pressures.
- Optimization algorithms find the best placement for sensors to capture leak signatures.
- Sectorization of a large water distribution system reduces the computational cost and increases the performance of leak quantification techniques.

5.1 INTRODUCTION AND MOTIVATION

Leaks in water distribution networks are significant economic and environmental problems worldwide that result in the loss of billions of liters of water per year. Aging water distribution infrastructures have contributed the most to these losses. A report from the Energy and Water Department of the World Bank stated that the

---

A Shiddiqi ✉

CRC for Water Sensitive Cities, University of Western Australia, AU,

and Institut Teknologi Sepuluh Nopember, ID

E-mail: ary.shiddiqi@research.uwa.edu.au

R Cardell-Oliver

CRC for Water Sensitive Cities, University of Western Australia, AU

A Datta

University of Western Australia, AU

global water loss per year is about 48 billion m^3 with estimated financial costs at US$14 billion [6]. Leak quantification systems aim to reduce water loss by detecting and locating leaks in water pipeline systems.

A simple understanding of leak quantification is described in the "water balance" law. This law states that the sum of water supplied into a water distribution system (WDS) must be equal to the sum of the demands. If the amount of water supplied is larger than the demand, then there is a leak in the WDS. Each leak position has specific influences on different parts of a WDS driven by the leak size. The bigger the leak size, the bigger the influence.

Detecting and localizing leaks can be performed by observing the leak influences and using the information to estimate the location and size of the leaks, as shown in Figure 5.1. Many leak quantification techniques have been developed, such as mathematical modelling and machine learning algorithms. Mathematical modelling uses the physical features of a WDS (e.g., the network structure) and known demands to simulate hydraulic behaviors. The model is compared to the real observations of the WDS. Features that differ significantly to the model are suspected to have leakage. Machine learning algorithms work by developing a model extracted from training data. Machine learning for leak quantification can be categorized into two methods: optimization and classification algorithms. Optimization algorithms have similar principles to mathematical modelling techniques; but with lower computational cost. The optimization algorithms work by repetitively finding a model that fits an operational WDS features by evolving the model's variables. Classification algorithms use a training model that contains classes to identify which class an observation profile belongs to. While mathematical modelling has built the foundation of leak quantification technique, the other techniques have opened up many research opportunities in this field.

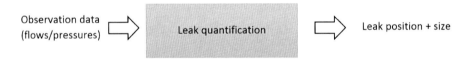

Figure 5.1: Leak quantification process.

Finding leaks through full observation of all WDS features (e.g., pipes) will give the best result. However, it is not feasible to install sensors on every pipe. One solution is to develop a sensor placement strategy that maximizes leak quantification accuracy given a fixed number of sensors. This can be performed through exploring possible sensor locations that give the highest leak quantification performance. Optimization algorithms are the most common means to find such locations. Some techniques use network structure information to enhance the original optimization algorithms.

In this chapter, we first present mathematical modelling for leak quantification. Then we review techniques to solve leak quantification using machine learning algorithms, sensor placement strategies, and sectorization of a large WDS.

5.2 MATHEMATICAL MODELLING BACKGROUND

Pudar and Liggett [10] were the first to investigate a leak quantification technique. Their technique used nonlinear equations to simulate hydraulic behavior in a WDS. The nonlinear equations used to solve the behavior of water in a WDS based on the water balance law as shown in Equation 5.1 [10].

$$\sum_{k=1}^{n} \text{sgn}(h_i - h_j) \left| \frac{h_i - h_j}{K_{ij}} \right|^{\frac{1}{n}} = D_i + q_i \tag{5.1}$$

where n = the velocity exponent in the friction equation; n_i = the number of pipes intersecting at node i; j = an adjacent node to i; h = head; and $\text{sgn}(h_i - h_j)$ is the signed difference between heads at nodes i and j with flows heading to a node is positive and vice versa; K_{ij} = a coefficient of resistance of pipes connecting nodes i to j; D = demand at node i; and q_i = leak at node i.

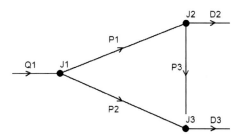

Figure 5.2: Small network example.

Consider for example, a WDS in Figure 5.2 with three nodes and three pipes. The network has one inlet ($Q1$), two demands ($D1$ & $D2$), and $D2 < D3$. The network requires three non-linear equations used to describe the network. When $n = 2$, then the three equations are:

$$\left| \frac{h_1 - h_2}{K_{12}} \right|^{\frac{1}{2}} + \left| \frac{h_1 - h_3}{K_{13}} \right|^{\frac{1}{2}} = D_2 + q_2 + D_3 + q_3 \tag{5.2}$$

$$\left| \frac{h_1 - h_2}{K_{12}} \right|^{\frac{1}{2}} - \left| \frac{h_2 - h_3}{K_{23}} \right|^{\frac{1}{2}} = D_2 + q_2 \tag{5.3}$$

$$\left| \frac{h_1 - h_3}{K_{13}} \right|^{\frac{1}{2}} + \left| \frac{h_2 - h_3}{K_{23}} \right|^{\frac{1}{2}} = D_3 + q_3 \tag{5.4}$$

The calibration process starts by setting initial values of the heads. Then, Newton-Raphson and Levenberg-Marquardt methods [9] are used to smoothly revise the head values. This process is applied iteratively until the WDS features converge resulting in a stable state of flows and heads. In a no-leak condition, the sum of a node's inflow is equal to the sum of the node's demand and outflow. As stated in the equations, a leak is considered as an additional demand at a node (q) that reduces the node's normal pressure. Leak detection is performed by comparing real measurements and simulated node pressures. A significant difference between a calculated and the real measurement indicates a leak at a node.

Leaks can occur as single or multiple events simultaneously. The more leaks occur, the harder they are to localize. Pudar and Liggett addressed this issue as an observation vs. leak event problem classified into three categories: even-determined, over-determined, and under-determined problems. Even-determined is where the observations are equal to the number of leaks, while over-determined is where the observations are more than the number of leaks, and under-determined is the opposite. Each method employs an inverse problem solver to estimate unknown heads from some observed heads of nodes. An inverse problem solver works by inverting an equation to find an unknown value of a variable. For example, suppose we have an equation $y = 2x + 1$. If we know the value $y = 5$, then the equation is reversed as $x = \frac{y-2}{2}$ to find x. The technique performed well in the first two categories (even-determined and over-determined), but for the under-determined problem. In general, a leak quantification technique performs best if the number of observations is equal to or more than the number of leaks.

Finding best locations for observations is crucial for the quality of leak quantification performance. Each location has a different ability to detect leaks affected by the structure of a WDS. Mathematical modelling uses a sensitivity matrix to measure the detection ability of each possible observation location. The sensitivity matrix is calculated using a forward problem solver using the nonlinear equations from known physical features and demands of a WDS. A forward problem solver works by finding unknowns from known values. For example, we know the value $x = 2$ and we want to find y. By using the example equation above, we find $y = 5$. The matrix contains the probability of locations that will be affected by each possible leak event. Hence, the matrix can be used to decide where to make measurements. The highly affected places are the best places to conduct observations. A good sensitivity matrix is one with no extreme values. The extreme values can lead to higher error in the head calculations. A well–structured WDS with adequate link redundancies will result in better leak quantification performance.

5.3 SIMULATING A WATER DISTRIBUTION SYSTEM

While Pudar and Liggett's technique has given a systematic foundation for the leak quantification problem, the mathematical modelling involved has a high computational cost to quantify leaks. In addition, finding a large number of real cases of leaks is difficult. Therefore, a more feasible approach is to use WDS simulators. Recent techniques have used WDS simulation tools that automate the nonlinear equations, such as EPANET [11]. The tools enable leak scenarios to be generated in leak quantification technique developments.

The general work flow of the simulators is shown in Figure 5.3. The simulators work by taking physical information and node demands of a WDS as the input. Then, the simulators run an algorithm to solve the nonlinear equations, explained in Section

5.2, to generate node pressures and pipe flows. An example of the algorithms is the Gradient algorithm used in EPANET. The Gradient algorithm begins by estimating flows in each pipe. Then, node heads are calculated at each iteration to rectify the pipe flows. A leak is simulated as an orifice represented by an emitter coefficient at a node. To obtain all possible leak data, leaks are simulated at all nodes by changing the emitter coefficient value at different sizes. This data is recorded and used for learning as well as the testing process in developing leak quantification techniques.

Figure 5.3: Modelling leaks for generating observation data using simulators.

5.4 MACHINE LEARNING ALGORITHMS FOR LEAK QUANTIFICATION

Machine learning algorithms are well-known tools for extracting the *unseen* model from a set of data [5]. Machine learning consists of two steps: learning and testing. The learning process extracts and generates a model from a set of data, while the testing process compares the learned model to observed WDS features. Analysis of the comparison classifies the leak location and predicts the leak magnitude. We focus our review on algorithms commonly used in leak quantification systems: optimization and classification algorithms.

5.4.1 OPTIMIZATION AND CLASSIFICATION ALGORITHMS

The goal of optimization algorithms is to find the "best" solution from a set of possible solutions using some criteria. The best solution can be the maximum or the minimum values depending on the purpose of a leak quantification technique. Each solution contains a WDS configuration, such as demands, pipe roughness, heads, flows, etc. For example, when a WDS has a leakage at a node, the other node pressures change from their normal conditions. The altered node pressures are the signatures of the leak. Optimization algorithms seek a WDS configuration that results in the least difference between the simulated and the altered node pressures. An altered node pressure with the highest difference from the simulated pressure is flagged as the leak source.

Classification algorithms extract a model from a dataset to describe classes. The dataset contains tuples and their associated class labels. This learning process is termed *supervised* learning. The model, called a classifier, uses a mapping function to predict a class label Y of an instance X. The mapping function can be classification rules, decision trees, or mathematical equations. In leak quantification systems, a binary classifier can be used to identify whether a WDS has a leak or not. The classifier can be extended to localize leaks at nodes/pipes by training the classifier using a multiclass dataset that contains leak locations as the class labels.

5.4.2 SINGLE AND MULTI-LEAKS QUANTIFICATION

Leaks can occur as single events or multiple events simultaneously. Quantifying multiple leaks is harder than a single leak and often comes with trade-offs, such as high computational cost or lower performance. The trade-offs are caused by the presence of more leak combination possibilities to cover for quantifying leaks. This has brought challenges in the development of both single and multi-leak quantification algorithms.

A. Genetic Algorithm

A genetic algorithm (GA) is an optimization algorithm that works based on a natural selection mechanism to find the best candidate of a population. In the population, a solution is encapsulated in a *chromosome*. For each generation, the chromosomes are evaluated using an objective function to select the best chromosomes of the generation. The selected chromosomes are modified by using mutation and crossover methods to form the next generation. Mutation means randomly altering genes of the selected chromosomes, while crossover means keeping similar genes of a pair of chromosomes and swapping the different genes. This process continues until certain criteria are met. Common stopping criteria are: if the objective function results are no longer changing or have only small changes, and if a fixed number of generations are reached. GA is more efficient than exhaustive search techniques, such as brute force, in terms of computational cost.

A GA for leak quantification works by finding the best combination of WDS features that produces the least difference with the currently observed ones [15]. The technique used two variables encapsulated in a chromosome to represent a solution, i.e., node identifier and emitter coefficient. Each solution is simulated to produce node pressures and pipe flows of the WDS. The technique observed the pressure and flow changes caused by different leak positions. Three objective functions were used to select the best candidates for each generation. The three objective functions work by minimizing the sum of difference squares, minimizing the sum of absolute differences, and minimizing the maximum absolute difference squares of the simulated and observed values of the two features.

The GA-based technique terminates when the maximum number of trials or the acceptable fitness values are reached. The technique was able to locate and predict size in a single or multiple leaks. Real data can be used to improve the GA model, such as time series data of flows / pressures, and hydrant discharges.

B. Support Vector Machine

Support vector machine (SVM) is a classification algorithm that transforms a set of data into a high-dimensional space using a function, called a *kernel*. The goal is to find as clear and as wide a gap as possible (called a *hyperplane*) to separate the data into their classes. The classes can be used to detect whether leaks are present or not.

Leak-driven pressure profiles of different leak positions and sizes are used to develop an SVM model and then tested against an observed pressure profile to classify a leak location [7]. The SVM-based technique used a multiclass which is an extended version of the original SVM that can only handle two classes. The technique classifies leaks of different sizes into three categories: low, medium, and high. Therefore, four SVM models were used for all of the leak classes and the no-leak condition. The multiclass SVM used distance measurement and classified leak signatures into *one* vs. *the others* categories. The closest distance was labeled as +1, and the others were -1. The classification process was recursively performed until the signatures were tested against all SVM models and then stored the classification results in a vector. The vector was compared to a *code matrix* that used as the translator from coded SVM outputs to leak details. For example, a no-leak condition is coded as [+1, -1, -1, -1], while a low leak is coded as [-1, +1, -1, -1], and so on. Leak signatures that match to a code matrix entry were labeled with the corresponding class.

Real WDS comprise various pipeline structures, such as size, edge, material, etc., that influence the WDS system. Discretized SVM outputs diminish information on the effect of the various structures. Instead of directly using the discrete classes, the SVM outputs were calibrated using Bayes' rule probability to produce probabilistic results. This approach had made the technique more informative for decision making. The technique had promising performances in leak quantification. The technique's performance can be maximized through further experiments using real data with multiparameter monitoring sensors. Another important aspect to enhance the technique is to develop a leak localization function.

C. Artificial Neural Networks

ANNs have been used in many applications for pattern recognition. The ANN works in a similar way to biological neural networks that use interconnected neurons to compute outputs from known inputs. The key is at the hidden neurons that sit between the input and output. The hidden neurons contain activation functions, a nonlinear equation, to represent the complex behavior of the neural net. Inputs are summed up at each hidden neuron, and the activation function decides to either "activate" the neuron or not to the next hidden layer/output computations. Optimization takes place at the learning process to obtain the best weight values for each hidden neuron to meet a range of outputs.

An example of ANNs for a leak quantification system used flow measurement delay signals taken from the inlet and outlet of pipes as the input of the ANN [2]. A delay signal is a scalar which is sampled over a specified period of time. The technique placed two sensors at the inlet and the outlet of a pipe and possible leaks are located between them. The ANN model was trained using the delay signals taken over a period of time to find optimized weights of the hidden neurons. The training process produced a model that fits into a range of leak classes. Outputs were calculated from observed delay signals using the optimized weights and activation functions at the hidden neurons. Each output represented a unique condition in the pipe. The technique used a pipe with three possible leak locations. The combinations of all possible leak events of the three locations and a no-leak condition were encoded using 8 unique outputs. A state decoder was used to translate an output of the ANN to the detail of a condition: no-leak or the leak location(s). For example, if the ANN output is 8, then leaks are located at positions 1, 2, and 3. Any other outputs will follow the same logic of the state decoder.

The technique performed well with low error rate even in the presence of noise. This means the technique has shown a promising prospect in leak quantification systems, particularly for multiple leaks. However, this technique has a high computational cost that limits the ability of ANN to quantify leaks to only a few observable pipes.

D. Rule Learning

Leak quantification techniques discussed in the previous sections use "black box" approaches that give no explanation of their decisions. Black box techniques take inputs and decide the output based on the distance of the inputs to a corresponding instance of the model. The opposite of this decision-making model is the "white box" approach that explains the logic of the model.

Rule-based classification is a white box machine learning algorithm where the logic of the model is interpretable. An example is the repeated incremental pruning to produce error reduction (RIPPER) which is a decision tree classifier based on association rule algorithms. The algorithm initially generates a large tree to accommodate all possible rules from a training dataset. Then, the large tree is repeatedly simplified by removing any single rule with the highest error reduction in the simplified tree. This process ends when removing a rule resulted in the increase of error on the simplified tree. Through this mechanism, RIPPER assures that there will be no rule conflicts as each tuple can only trigger one rule.

RIPPER presents its logic as a set of rules learned from a database of observed WDS features [3]. To illustrate how the technique works, let us consider a pipeline network structure with one reservoir, two demand nodes, and pipes that channeled water from the reservoir to the demands. Leaks of different sizes were simulated at all possible locations at the actual and virtual nodes on pipes. The simulation results were different flows and pressures on pipes and nodes respectively stored in the same database at each corresponding leak scenario. Flow data is used to train RIPPER to develop a tree-based classification model. An example of the model logic of the RIPPER-based technique extracted from the learning process is given as:

FOR a running WDS

IF (p1 $\geq$ 7.5 L/s) AND (p2 $\leq$ 5.4 L/s)

THEN LeakSize = 0.3 L/s AND LeakPosition = Node 8

Confidence 15/1

The rule states that if two conditions are met, i.e., flow in pipe 1 $\geq$ 7.5 L/s and flow in pipe 2 $\leq$ 5.4 L/s, then a leak is detected with the size of 0.3 L/s located at node 8. The confidence level was determined in the training process. If we set the confidence level high, then detailed rules will be generated. This type of rule aims to cover all possible leaks in a WDS. However, the detailed rules are susceptible to over-fitting to the training data. On the other hand, if the rules are too loose, then the classification might end up in the same class for all testing data. Therefore, tuning the RIPPER parameters is critical to achieving the best model.

The rule-based technique has demonstrated good performance for the leak quantification problem. The advantage of the technique is that it learns and presents interpretable rules to explain the leak quantification logic. However, a real implementation of data gathering using a systematic method for sensor placement is still open for further investigation.

5.5 SENSOR PLACEMENT STRATEGIES

Installing sensors on every pipe is the best method to capture leak signatures. However, this approach is not feasible due to provision and maintenance costs. A strategy to find the best sensor placement is needed that maximizes the accuracy of a leak quantification system given a fixed number of sensors. Optimization algorithms can be used for finding the best sensor positions. The optimization algorithms work by comparing all possible sensor places and choosing positions with the highest detection and localization abilities. However, conventional optimization techniques, such as brute force, have a high computational cost. Heuristic optimization algorithms can overcome this shortcoming.

5.5.1 STATE SPACE SEARCH

A state space is a set of values as a result of a process. In sensor placement strategy, the state space is the set of all possible configurations of sensors. Configuring optimal sensor locations can be regarded as finding the best state from an initial state. Each transition is constrained by the leak quantification goals: detection and localization of leaks.

Depth-first branch and bound algorithm searches for a desired state by exploring all possible branches that meet certain properties [12]. The technique searches the best sensor configuration based on maximum detection and leak isolation. Maximum detection measures the leak detection coverage while fault isolation measures the localization ability when a set of sensors is installed. The state-search-based technique started the sensor placement strategy from an over-determined model. At this stage, all leaks can be detected and localized. The depth-first branch and bound search algorithm is then run to find best positions for a given number of sensors. At the root, an initial candidate sensor set for installation is set equal to all possible observation locations, the best sensor configuration is set to empty, and both overall minimum detection and localization scores are set to 0. For each branching step, a sensor is randomly removed from all possible sensor locations and the branch's minimum detection and localization scores are evaluated. If the current branch's minimum detection and localization scores are equal to or more than overall minimum detection and localization scores, then the best sensor configuration is updated to the current sensor configuration. Otherwise, the branch is discharged.

This process is recursively performed until certain conditions at the parent or child were met. At the parent node, the process stopped when all ancestors of some m-sensor configuration had been visited and there was a solution from a child node with duplicate isolation score to the parent node. On the other hand, the branch and bound process stopped if: the branch's leak isolation score does not improve the current best one, the current m-sensor configuration does not cover all leaks, and the number of m-sensor had been reached. The last criterion means the subnode branches would have fewer sensors than m.

Providing m-sensors for a WDS sometimes does not give the best performance of a leak quantification system. This means there are some undetected leaks that result in constant water loss. The more undetected leaks, the more losses to bear. Having an approximation of an "ideal" m-sensor configuration of a WDS will be useful from the management's point of view. The technique was able to suggest the ideal m-sensors and the placement in a WDS. Parameter sweeping method was used to find the ideal m-sensors by setting m from 1 to the maximum sensors that can be placed in the WDS. The ideal number of m-sensors is reached when minimum detection and localization scores have reached a steady state indicated by insignificant increases in both scores.

Although the technique is efficient for sensor placement strategy, it may find suboptimal solutions. The tree-based optimization method has limited the technique from exploring all possible solutions. Further investigations are open to solve this problem. Implementing a cost function for sensor provision that constrains the selection of m sensor configuration is another future work to enhance the technique.

5.5.2 MINIMUM SET COVER

A leak event triggers unique changes in pressures and flows in a WDS. This is known as the leak signature. For example, suppose a WDS has n nodes and m pipes. A leak event at pipe 1 (x_1) causes the pressure changes at nodes $\{y_{x1-1},...,y_{x1-n}\}$. If we simulate all possible leak events on pipes (m), we will have a set of leak events $X = \{x_1,...,x_m\}$ and the set of signatures is $Y = \{\{y_{x1-1},...,y_{x1-n}\},...,\{y_{xm-1},...,y_{xm-n}\}\}$. The relations between the leak event and the pressures can be mapped into a matrix of X and Y, called the *influence matrix*.

Leak localization can be performed by finding a set of sensors that can uniquely identify each leak event. This problem is known as the minimum test cover (MTC), an NP-hard problem. One solution to avoid the NP-hard problem is to transform the MTC problem into an NP-complete problem, such as minimum set cover (MSC) problem [1]. The technique transformed the influence matrix into a new representation called *identification matrix* where the rows are *pair-wise* leak events and the

Sensing-Based Leak Quantification Techniques in Water Distribution Systems **139**

columns are the result of *XOR* operation product of the *pair-wise* event's individual sensor readings. The intuition of this transformation is that a sensor that produces different sensing outputs of a pair-wise leak event can distinguish the pair-wise individual leak signatures. Then, a minimum set of sensors is selected to cover all pair-wise events in the identification matrix. A simple way to find the minimum set of sensors is to use the brute force method. However, this approach is computationally expensive. The MST-based technique uses a heuristic method to select the minimum set of sensors based on two criteria: the number of pair-wise leaks that are not detected by the current set of sensors that can be detected by the additional sensor, and the number of pair-wise leaks that are detected by current sensors that can be uniquely identified by the additional sensor. The number of sensors giving the highest sum of the scores is selected as the solution.

Recently, sensor manufacturers have developed advanced sensors that are capable of producing high-resolution readings. For example, a WDS simulation is run in a period of [0 T] time. The MST-based technique divided the running time into two subintervals $[0\, t_1)$ and $[t_1\ T]$. This division results in three sensor states to recognize a leak event: [0 0] if an event is not detected at all, [1 0] if an event is detected in $0 < t_1$, and [0 1] if an event is detected in $t_1 \leq T$. Through this approach, leak detection and localization performance can be improved compared to single level sensors. For example, two leak events occur in different intervals identified by two single-level sensing sensors $S_1 \& S_2$ as [0 1] & [1 0], can be sensed by one multilevel sensor. Consequently, the number of sensors required is reduced. However, most sensors in the market have a certain range of inaccuracy that was not addressed in this method [1].

5.5.3 GENETIC ALGORITHMS

One technique that used GA for finding best sensor locations was by finding the maximum difference of node pressures between a no-leak and leak conditions of a WDS [4]. The technique modeled pressure data using a residual vector and a sensitivity matrix. A residual vector contains the difference of node pressures between the leak and no-leak conditions, while a sensitivity matrix contains each node pressure difference between the leak and no-leak conditions for all possible leak positions. The values in the residual vector and sensitivity matrix are affected by the location of each leak. The variation is normalized by projecting a leak's pressure residuals to the sensitivity matrix and stored in a projection vector. The highest projection value in the vector indicates the leak position.

The quality of sensor placement for a given number of sensors is determined by the error score of each possible set of sensor configuration. To illustrate how the error scoring works, let us call q_i the leak size at i and the highest projection value in the projection vector is ψ. If $q_i = \psi$, then the error score is 0, otherwise the error score is 1. This calculation is performed for all possible leaks. A perfect error score is 0 and the worst score is m. The perfect condition is mostly hard to reach, therefore the optimal sensor configuration is the one with the lowest error score.

The GA-based technique will reach the maximum performance if the leak size is close to the residual value. In reality, this ideal condition sometimes does not happen. The non-ideal condition causes various sensor configurations that result in higher error score in the discrete scoring approach. Three improvements to the error measurement function were developed in [4] to address this issue by:

- Using time series data to calculate the projection vectors. Each snapshot of the time series data results in a different projection vector. The highest mean of a similar position in the projection vectors was selected as the representation of the leak node.
- Replacing leak localization binary scoring with a distance-based scoring. Particularly for incorrect scores, rather than a straight penalty of 1, the new scoring approach measured the topological distance between an actual leak node i and the maximum projection value j by a threshold, i.e., the closest cut-off distance. Through this approach, the error score will not increase sharply.
- Generating all combinations of residual and sensitivity matrices from a range of possible leak sizes. Then, discharge all couples with duplicate error scores. The new error score was computed by averaging the error scores of the combinations.

By applying the three approaches, the technique becomes more robust for leaks at different sizes. However, this research did not address the problem of sensor inaccuracies where sensor readings might differ from the actual conditions.

5.5.4 ONE SENSOR TO FIND ALL LEAKS

A leak is analogous to an electrical "short circuit" that attracts all currents to the circuit short location. A similar trend happens if a leak is big enough; the leak source becomes a "sink" that attracts flows. Different leak positions have different leak signatures indicated by different excess flows at the inlet. The difference is caused by hydraulic resistances of accumulated individual links from a WDS inlet to the leak position. This indication fits into the *current-flow centrality* concept that measures the importance of a link in a network by measuring the average amount of current that passes through a link when all possible node pairs' currents in an electrical network are considered. The current-flow centrality concept was adopted for modelling a map to trace leak signatures [8]. The technique used one meter at the inlet of the WDS to observe flows and predict node heads to locate a leak.

A. Leak Detection and Localization

The technique generated an all node head pressures database based on WDS periodic demands. A leak event causes node heads to change resulting in the difference between the predicted (in the database) and the actual heads at nodes.

The current-flow centrality based technique modeled a WDS as a directed acyclic graph (DAG) to characterize each leak signature. This approach was used due to the similar characteristics between a DAG and the non-cyclic water flows in a WDS: excess water channels from the reservoir/the inlet all the way to the sink (leak source).

This means that the leak source is a node where the head difference matches with the excess water at the inlet.

A leak occurs when the leak pressure is at least equal to the static head from the outside environment. The outside head is due to soil weight, depth of leak, and the leak's elevation from mean sea level. The current-flow centrality based technique finds a leak by predicting initial node heads and performing iterative refinements until the system is converged. If a node's profile equals a leak signature detected at the WDS inlet, then the node is flagged as leaky. To illustrate, a leak at i is notated as $b'_i = \delta + d_i$, where δ is the difference of the sum of all demands in no-leak and leak conditions, and d_i is the actual demand at node i. The head at node i to penetrate the soil above the node is H_i and the head caused by the soil mass above the node is h_0 (Figure 5.4). The algorithm calibrates δ in order to make H_i equal to h_0. When both heads are equal, the signature of the leak at i is $b'_i - d_i$ detected at the inlet.

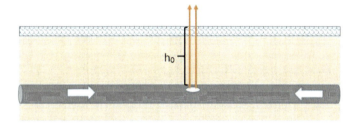

Figure 5.4: Burst penetrating soil above a leaky pipe.

B. Leak Signature Database

The current-flow centrality based technique generated a database of leak signatures for all possible leak positions from each snapshot of time. The signatures are generated by predicting all node heads from WDS demands and physical information such as network connectivity, diameters and lengths of pipes, rated power of pumps, and height and capacity of tanks. Each incoming leak signature is compared to the database instances. When the leak signature matches one of the instances, then a leak flag is raised along with the leak size and position. On time series data, the suspected location might be more than one that results in a set of suspected locations. The current-flow centrality based technique used a proximity threshold (5%) applied to the heads of the suspected leak locations and the excess flow detected at the inlet. This approach resulted in an area of suspected leaks. Then, a manual scanning method using an acoustic sensor is performed to find the actual leak position.

So far, the current-flow centrality based technique is the most efficient leak quantification technique in terms of the number of sensors used: just one sensor at the inlet. However, the method is most suitable for big leaks, not small leaks. In practice, small leaks contribute significantly to losses because they are undetected for a long time.

5.5.5 SMALL LEAK QUANTIFICATION

Excess flows of all possible leaks detected at the inlet that split to all possible paths heading towards a leak source can be used to characterize leak signatures [13]. Each leak position has a different flow change pattern, i.e., zero, positive, or even negative. Pipes with zero changes cannot detect leaks, while positive and negative changes can. The technique used the non-zero flow changes to distinguish leak signatures that form a DAG, a subset of a WDS, called the *lean graph*. A lean graph consists of a set of influenced pipes with flow changes heading to a leak source. For example, the signatures of leaks at J55 and J246 are indicated by the bold arrows as shown in Figures 5.5 and 5.6. The figures show that the lean graphs of leaks at different positions are unique as indicated by the different set of edges and flow directions. The lean graph is driven by leak sizes. The bigger the leak, the bigger the lean graph. If the lean graph is too big, the leak signatures overlap and become indistinguishable. The lean graph-based technique used a *minimum common leak size* for each snapshot of a WDS to avoid this problem.

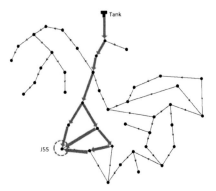

Figure 5.5: J55-Lean graph.

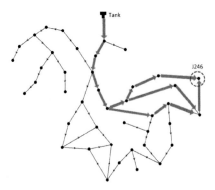

Figure 5.6: J426-Lean graph.

Leak localization was performed based on a clustering approach. The intuition is that if a leak is detectable at a WDS inlet, then it will also be detected at a cluster inlet. This is because essentially a cluster is a miniature of a WDS. Finding best separation of the clusters is the key to gain the best leak localization performance. The lean graph-based technique used *Jaccard Index (JI)* to measure the similarity of lean graphs. *JI* is defined as the measurement of similarity and diversity of a couple of lean graphs. This comparison was applied to all lean graph pairs and the results are stored in a table. Then, clustering a WDS started from finding a centroid. An edge that strongly connects two nodes, indicated by a high *JI*, was selected as a centroid of a cluster. A threshold (τ) was then set to define the boundary of the cluster. All nodes within the boundary were included in the cluster. This process was performed iteratively until the remaining nodes were assigned to a cluster.

Finding the inlets was based on selection of places with the most distinctive sensor readings. This approach was performed for all of the clusters formed through a selection process from a set of candidates. The distinctiveness of the inlets was measured using *cosine similarity* (*CS*) function. The *CS* function sums the dot products of each flow change recorded by pair-wised sensors at the inlets of two clusters and normalized by the dot product of the individual sensor readings squares. The lower *CS* score shows the more distinctive inlet readings, which means better inlet selection.

The lean graph-based technique modified τ from 0.1 to 0.9 to find the closest number of clusters (n) to a given k sensors. If the number of clusters was not equal to k, then post-partitioning adjustment was performed by merging ($n>k$) or partitioning a cluster ($n<k$). This approach is effective in finding the best sensor locations for a predetermined k sensors. However, the τ sweeping process is inefficient. A better method to find initial partitions closest to k is needed.

5.6 LEAK QUANTIFICATION IN LARGE-SCALE WATER DISTRIBUTION SYSTEMS

Large-scale pipeline networks can be challenging for leak quantification. This is because the size of a WDS contributes to the computational cost of a leak quantification system. The complexity of the computational cost increases along with the size of the network that can be linear, quadratic, or even exponential. A *sectorization* approach called district metered area (DMA) can be a feasible method for leak quantification in large-scale WDSs. The approach defines smaller areas or zones from the larger-scale WDS. An example of the approach is shown in Figure 5.7 that divides a WDS into five DMAs with sensors located at the inlet of each DMA [14].

A DMA–based leak quantification technique works hierarchically dividing a large WDS into smaller areas and sub-areas. When a DMA of a large WDS is suspected to have a leak, a leak quantification technique is performed in the suspected DMA to find which node(s) is the leak source [7]. The technique used the multiclass SVM to recognize a leak's signatures in a large WDS and label a DMA's profile that matches the leak signatures with (+1) while the other DMAs were labeled with (-1). Then, an SVM classification was performed in the suspected DMA.

The size of a DMA can be varied, depending on the scalability of the leak quantification systems. A big DMA can have hundreds of nodes and pipes. However, some DMAs may only contain fewer than one hundred nodes and pipes. A similar initial WDS partitioning approach with [14] was performed in [12]. The later technique represented a network as a directed graph and used the branch and bound algorithm to find positions with the highest leak detection and isolability for a given m-sensor configuration. Locations with the highest scores on both measurements were selected to place sensors.

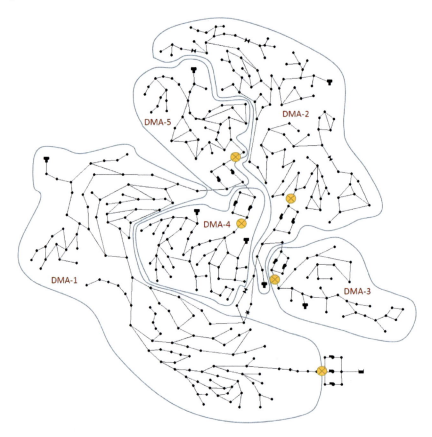

Figure 5.7: Sectorization in a large WDS.

A GA optimization algorithm is also a promising sensor placement technique for a large WDS [14]. C-Town network was used as a case study and divided the network into five DMAs (Figure 5.7). The technique used GA to produce combinations of pipe diameters or duplication of existing pipes in the network, tank capacities diameters modification, old pumps changing, and closing of existing pipes. The features were encapsulated in chromosomes. At each generation, chromosomes were evaluated using three objective functions to minimize: annual costs for operational (energy

+ water loss), pressure reduction valve (PRV), and physical component replacement (pump, tank, and pipe). The results were compared with the previous generation. Chromosomes that satisfy the evaluation functions were selected and used to generate the next generation. The next generation chromosomes were expected to produce the best or nearly optimal results (Pareto-optimal solutions).

The DMA-based strategy has been widely used in research for large-scale leak quantification systems. This is due to its practicality in localizing leaks into smaller clusters. The key to this method is defining an inlet for each DMA. Pipes with high leak detection ability are good candidates for a DMA's inlet.

Table 5.1

Summary of leak quantification techniques

LQ Technique	Data	Number of leaks		Fault tolerance	Large network	Localization	Leak size		Sensor plcmt. strat.
		Single	Multi				Small	Large	
Genetic algorithm	EPANET + real data	v				v		v	v
Support vector machine	EPANET + real data	v			v		v	v	
Artificial neural network	Flow observations		v			v	v	v	
Rule based learning	EPANET	v		v		v	v	v	
Search state space	EPANET + real data	v				v		v	v
Minimum set cover	HAMMER	v				v		v	v
Current-flow centrality	EPANET+ real data	v				v		v	v
Lean graph	EPANET	v			v	v	v		v

5.7 CONCLUDING REMARKS AND FUTURE DIRECTIONS

Although there have been advanced techniques developed, research opportunities in this field are still wide open. Table 5.1 summarizes the aspects covered by the leak quantification techniques reviewed in this chapter as well as some possible future developments. Among the future works are small and multi-leak quantification, sensor faults and inaccuracy, cost-sensitive clustering and machine learning for sensor placement strategies in large-scale networks. Leak quantification systems that address these problems would accelerate science discovery in this field and improve business and financial performance of water utilities.

5.8 REFERENCES

1. W Abbas, L S Perelman, S Amin, and X Koutsoukos. An efficient approach to fault identification in urban water networks using multi-level sensing. In *Proceedings of the 2Nd ACM International Conference on Embedded Systems for Energy-Efficient Built Environments*, BuildSys '15, pages 147–156, New York, NY, USA, 2015. ACM.
2. I Barradas, L E Garza, R Morales-Menendez, and A Vargas-Martínez. Leaks Detection in a Pipeline Using Artificial Neural Networks. In *Progress in Pattern Recognition, Image Analysis, Computer Vision, and Applications*, 2009, pages 637–644. Springer Berlin Heidelberg, Berlin, Heidelberg, 2009.
3. R Cardell-Oliver, V Scott, T Chapman, J Morgan, and A Simpson. Designing sensor networks for leak detection in water pipeline systems. In *2015 IEEE Tenth International Conference on Intelligent Sensors, Sensor Networks and Information Processing (ISSNIP)*, pages 1–6, April 2015.
4. M V Casillas, V Puig, L E Garza-Castañón, and A Rosich. Optimal sensor placement for leak location in water distribution networks using genetic algorithms. *Sensors (Basel, Switzerland)*, 13(11):14984–5005, January 2013.
5. J Han, M Kamber, and J Pei. Introduction. In *Data Mining (Third Edition)*, The Morgan Kaufmann Series in Data Management Systems. Morgan Kaufmann, Boston, 2012.
6. B Kingdom, R Liemberger, and P Marin. The challenge of reducing non-revenue (nrw) water in developing countries. How the private sector can help: A look at performance-based service contracting. Technical report, Water Supply and Sanitation Sector Board discussion paper series, 2006.
7. T G Mamo. Virtual DMA municipal water supply pipeline leak detection and classification using advance pattern recognizer multi-class SVM. *Journal of Pattern Recognition Research*, 1:25–42, 2014.
8. I Narayanan, A Vasan, V Sarangan, and A Sivasubramaniam. One meter to find them all-water network leak localization using a single flow meter. In *Information Processing in Sensor Networks, IPSN-14 Proceedings of the 13th International Symposium on*, pages 47–58, April 2014.
9. W H Press. *Numerical recipes 3rd edition: The art of scientific computing*. Cambridge University Press, 2007.
10. R S Pudar and J A Liggett. Leaks in pipe networks. *Journal of Hydraulic Engineering*, 118(7):1031–1046, 1992.
11. L A Rossman et al. Epanet 2: Users manual. 2000.
12. R Sarrate, F Nejjari, and A Rosich. Sensor placement for fault diagnosis performance maximization in distribution networks. In *20th Mediterranean Conference on Control & Automation (MED)*, pages 110–115. IEEE, 2012.
13. A M Shiddiqi, R Cardell-Oliver, and A Datta. Sensor placement strategy for locating leaks using lean graphs. In *Proceedings of the 3rd International Workshop on Cyber-Physical Systems for Smart Water Networks*, CySWATER2017, Pittsburgh, Pennsylvania, USA, 2017. ACM.

14. B A Tolson and A Khedr. Battle of background leakage assessment for water networks (BBLAWN): An incremental savings approach. *Procedia Engineering*, 89:69–77, 2014.
15. Z Y Wu and P Sage. Water loss detection via genetic algorithm optimization-based model calibration. *Water Distribution Systems Analysis Symposium 2006*, 247(2005):180, 2006.

Part II

Applications and Emerging Topics: Water Infrastructures and Urban Environments

6 An Agent-Based Storm Water Management System

LUIS A MONTESTRUQUE

CHAPTER HIGHLIGHTS

- We present the case study of a city using a cyber physical system to manage storm water in sewer systems. The system, called CSOnet, was designed to minimize pollution to a nearby river which results from excess storm water entering the sewer system. CSOnet is composed of 150 wireless sensor nodes and 12 actuator nodes.
- The case study describes CSOnet hardware, communication algorithms, control algorithms, implementation, and results. In particular, an agent-based control system which emulates a free market economy was implemented. The agent-based control implements buyers and sellers of hydraulic capacity with supply curves designed to more effectively utilize existing infrastructure and prevent overflows.
- The implementation of CSOnet resulted in effective reductions of wet weather induced sewer overflows. A secondary but important effect of implementing the system, CSOnet also provides visibility into the sewer infrastructure of the city that enables data-driven operation and maintenance activities.

6.1 INTRODUCTION AND MOTIVATION

More than 700 cities in the U.S. have sewer systems that transport both sanitary and storm water flows in the same system [2]. During rain storms, the combined sanitary and storm water flows can easily overload these sewer systems. To prevent excessive surcharging of the sewer lines and the overwhelming of the wastewater treatment plant (WWTP), a series of relief structures are designed to overflow excess water into the nearest water body. The discharge is called a combined sewer overflow (CSO) event. The discharged water is highly impacted with biological and chemical contaminants, thereby creating a major environmental and public health

---

L A Montestruque ✉

EmNet, LLC, US. E-mail: lmontest@emnet.net

151

hazard. Under the provisions of the 1972 Clean Water Act [4], the United States Environmental Protection Agency (USEPA) has compelled wastewater utilities to eliminate CSOs [1].

Traditional solutions to mitigate or eliminate CSOs typically involve large and expensive construction projects. For example, sewer separation projects consist of building an entire new sewer system to separate the storm water flows from the sanitary water flows. This process is not only highly disruptive to the public as it involves heavy construction in virtually every street at a greal expense, but it also does not provide the best environmental outcome as storm water runoff is also typically highly contaminated. Furthermore, keeping storm water entirely off sanitary sewer systems is not possible and, in many cases, leads to separate sewer overflows (SSOs).

The construction of large storage facilities is another traditional approach to CSO mitigation. In many cases these projects involve the construction of a large deep-rock tunnel to store and convey wastewater during rain events. After the rain subsides, the water is pumped out of the tunnel and into the WWTP for treatment. These types of projects reduce disruption to citizens as most of the construction is done underground but they represent a significant expense to rate payers.

Lately, green infrastructure has also been applied to reduce the amount of storm water runoff that leads to CSOs. Green infrastructure typically consists of rain gardens, bio swales, permeable pavement, and other structures that allow storm water to infiltrate the soil. While green infrastructure provides benefits to citizens by improving the urban landscape it is not always possible to achieve the level of performance required by environmental regulations.

An alternative method to eliminate or reduce CSOs is to dynamically control the hydraulics in the combined sewer system with the use of a sensor network. This approach uses control structures such as actuated gates, movable weirs, pump stations and other mechanisms that can be actuated to change the hydraulic behavior of the sewer system. Through dynamic control it is possible to take advantage of the spatial and temporal distribution of rainfall as well as concentration time differences among sub-catchments with diverse land use. By actively controlling the hydraulics in the sewer system it is possible to hold flows in areas that have available storage while dewatering areas that have excess flows effectively maximizing the infrastructure utilization [9, 19]. The cost of installing the infrastructure needed for dynamic control (actuators, sensors, and communication) is typically lower than traditional, capital-intensive, and heavy construction solutions.

The dynamic control of sewer systems for CSO mitigation is not new [6, 11]. Typically, model predictive control has been utilized in conjunction with a hydraulic model and weather forecast to dynamically compute setpoints for actuated controls [5, 17]. Significant challenges have been encountered in past implementations. Some challenges are related to the complexity of the hydraulic dynamics in sewer systems which typically require the linearization of 1-D Saint Venant equations. The model predictive control approach attempts to identify the optimal set points for control for a predetermined time horizon (usually between 3 and 8 hours) by utilizing weather forecasts and the hydraulic model. The optimal setpoints identified for the

An Agent-Based Storm Water Management System **153**

immediate time interval are used. Optimal setpoints for the rest of the time horizon are discarded. The process requires significant computational requirements as it needs to be repeated at each time interval (typically 5 to 15 minutes).

Additionally, weather forecasts introduce uncertainty that lowers the level of confidence that the system will be able to effectively mitigate CSOs while keeping safe hydraulic conditions in the sewer system. This chapter presents a framework for dynamic control of combined sewer systems called CSOnet, previously introduced in [13, 14]. The CSOnet framework is based on an extensive sensor network coupled with an agent-based control scheme that is both scalable and decreases the computational complexity and uncertainty associated with weather forecasting. This chapter describes the CSOnet framework as implemented in South Bend, Indiana, U.S.A.

6.2 APPROACH

The CSOnet framework attempts to take advantage of existing conveyance and storage capacity of the collection system as storm events present high temporal and spatial variability literally leaving some pipes half empty while others are overflowing. By doing so, CSOnet is able to redirect flows from areas that are overwhelmed to areas with lesser hydraulic loads. Instead of utilizing a model predictive approach that mostly relies on a hydraulic model and weather forecasts, CSOnet relies more on actual sensor measurements from the sewer system. The sensor measurements as distributed throughout the sewer-shed eliminate the uncertainty associated with the hydraulic model. By monitoring the sewer-shed sufficiently upstream of the controlled assets CSOnet also decreases its dependence on weather forecast input and its associated uncertainty.

At the core of the CSOnet framework is an agent based control methodology in which each actuated control asset has at a minimum two choices to redirect flows (or store them). An asset that can optionally convey water (or store it) is called a selling agent and sells conveyance capacity (or storage capacity). The agent that controls which selling agent gets to convey (or store) water is called the buyer agent. The buyer agent buys conveyance (or storage) capacity from at least two offering selling agents. Each selling agent is preprogrammed with a supply curve that generally matches a decreasing available capacity of the corresponding conveyance (or storage) asset with an increasing virtual price. The buyer agent is then confronted with a simple optimization problem to minimize its cost by balancing flows between the two or more selling agents.

The CSO framework can scale easily as a seller agent can also be a buyer agent in the network. The approach also reduces in-line computational requirements as the supply curves are designed offline through extensive model simulation prior to implementation. The CSOnet framework is distributed in nature as each agent can be programmed and implemented in the field at the device level. This presents desirable robustness properties as an agent failure does not prevent other agents to continue executing control. Because the CSOnet framework requires constant updates on the hydraulic status of selling agents, it requires a sensor network capable of providing frequent information on the levels or flows at each selling agent. When sensors are installed upstream enough of selling agents they can also provide forecasting capabilities that can be used to set prices based of future flows.

6.3 ARCHITECTURE

CSOnet was designed as a dense wireless sensor actuator network that could reliably and economically monitor the collection system and implement a distributed control system. CSOnet consists of three main components:

6.3.1 MONITORING HARDWARE

The CSOnet hardware component is composed of the wireless sensor-actuator network installed in the sewer and that is in charge of acquiring and processing sensor data, transmitting the data, and, in some cases, adjusting the position of a valve, gate, or movable weir through an electrical actuator. There exist three types of devices: sensor nodes deployed in the sewer which are in charge of measuring hydraulic conditions such as water levels or flows, gateways which are mounted above ground structures such as traffic poles or lighting poles and are in charge of managing and collecting information from the sensor nodes and, using a wide area network (WAN) connection, retransmitting such data to a database, and repeaters which fill in the spatial gaps when sensor nodes are distant from a gateway.

Sewer systems present challenging conditions for the deployment of electronic systems. Expected issues such as 100% humidity, submerged conditions, and large debris travelling at 5 meters per second are common in sewer systems. In addition, hydrogen sulfite gases can easily corrode metal, methane gases can ignite easily in case of a short circuit, and biofilm can both corrode and impact sensors. Furthermore, since sewer lines are usually installed under city roads, maintenance requires traffic control in potentially dangerous conditions. As a result, the requirements for a sensor node are significant: it must have an enclosure that is able to protect the electronics from the physical and chemical environmental conditions in the sewer and it must be able to operate unattended for a minimum of 12 months.

The sensor node is based on the original Mica2 mote described in [8]. The new sensor node, called the Chasqui mote, was augmented with the necessary components to implement low power operation, connect to a variety of off-the-shelf sensors and communicate in an urban scale network. To allow interfacing with commercially available sensors, the Chasqui mote is outfitted with industrial hardened inputs for up to four 4-to-20mA analog signals and one digital MODBUS RS232 input. To reduce power consumption, each input has its own separate power control mechanism where sensor power can be entirely removed. An 8Mbyte non-volatile flash memory provides storage for sensor data. A highly accurate real time clock (RTC) allows precise time tracking with low temperature drifting. This feature is particularly important as temperatures cannot only swing dramatically during the day but can be dramatically different between devices that must be synchronized.

Extended operation is an important consideration. Several power sources and battery configurations were considered. Rechargeable batteries are desirable because

An Agent-Based Storm Water Management System

they can be reused. However, in the midst of the winter season, temperatures can drop below operation tolerances. The design opted for a lithium non-rechargeable battery which can operate at extreme temperatures and for an extended period of time with little impact on its capacity. A series of switching power supplies form the power subsystem of the Chasqui mote. The microprocessor is able to independently remove power from other subsystems to conserve power. The Chasqui board draws from the 13V lithium battery currents up to 1A during data transmission but ramps down to less than 30μA during its sleep mode. While the sensor nodes can accommodate a virtually unlimited number of batteries, a battery pack consisting of 4 D-sized lithium batteries in series provides about 12 months of uninterrupted operation when sampling and transmitting at 5 minute intervals.

Other improvements over the original Mica2 board include the use of a 1 watt spread spectrum radio operating at 900MHz which allows it to communicate up to 300 meters in highly urban environments. Later, with the reductions in cost of cellular data, the device was provided with a way to also use a cellular modem for data transmission. The use of cellular telemetry at the sensor node level provides an inexpensive way of sending data directly to the systems database. However, cellular systems take a significantly longer amount of time to register with the carrier network. As a result, to achieve the same 12 months of battery life the 900MHz system is able to achieve, the sensor node must transmit its data only every 3 hours. On each transmission the sensor node is able to transmit all its data which continues to be sampled every 5 minutes. An internal programmable threshold places the sensor node in alert mode during abnormal conditions. In alert mode the cellular sensor node transmits every 5 minutes. Because this mode can rapidly deplete the battery, it is used sparingly.

Each sensor node consists of two parts: a telemetry part and a sensor part. The telemetry part is located at street level and is built under a composite manhole cover to allow for surface level radio transmissions (see Figure 6.1). An explosion-proof box (to comply with deployment of electronics in an explosive environment) houses the Chasqui mote and its battery pack. The antenna located on the outside of the explosion proof box and mounted under the composite material manhole cover is a slot antenna designed to broadcast radiofrequency signals out of the manhole. The sensor part is typically located in the sewer pipe and consists of third party available level or flow sensors. Typical level sensors include pressure transducers or ultrasonic transducers to measure water level and ultrasonic sensors that measure flow by using the Doppler effect to measure water velocity.

A locking mechanism keeps the otherwise light composite manhole cover secured to the manhole frame. This prevents unauthorized entry or tampering as well as preventing the cover from springing off the manhole frame when high speed and heavy vehicles run over the cover.

Sensor data is typically sampled at 5 minute intervals. The Chasqui mote is periodically woken up by its real time clock to take a measurement and communicate. To take a measurement, the Chasqui mote powers the corresponding sensor, waits for the sensor to take its measurement, acquires the sensor measurement, stores and

Figure 6.1: Underneath a sensor node. Visible are the composite manhole cover, the explosion proof box containing all the electronics, a cam-lock system to maintain the lightweight cover in place, and the antenna (under the explosion proof box).

communicates the sensor measurement and finally removes the power from the sensor. The gateway consists of an embedded PC running a Linux-based operating system, a 900MHz radio to communicate with the sensor nodes, and a WAN connection (typically a cellular modem). The gateway is usually permanently powered through a connection to the electrical grid or through a solar cell system. The gateway also contains interface systems to control sewer infrastructure. For example 4 to 20 mA outputs or RS232 Modbus protocols are used to open and close gates or to set pump stations to a speed setpoint.

The repeater consists of a similar architecture to a sensor node, however, since the repeater is typically installed above ground, the repeater does not require a cover, sensors, or an explosion proof box. Repeaters are either battery powered or permanently powered.

6.3.2 COMMUNICATION SUBSYSTEM

CSOnet was initially designed to exclusively use its 900MHz radio to transmit data by establishing an ad-hoc, multi-hop wireless network (see Figure 6.2). In this scheme, data from various sensors is aggregated to a gateway device with wide area network (WAN) connectivity. The permanently powered gateway is in charge of managing a set of sensor and repeater nodes on its vicinity. To do this, the gateway uses its WAN connection to synchronize 4 times a day its internal clock to a National Institute of Standards and Technology (NIST) time server. A process sends a synchronization message to all nodes in the sensor network every 24 hours. The synchronization message is transmitted using a multihop broadcast flooding protocol. Each node in the sensor network corresponding to the gateway that sent the signal resets its internal clock. The synchronization signal is also used to inform each node in the sensor network with its relative distance to the gateway as indicated by the number of hops the synchronization message needed to take to arrive at the particular node.

To conserve power, the synchronization function allows the network to sleep and be awake at the same time to allow for multi-hop interconnectivity. Every sensor node in the sensor network wakes up every 5 minutes to take and record a sensor measurement. It then transmits the sensor measurement to all its neighboring nodes. Nodes that are closer to the gateway as measured by the number of hops first acknowledge receipt of the data and then rebroadcast the message until it is received by the gateway. Nodes that do not hear an acknowledgment from a node that has "picked up" the message, retransmit the sensor measurement up to 3 times. All data transmissions happen in a period of approximately 3 seconds after which all nodes go to low power mode.

The gateway receives sensor measurements from its sensor network every 5 minutes. The gateway processes the data, records the sensor measurements, and using its WAN connection transmits the data to a centralized server for visualization. The gateways also use their WAN connection to exchange messages. The city has several gateways (32 gateways) including a gateway at each actuator. To allow for distributed control where each gateway computes its control action, a peer-to-peer communication protocol allows a sensor measurement collected by a gateway to also be transmitted to a different gateway. In this manner, gateways connected to actuators can have access to the sensor data of any sensor in the city. The peer-to-peer communication protocol enables robust operation in case of server failure.

The 5 minute sampling time is designed to capture hydraulic transients in the sewer system. While transients with a time constant of less than 5 minutes do occur, they are rare. Hydraulic modelling software is typically used to design the system. Several years worth of rainfall are simulated in the modelling software to verify that faster sampling times are not required.

As previously mentioned, the advent of machine-to-machine applications has significantly reduced cellular hardware costs and data rates making it possible to have a cellular communication system at the sensor node level. This effectively eliminates the need for gateways and repeaters. However, due to power constraints, 5 minute sampling and transmission has to be limited to extreme circumstances. This communication methodology is usually reserved for sites where data for control is not needed but where system operation can be analyzed for maintenance purposes.

All processes in the gateway are multithreaded non-blocking processes where each thread is monitored by a watchdog. A process is restarted if it is detected to be stalled. This is especially useful when WAN network glitches prevent a process from successfully completing a handshake.

Security of communication networking protocols are of utmost importance for the operation of critical infrastructure. The system security is handled by a combination of encryption protocols, security keys, and operation behind a firewalled private network.

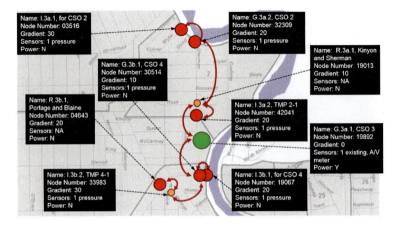

Figure 6.2: Example of a deployed network. The red circles represent the sensor nodes, the orange circles represent repeaters, the green circle represents a gateway.

6.3.3 DISTRIBUTED CONTROL SUBSYSTEM

The control of the collection system is performed by dynamically controlling actuators throughout the collection system. Actuators typically utilized in collection systems are movable weirs, valves, gates, and pump stations that are electrically controlled. Some of these actuators may already be part of the existing infrastructure (e.g., pump stations). Other actuators had to be installed as part of the implementation of CSOnet (e.g., gates). By changing the set points on these actuators, CSOnet is able to actively reconfigure the sewer system to better accommodate the time-varying hydraulic loads in the sewer system. As a storm moves through the city, these actuators mirror the precipitation patterns by balancing the hydraulic loads of different parts of the sewer network.

Given the scale of the collection system, the unreliability of actuator systems and communication systems (especially during storm events), all control algorithms were required to have a distributed architecture. A distributed control system architecture implements the control algorithm for each actuator at the location of the actuator (gateway). Each actuator-enabled gateway is able to directly communicate with devices upstream and downstream. The distributed control architecture contrasts with a centralized control architecture where the control logic for all actuators is implemented at a control center. By distributing the control logic, the impact of a communication outage, a problem with the control center, or with any particular actuator is minimized. It is assumed that operation with a disabled component in the system will be a common occurrence.

One of the early opportunities to improve the operation of the combined sewer system was to dynamically balance flows along the city's single interceptor line. An interceptor line is a sewer line that collects water from trunk lines at each CSO original discharge point and takes it to the WWTP. A CSO diversion structure allows

An Agent-Based Storm Water Management System

a maximum amount of flow to go from the trunk line to the interceptor line. The maximum allowable flow to enter the interceptor is determined by the throttle line diameter. A diagram of a typical CSO diversion structure is shown in Figure 6.3 and a diagram showing how CSO diversion structures are connected to the interceptor line is shown in Figure 6.4.

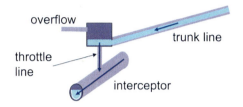

Figure 6.3: Diagram of a CSO diversion structure.

Each CSO diversion structure was designed such that the sum of the maximum flows allowed to enter the interceptor equaled the capacity of the WWTP. That is:

$$Q_{WWTP} = \sum_{i=0}^{N} Q_{Ii} \quad (6.1)$$

$$\bar{Q}_{WWTP} = \sum_{i=0}^{N} \bar{Q}_{Ii} \quad (6.2)$$

$$Q_{CSOi} = \begin{cases} Q_{Ti} - \bar{Q}_{Ii} & \text{if } Q_{Ti} \geq \bar{Q}_{Ii}; \\ 0 & \text{if } Q_{Ti} < \bar{Q}_{Ii}. \end{cases} \quad (6.3)$$

Where Q_{WWTP} is the flow to the treatment plant, N is the total number of CSO diversion structures, Q_{Ii} is the flow diverted to the interceptor at CSO diversion structure i, $\bar{Q}_{WWTP}$ is the maximum allowable flow to the treatment plant, $\bar{Q}_{Ii}$ is the maximum flow that is allowed to be diverted to the interceptor at CSO diversion structure i, Q_{CSOi} is the flow released to the environment at CSO diversion structure i, and Q_{Ti} is the flow that is carried by the trunk line. To set the maximum flow to the interceptor $\bar{Q}_{Ii}$, a weir diverts flows from the trunk line to the interceptor through a narrow pipe called a throttle line. The height of the weir and the diameter of the throttle line determine $\bar{Q}_{Ii}$.

The opportunity arises because rainfall does not fall evenly throughout the city and because different land uses produce different concentration times. An example showing the variability of rainfall intensity is shown in Figure 6.5. As a result, not all the CSO diversion points overflow or are about to overflow at the same time. This implies that the WWTP and the interceptor are not at their maximum capacity when some CSO diversion points are overflowing. This is particularly true at the beginning of a storm event and at the end of the storm event.

To maximize the WWTP and interceptor line capacity, CSOnet must be able to dynamically control $\bar{Q}_{Ii}$. This can be achieved by installing an oversized throttle

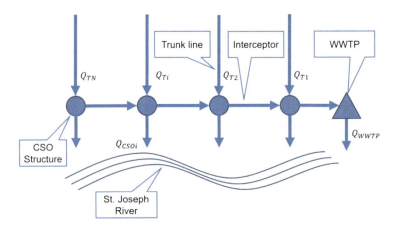

Figure 6.4: Diagram showing the trunk lines, CSO diversion structures, interceptor line and WWTP in South Bend, Indiana. This arrangement is typical of CSO systems in the United States.

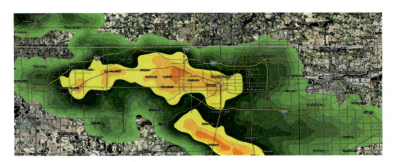

Figure 6.5: Example of typical storm event over South Bend, Indiana. Significant rainfall spatial variability in space can be observed.

line augmented with an electrically actuated valve (see Figure 6.6). When capacity exists in the interceptor line, the valve would open to allow more water to enter the interceptor thus reducing overflows. When capacity in the interceptor is depleted, the valve would close to prevent too much water from entering the interceptor thus protecting the interceptor and WWTP from excessive flows.

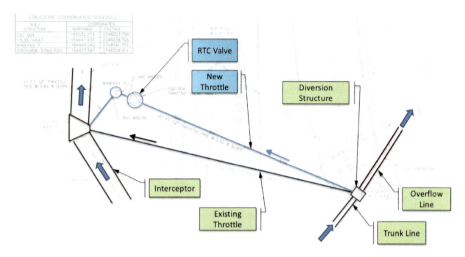

Figure 6.6: Sewer infrastructure for balancing flow in the interceptor. The blue lines (new throttle line and RTC valve) were added as part of this project.

The control strategy is implemented on gateway nodes located where the throttle valves are installed. The gateway nodes are able to communicate with each other through the use of a peer-to-peer communication system implemented at the WAN level.

Several distributed control methodologies have been considered in the past. For example, in [18] a distributed control used Pontryagin maximum principle to obtain the optimal setpoints for each valve. Other methodologies use economic principles and game theory for the control of sewer systems [3, 7, 15]. The distributed control methodology chosen for implementation was an agent-based control mechanism that emulates a free market economy [10]. Under this methodology two kinds of agents exist: selling agents and buying agents. A selling agent is associated with a conveyance, storage, or treatment asset with a finite capacity. A buying agent is associated with a controllable asset which has the ability to choose from two or more selling agents to take its load of wastewater flows. Each selling agent determines a virtual cost per unit of flow that is, generally, proportional to its utilized capacity. The relationship between the virtual cost and the selling asset capacity is referred to as the supply curve of the agent. As the assets capacity is depleted, the corresponding selling agent increases its own virtual cost. The buying agent on the other hand must decide how to divert flows to each selling agent such that its total cost to "buy capacity" from the selling agents is minimized.

In the case of the South Bend example, the buying agents are the gates located at the overflow diversion points. An analysis was conducted to determine the locations where a CSO diversion point could be automated. Each automated CSO diversion point has the option of opening the valve to buy capacity from interceptor line or closing the valve and buy capacity from the CSO discharge point.

A set of selling agents monitors the interceptor line at several locations throughout its length. This is important as the interceptor line, which has varying diameters as it goes through the city, experiences capacity limits at different locations and at different times. For convenience, all CSO diversion structures are numbered sequentially starting with the CSO diversion point closest to the WWTP. Also for convenience, the monitoring points along the interceptor are numbered according to the closest CSO diversion structure immediately upstream from the monitoring point. There are no more than 1 interceptor monitor between two consecutive CSO diversion points. Each monitor is associated with an interceptor selling agent where the supply cost curve $C_{Ii}(L_{Ii})$ which is generally a linear relationship based on the level at the interceptor. Typically, the cost is zero if the level in the interceptor is less than the levels experienced during dry weather. From there the cost increases linearly to 1 when the interceptor line is full and thus at full capacity. The WWTP also has a supply cost curve $C_{WWTP}(Q_{WWTP})$ which starts at a zero cost when the WWTP flow is less than the flows experienced during dry weather conditions and increases linearly to 1 when the WWTP is at full capacity $\bar{Q}_{WWTP}$. An example of supply cost curves is shown in Figure 6.7.

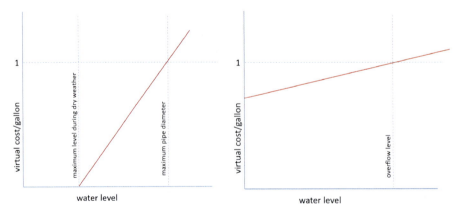

Figure 6.7: Left: Typical supply cost curve at an interceptor selling agent. Right: Typical supply cost curve at CSO diversion line.

Similarly, a supply cost is assigned to the CSO diversion point $C_{CSOi}(L_{CSOi})$. Here the supply cost is related to how close the CSO diversion point is from overflowing. Typically, this supply cost starts at a value slightly less than 1 and then linearly increases reaching a value of 1 at the point at which the CSO diversion point starts overflowing.

The buying agent *i* located at the throttle line attempts to minimize its cost given

An Agent-Based Storm Water Management System **163**

by:

$$C_{Ti} = C_{CSOi}(L_{CSOi}) + \max_{j\in[1,i]} (C_{Ij}(L_{Ij}), C_{WWTP}(Q_{WWTP})) \tag{6.4}$$

This results in a buying agent which generally keeps the throttle line open to divert water to the interceptor as long as the interceptor has capacity. As the storm hits the city and the interceptor fills up and starts getting close to full, the throttle line will start closing to prevent surcharging. Eventually the storm recedes and the system returns to dry weather flow conditions, the interceptor capacity becomes available and the throttle line gate starts opening until it diverts all flows to the interceptor.

Generally, this type of control results in a type of proportional gain controller which operates based on the levels throughout the interceptor and CSO diversion point. Search algorithms are used to tune the supply curves using a simulation of the sewer system behavior under different storm event conditions. Other elements require more complex control algorithms. For example, water retention basins that can store storm water before it enters the sewer system are sometimes located several kilometers upstream of the nearest CSO diversion structure. In these cases, a hydraulic model allows the agents to determine their costs based not on the current water levels in the system but on predicted levels. Also, wastewater tanks may have requirements on the length of time that they are allowed to store water resulting in time varying cost curves.

6.4 IMPLEMENTATION

The CSOnet monitoring system was installed in 2008 and its control subsystem was implemented in 2010. Currently, the CSOnet system consists of 150 wireless nodes monitoring 111 locations. These locations include 36 discharge locations, 28 sewer interceptor points, 42 trunk line points, and 5 basins (see Figure 6.8). The system is divided into 32 subnets interconnected through the use of a cellular data network. The CSOnet system controls 9 valves and 3 movable weirs. Figure 6.9 shows the installation of a valve.

Originally the system was designed with the objective of reducing overflows. However, data analytics proved useful for other purposes as well. For example, the sensor network is able to detect areas that operate outside of the normal regime observed historically. Sewer cleaning crews have used this information to pinpoint and troubleshoot issues. As a result, city workers routinely pull large amounts of gravel or other debris that otherwise limits the conveyance capacity of the sewer system.

In Figure 6.10, an example of data analytics is presented. The analysis compares the level and velocity of the water at a particular location and compares it to the theoretical relationship as given by the Mannings equation. In this case, because no backwater effects occur, it is relatively straightforward to make the comparison. The example shows that the 610 mm (2 feet) diameter pipe has a severe flow restriction downstream which keeps its maximum flow below 80 liters per second (2.8 cubic feet per second) or about 1/8th of the capacity of a pipe of that diameter and characteristics. Using these tools the utility has been able to implement a data-driven

Figure 6.8: Location of the CSOnet monitoring sites in South Bend, Indiana.

Figure 6.9: CSOnet throttle line installation. Left: The pinch valve installation. Center: Pinch valve and electrical actuator installed. Right: Control cabinet for the valve with the agent-based control implementation.

An Agent-Based Storm Water Management System

maintenance and cleaning program which has increased the hydraulic performance of the sewer system.

Ultimately, a dashboard with data from all the sensor network locations shows operators the status of the interceptor line and CSO outfalls (see Figure 6.11). This dashboard also provides the capability to perform forensic analysis of the performance of CSOnet and the sewer system in general during storm events. These tools improve the ability for the utility staff to explore and understand the operation of the system and thus improve the chances that the project is successful. The utility staff buy-in is considered one of the biggest hurdles in the success of dynamic control of sewer systems.

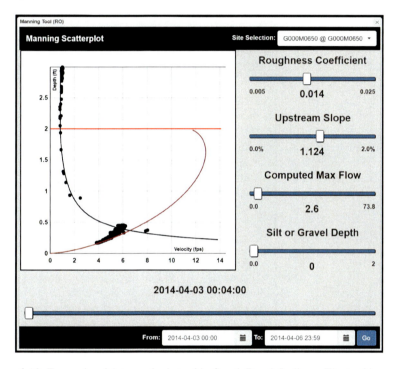

Figure 6.10: Example of data analysis tool in South Bend, Indiana. The tool is used to compare relationship between velocity and depth in a 610 mm (2 ft.) pipe as collected by a flow meter (black dots) versus its theoretical relationship as given by Manning equation (brown line).

The implementation of the control algorithm has been done in stages to gain the trust of the utility staff [16]. The initial state of the implementation of the control algorithm was done by verifying the effectiveness of the system via hydraulic simulations. On the field, increasing levels of control aggressiveness were achieved by limiting the actuator dynamical range.

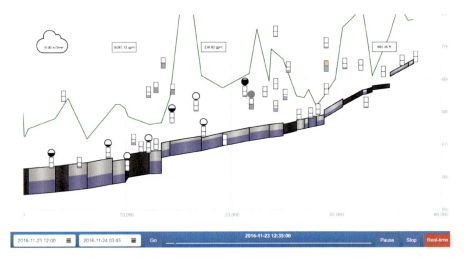

Figure 6.11: Dashboard showing, at scale, the hydraulic status of South Bend's main interceptor line. The interceptor line runs from the top right part of the dashboard to the bottom left side. It increases its diameter as it collects water from several trunk lines. The blue portions of the dashboard show the water level along the interceptor. The green line represents the ground surface elevation along the interceptor. The rectangles immediately above the interceptor show the water levels at the diverse CSO diversion structures. The circles show the position of the different CSOnet valves.

6.5 RESULTS

Computer simulations showed a 25% reduction of overflows using synthetic rainfall based on precipitation measured in 1995 (called "typical year" precipitation for regulatory purposes). Actual measurements indicate a more drastic reduction of overflows, although it is worth noting that other improvements to the sewer system have been done over the same period. A full attribution analysis to isolate the contribution of each project is currently being considered.

Table 6.1 shows a performance index that relates the total yearly CSO volume to the total yearly amount of precipitation. Before CSOnet was commissioned (2006-2008) the performance index hovered around $6m^3 \times 10^3$/mm. After CSOnet was fully installed (2012-2016) the performance index has been reduced by a factor 3.

6.6 CONCLUDING REMARKS AND FUTURE DIRECTIONS

The CSOnet system has successfully mitigated CSO events over the last 7 years and demonstrated the use of sensor networks to understand and control sewer-sheds in a smart city environment. Using the CSOnet framework, similar systems have been successfully designed and implemented in large sewer systems, including Buffalo, New York, Kansas City, Missouri, and Columbus, Ohio. Control parameters such as supply cost curves were derived heuristically in all cases, but mathematical

Table 6.1
CSOnet performance

Year	Total Precipitation (mm)	Total Overflow (m^3x10^6)	Performance Index (m^3x10^3/mm)
2006	1143	7.991	6.99
2007	1067	5.421	5.08
2008	1219	7.775	6.38
2009	927	4.539	4.90
2010	749	3.043	4.06
2011	1064	3.369	3.17
2012	747	1.685	2.26
2013	909	2.707	2.98
2014	1013	1.734	1.71
2015	851	1.510	1.78
2016	1455	2.669	1.83

frameworks aimed at synthesizing these parameters are being investigated. Also, the robustness, stability and optimality of these algorithms are currently being studied in the context of stochastic inputs, specifically rainfall measurements and weather forecasts. While literature on methodologies that address these problems may seem generally abundant, they are generally not applicable due to the highly nonlinear aspects of sewer networks, which may contain many (100s to 10000s) nonlinear and uncertain elements. Currently CSOnet control algorithms are being tested to incorporate probabilistic precipitation forecast as an input [12]. Also, CSOnet infrastructure is being leveraged to test other applications such as optimization of drinking water facilities, optimization of salt application during winter time, and monitoring river streams.

REFERENCES

1. United States Environmental Protection Agency. Combined sewer overflow control policy, 1994.
2. United States Environmental Protection Agency. Report to Congress: Implementation and enforcement of cso control policy, 2001.
3. J Barreiro Gomez, G Obando, G Riano-Briceno, N Quijano, and C Ocampo-Martinez. Decentralized control for urban drainage systems via population dynamics: Bogotá case study. In *Control Conference (ECC), 2015 European*, pages 2426–2431. IEEE, 2015.
4. United States Congress. Federal water pollution control act, 1972.
5. O Fradet, M Pleau, A Desbiens, and H Colas. Theoretical and field validation of solutions based on simplified hydraulic models for the real-time control of sewer networks. *NOVATECH 2010*, 2010.

6. L García, J Barreiro-Gomez, E Escobar, D Téllez, N Quijano, and C Ocampo-Martínez. Modelling and real-time control of urban drainage systems: A review. *Advances in Water Resources*, 85:120–132, 2015.

7. J Grosso, C Ocampo-Martinez, and V Puig. Reliability–based economic model predictive control for generalised flow–based networks including actuators health–aware capabilities. *International Journal of Applied Mathematics and Computer Science*, 26(3):641–654, 2016.

8. J Hill. A software architecture supporting networked sensors. Master's thesis, UC Berkeley Department of Electrical Engineering and Computer Sciences, 2000.

9. B Kerkez, C Gruden, M Lewis, L Montestruque, M Quigley, B Wong, A Bedig, R Kertesz, T Braun, O Cadwalader, P Aaron, and C Pak. Smarter storm water systems. *Environmental Science & Technology*, 50:7267–7273, 2016.

10. R Kertesz, B E McDonnell, and L A Montestruque. Stealing from the market: A novel application of market based principles for real time collection system coordination. In *Proceedings of Conference on Water Management Modelling*, 2017.

11. M Marinaki and M Papageorgiou. *Optimal real-time control of sewer networks*. Springer Science & Business Media, 2005.

12. L A Montestruque, B Kerkez, R Kertesz, and B P Wong. The role of weather uncertainty in the real-time operation of sewer and storm water systems. In *Proceedings of the Environmental & Water Resources Institute Congress*, 2016.

13. L Montestruque and M Lemmon. Csonet: A metropolitan scale wireless sensor-actuator network. In *International Workshop on Mobile Device and Urban Sensing (MODUS)*, 2008.

14. L Montestruque and M Lemmon. Globally coordinated distributed storm water management system. In *Proceedings of the 1st ACM International Workshop on Cyber-Physical Systems for Smart Water Networks*, page 10. ACM, 2015.

15. C Ocampo-Martinez and N Quijano. Game-theoretical methods in control of engineering systems: An introduction to the special issue. *IEEE Control Systems*, 37(1):30–32, 2017.

16. T P Ruggaber, J W Talley, and L A Montestruque. Using embedded sensor networks to monitor, control, and reduce cso events: A pilot study. *Environmental Engineering Science*, 24(2):172–182, 2007.

17. L Vezzaro and M Grum. A generalized dynamic overflow risk assessment (dora) for urban drainage rtc. In *9th International Conference on Urban Drainage Modelling*, pages 3–7, 2012.

18. P Wan and M Lemmon. Distributed flow control using embedded sensor-actuator networks for the reduction of combined sewer overflow (cso) events. In *Decision and Control, 2007 46th IEEE Conference on*, pages 1529–1534. IEEE, 2007.

19. B P Wong and B Kerkez. Real-time environmental sensor data: An application to water quality using web services. *Environmental Modelling & Software*, 84:505–517, 2016.

7 Gamified Approaches for Water Management Systems: An Overview

ANDREA CASTELLETTI, ANDREA COMINOLA, ALESSANDRO FACCHINI, MATTEO GIULIANI, PIERO FRATERNALI, SERGIO HERRERA, MARK MELENHORST, ISABEL MICHEEL, JASMINKO NOVAK, CHIARA PASINI, ANDREA-EMILIO RIZZOLI, CRISTINA ROTTONDI

CHAPTER HIGHLIGHTS

- The chapter focuses on behavioral interventions and educational/awareness tools related to water issues and on gamified approaches and persuasive technologies for water management.
- It provides background notions on the enabling technologies for water management systems, i.e., smart metering infrastructures, big data analytics, and persuasive technologies.
- It reviews some research and commercial gamified water management applications.
- It describes in detail the gamified platform for water management developed as part of the EU-funded SmartH2O project [62].

7.1 INTRODUCTION AND MOTIVATION

Water demand is growing worldwide, especially in densely populated areas, as a consequence of population growth [38] and urbanization [74]. The spatial concen-

---

A Castelletti • A Cominola • M Giuliani • P Fraternali • S Herrera • C Pasini

Politecnico di Milano, IT.

M Melenhorst • I Micheel • J Novak

European Institute for Participatory Media, Berlin, DE.

A Facchini • A Rizzoli • C Rottondi ⊠

Dalle Molle Institute for Artificial intelligence (IDSIA), CH. E-mail: cristina.rottondi@supsi.ch

tration of water demand in urban areas is impacting demand magnitude, peak intensity, share between use sectors, and indoor and outdoor usage [50]. This, coupled with climate change and land use change, is intensifying the stress on finite water resources, creating both operational and environmental challenges to water supply. To cope with this evolving context, traditional supply and management schemes need to be adapted to meet future demand while preventing unsustainable resources exploitation. In this context, taking into account geographical constraints and increasing marginal cost often limiting capacity expansion through infrastructural interventions [39], demand-side management strategies are key to complement supply-side interventions for securing reliable water supply as well as reducing shortages and overall utilities' costs [16]. The potential of demand-side management interventions has been demonstrated in a number of works (see, for a review, [29] and references therein), especially in the last two decades, fostered by the promotion of several water saving programs worldwide, particularly in areas affected by prolonged droughts (e.g., California [23]) or low-recharge periods (e.g., Australia [23]).

Within the broad portfolio of demand-side management interventions that can be implemented (i.e., technological, financial, legislative, maintenance, and educational), many recent works (e.g., [80] [37] [78] [73]) devoted attention to exploring the effect of behavioral change programs aiming at promoting water saving practices and increasing water awareness among users. Detailed investigation regarding behavioral interventions has been possible especially since the late 1990s, with the development of smart water meters [49], which enabled the monitoring of water consumption data with very high spatial (household) and temporal (from several minutes up to few seconds) sampling resolution [25]. Smart meters provide essential data to characterize the behavior of individual users and support the development of customized demand-side management interventions targeted on specific groups of users, and tailored to act on the drivers of their water consumption behavior. Indeed, a major role in forming the water demand is played by the behavior of water consumers, each potentially driven by diverse social and individual motivations and triggers. The achievement of an efficient water demand management depends on the aggregate effect of actions undertaken by several, diverse, individuals. This suggests that behavioral change programs built on customized demand-side management interventions can act on the specific drivers of users' demand and promote water and environmental friendly behaviors ([64] [35]).

In this chapter, we focus on behavioral interventions and educational/awareness tools related to water issues and on gamified approaches and persuasive technologies for water management, which aim at empowering users' knowledge and incentivize virtuous attitudes towards water conservation based on gaming mechanisms [67]. Computer and social sciences offer a wide set of gamified tools to build or raise environmental awareness in a variety of sectors (e.g., [19] [26] [54]) and most recently also in the water domain. According to a recent review on serious games on environmental management [47], water management games constitute a relevant portion of the games belonging to the more general category of serious games for environmental management (more than 20% of the 25 games considered in the study and developed between 1994 and 2013). These games represent promising tools both for learning purposes (game based learning), because they usually increase motivation and foster engagement, and for facilitating decision making processes, especially when a number of stakeholders with conflicting interests are involved, or validating behavioral models, such as agent-based models. Therefore, a variety of users, ranging from students and people involved in learning processes, to professionals, local stakeholders and decision makers can potentially benefit from the use of such tools. Yet, state-of-the-art literature often lacks exhaustive descriptions of some gamified approaches to environmental and water problems, as well as quantitative measures enabling a fair assessment of the effectiveness of such tools for education and awareness purposes, or a comparison of different games under common criteria [47]. Our goal is therefore to review and critically analyze the most relevant research and industrial gamified approaches to water management, with a focus on water awareness and behavioral change. Additionally, we present some preliminary results in terms of user response and water conservation achievements obtained with the real-world application of a gamified approach.

The chapter is organized as follows. In the first part, we introduce some background notions on the enabling technologies for water management systems, i.e., smart metering infrastructures, big data analytics, and persuasive technologies. In Section 7.3, we review in detail some research and commercial gamified water management applications, especially developed within EU-funded projects belonging to the ICT4Water cluster.[1] Finally, in Section 7.4, we present the gamified platform we developed as part of the EU-funded SmartH2O project [62], highlighting the current strengths and weaknesses identified during its application in two real-world case studies, in Switzerland and in Spain.

7.2 TECHNICAL CONCEPTS AND METHODOLOGIES

Water utilities need support tools to match water demand with supply in a sustainable way. To this aim, it is first necessary to understand the factors that drive the demand by mining water consumption data collected at high frequency. This is now possible thanks to the development of smart metering technologies and Internet of Things (IoT) protocols [13] that support real-time metering data collection, whereas cloud computing and big data infrastructures allow utilities to store and process meter data, and machine learning and pattern analysis enable the extraction of useful information such as identification of leakages and disaggregation of end-uses.

Water utilities deal with other important issues such as providing the consumers with educational materials on efficient water use, and designing adequate campaigns to influence their behavioral habits. Gamification and persuasive interfaces offer a viable solution to promote behavioral change, e.g., by providing consumers with visualization interfaces to increase their awareness on the amount of resources they use, and integrating incentives and gamified activities aimed at influencing their behavior towards more sustainable attitudes.

---

[1] http://www.ict4water.eu/

Finally, water utilities have to deal with fast changing policies, extreme weather effects and other factors that impact their revenues, therefore they require simulation tools to forecast the effects of different pricing policies and observe the expected reaction of the consumers: in this case machine learning techniques in combination with game theory can be applied to tackle the problem.

Therefore, in this section, we will provide basic notions on smart metering technologies, big data analytics for end use disaggregation and profiling, persuasive technologies such as visualization interfaces and gamification.

7.2.1 SMART METERING TECHNOLOGIES

Smart meters are electronic measurement devices installed at the users' premises to collect detailed household consumption data and provide them to the utilities for billing and monitoring purposes. Additionally, smart meters support bidirectional communications, thus enabling remote control and monitoring of the overall performance of the grid and helping to detect failures and leaks [20] [17]. To do so, smart metering systems rely on a local area network (LAN) to transmit individual consumption readings to a local data collector. This transmission can occur with different frequencies (e.g., daily or hourly), according to the capabilities of the meter and the specific usage of the collected data. The collector gathers the data and possibly operates some preprocessing. Data is then transmitted via a wide area network (WAN) to the utility for processing and use by business applications.

On the customer side, one of the main benefits of smart metering is that it enables real-time or close to real-time consumption monitoring [71], which in turn raises the understanding and awareness of consumption and possibly leads to more efficient water usage. Furthermore, smart metering enables utility companies to offer better customer service by allowing for on demand readings, reductions of meter reading service costs, reductions of the response time to customer enquiries, more accurate billing, and reduction of the volume of non-revenue water losses thanks to a more accurate leaks detection [21]. Finally, by processing the huge amount of collected consumption data with statistical tools, companies are provided with a powerful tool for planning and forecasting the production and distribution of the resources based on real-time data.

Despite the benefits of smart metering systems, there are issues related to their implementation and perception that have hindered their adoption. One of the main concerns is privacy and security of the consumption data. Though secure design approaches have been applied to every component of the metering system, including level access control, data validation, and error checking as well as the encryption of data, the widespread usage of the utility backhaul communications and of public wireless networks has increased the exposure to potential security intrusions. Utilities are not only concerned about unauthorized access to user data, but also by potential unauthorized data manipulation to reduce billing [57].

Another issue is the meter accuracy: even though smart meters are very accurate and precise devices, individual meters are still prone to hardware/software failures. Therefore, utilities need to develop monitoring strategies and control the operational

Gamified Approaches for Water Management Systems: An Overview **173**

processes to identify anomalies and errors and correct them before the billing process is executed.

Finally, the biggest challenge is the initial roll-out time scale and cost of creating the smart metering infrastructure: the roll-out phase might take from months to years and should be managed with adequate planning strategies to minimize costs, technical and logistic issues [57]. There are also some hidden costs for the companies, e.g., the management of the infrastructure supporting the storage of big data, the acquisition or development of tools to perform analytics, and the adaptation of the currently deployed system to enable the coexistence with the new infrastructure [75].

7.2.2 BIG DATA ANALYTICS FOR END USE DISAGGREGATION AND USER PROFILING

In recent years, the decreasing cost of storage media and the network bandwidth increase have made data storage and transmission relatively easy and inexpensive. This has pushed a number of companies from diverse business sectors to collect huge volumes of data and extract useful information to support their business by means of "big data" mining and analysis techniques. The term "big data" refers to enormous amounts of heterogeneous unstructured data produced at high bit-rate by a wide family of application scenarios [31], e.g., social networks, data streams from sensors, and scientific and enterprise applications. In turn, big data analytics are a set of knowledge discovery and data mining techniques adopted to uncover structures, trends, and patterns in big datasets. Such techniques are used for a wide range of applications, e.g., medicine, scientific research, computer performance enhancement, financial market prediction, insurance, marketing customer profiling, and resource optimization.

In the field of water management, big data and big data analytics are becoming more and more relevant since the advent of smart metering technologies in the late 1990s. Such data potentially contain detailed information about the behavior of individual consumers in terms of water consumption habits and water demand patterns, which can be employed for designing and evaluating alternative portfolios of consumer-tailored water demand management strategies. The need for automatic tools able to extract relevant information supporting the demand management out of large smart metered datasets has been fostering the development of data analytics techniques and mathematical models to accurately characterize sub-daily water consumption patterns, as well as end-use water consumption profiles.

As highlighted in a recent paper reviewing comprehensively several studies on high resolution residential water demand modelling and management [29], data analysis and modelling have a key role in uncovering water users' behavior, especially for the purposes of the so-called (i) *automatic water end-use characterization* and (ii) *user modelling*. More specifically, water end-use characterization consists in the disaggregation of the total water consumption, metered at the household level, into the different end-use categories (e.g., shower, toilet flush, irrigation, etc.). Automatic and scalable algorithms able to perform water end-use characterization are needed because the direct monitoring of end uses is not viable, as it would imply high costs for the installation of multiple sensors, as well as scarce acceptability by water users because of its intrusive nature. End use disaggregation is still an open research issue and several algorithms have been proposed in the literature [82]. Most of them consist of two main phases:

- *Observation phase*: the actual consumption of the appliances needs to be measured (in an intrusive way) during a given period in order to obtain reference consumption patterns of the single end uses, typically named as "signatures." This information is used to create the so-called "consumption base" or "baseline" and calibrate the end-use disaggregation algorithms.
- *Classification phase*: In this phase, machine learning techniques, usually based on optimization and pattern matching methods, are applied to break down the aggregate consumption data into single end-use contributions, after algorithm training. The most common algorithms used in disaggregation problems are decision trees, classifiers, and clustering algorithms (for more information see [29] and references therein).

In turn, *user modelling* adopts data mining techniques to discover useful features in water consumption data to model heterogeneous water demand profiles at the household level, possibly as determined by natural and socio-psychographic factors, as well as by the users' response to different demand management interventions. More precisely, [29] suggests that user modelling includes two categories of models, i.e., *descriptive* and *predictive* models, whose goal is characterizing the patterns and trends of observed historical water demand time series, and forecasting future demands on the basis of natural (e.g., temperature and seasonality) and socio-psychographic determinants (i.e., variables characterizing user demography, as well as values, attitudes, interests, or lifestyles, such as age, number of house occupants, income level, conservation attitude, etc.), respectively. Data-driven techniques, multivariate and correlation analysis, data dimension reduction techniques (e.g., principal component analysis [30]), as well as feature extraction and clustering methods are commonly used for the above purposes. Thus, as it traditionally happened mainly for a number of marketing purposes, the development of big data analytics techniques in the water sector is becoming fundamental to understanding and characterizing the behavior of the heterogeneous consumers, ultimately supporting the implementation of tailored demand management strategies and customized feedback supporting water conservation and management programs.

7.2.3 PERSUASIVE TECHNOLOGIES

Persuasive technologies attempt to change the user's attitudes or behaviors without coercion or deception [36] and find applications in visualization interfaces, video games, mobile devices, and social media. We can find examples of persuasive technologies in online retail websites, health care, education, resource management, and collaboration tools. Persuasive applications aim at integrating within the habits of the user, and operate in a loop consisting of *triggers*, i.e., stimuli that alert the user and draw her attention, *actions*, which are the response of the user to a trigger, and *variable rewards*, which acknowledge the user's action in a way that surprises and stirs the desire to repeat the experience.

Using persuasive technologies provides some advantages over traditional media, or human persuaders. The most important is ubiquity: thanks to the proliferation of mobile and wearable devices, smart meters, and home-devices interconnected by means of IoT technologies, people can be targeted at any place and at any time. The diversity of modalities is another important advantage: computers and electronic devices can deliver messages in a huge variety of modalities including text, audio, video, graphs, animations with a higher persuasion impact. Finally, computers can store and access huge amounts of information that enable them to present new facts, statistics, and references as no other media can.

7.2.4 VISUALIZATION INTERFACES

Visualization has recently become a popular topic in computer science: it is formally defined as the use of computer-supported, interactive, visual representations of abstract data to amplify cognition [24]. In other words, the visualization of information deals with the design and aesthetic aspects of computer interfaces to represent complex abstract data in a clear and comprehensible way, in order to enable users to analyze them with visual tools. Data visualization leverages technical backgrounds like information retrieval and data mining, computer graphics, and human-computer interaction.

Other interesting features of information visualization are the persuasion capabilities that it can have on the user: it has been shown [59] [12] that a visual representation of some data can influence personal opinions about a topic by making people able to interpret the data, thus receiving feedbacks that showed them that their initial beliefs were wrong. Although ethical implications about the misuse of visualization to deceive or manipulate are currently under discussion, the persuasive capabilities of visualization can be positively used to influence users' behaviors.

7.2.5 GAMIFICATION

The concept of gamification has become increasingly popular in research and commercial applications. Gamification is defined as the use of game design elements in non-game contexts [65]. Some other terms like "serious games" and "games with a purpose" have also recently emerged. Although the two share many characteristics with gamification, they are fundamentally different: serious games are full functional games with a specific purpose like the education of a subject or the training on a skill, whereas gamified applications are enhancements of existing systems, services, tasks or processes, which incorporate game-based mechanisms but do not necessarily require them to function or be performed [40].

The main goals of gamification are the engagement of users and the enhancement of their experience, with the aim of increasing users' participation in the operations or social interaction aspects, or of raising users' productivity [41]. Gamification can also be used to set goals for the users in different contexts (e.g., teaching new skills): challenging users to meet an established requirement is known to be an effective motivational strategy [45].

In order to design a gamified application, it is important to characterize the people for which the game is intended and understand their motivations to play games. Bartle [15] proposed a player taxonomy based on the user motivations and interactions with the game and other players, which consists of four categories:

- *Achievers*: they are competitive and enjoy beating difficult challenges either set by the game or by themselves. The more challenging the goal, the most rewarded they tend to feel. Most children like to see progress in terms of points, clearing a level or a similar sense of progression.
- *Explorers*: they like to explore the world, not just its geography but also the finer details of the game mechanics. They try to learn all the mechanics, short-cuts, tricks, and glitches present in the game, and they are more likely to play a game several times in order to do that.
- *Socializers*: they are often more interested in having relationships with the other players than in playing the game itself. The game is merely a tool they use to meet other individuals or to hang out with their friends.
- *Killers*: They are driven by competition with other people and prefer them over artificial intelligence controlled by a computer.

Figure 7.1 shows Bartle's players' taxonomy: the horizontal axis represents the preference of the user to interact with other players or the environment, the vertical axis emphasizes the preference of the user to act or interact with or on something.

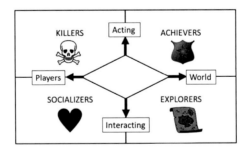

Figure 7.1: Bartle's player taxonomy [44].

Based on the target type of player, the gamification designer can think about how to design the game mechanisms in order to engage the players. For example, when targeting an explorer the designer has to think about the size of the world and the diversity of geographies and interactions. Differently, if the target is the group of socializers, the focus should be on personalization features of the game and on the communication and cooperation systems.

There are four necessary elements to design a game. The first is the mechanism that defines the procedures and rules of the game: the mechanics of a game describe the goal of the game, how to achieve it, and what happens when players do; choosing the mechanics defines the gameplay. There are several mechanisms that are widely used for gamifying everyday tasks [67]:

- *Achievements and Badges*: they are typically awarded if the player reaches a certain number of points in the game. They may require the user to perform tasks several times, master a skill, or explore the environment.
- *Progress Bar*: it indicates how close a user has come towards the achievement of a given goal. When the player completes a task or an activity, the progress bar is filled to indicate the progress of getting closer to a goal. How much the progress bar is moved is often determined by the severity of a task or by using points. The progress bar may often be combined with experience points, where the experience points collected determine the movement of the progress bar.
- *Experience Points*: experience points are an indicator of how much experience the player has gained within a game or setting. These points may be awarded by completing tasks, exploring areas and features or other similar activities. Experience points are usually combined with a leveling system, for example reaching certain amounts unlocks new levels, new rewards, or new features. Experience points are also often combined with leaderboards.
- *Leaderboards*: they consist of lists of the players ordered by their collected points, completed activities, or any other predefined metric. Each user has a score defined by rules set before a competition started, this score is compared to the one of the other players and players are ranked based on the scores. Leaderboards may be fully dynamic or state-based (i.e., a new order is recalculated after a certain period of time).
- *Contests*: they are categorized in duel-like head-to-head contests or free-for-all contests with no limit on the capacity of players. A duel-like contest may be a knockout style of competition where players compete to get the highest amount of points within a given time period or a similar type of a goal. In a free-for-all contest, the winner is whoever fulfills a specific goal to the best degree. The goal which players try to achieve is set be a specific set of rules determined before the start of the contest.
- *Real-World Rewards*: used together with leaderboards and contests, real-world awards may be given to some of the best players of the game. Usually the rewards are related with the topic of the application.
- *Social Networking*: it provides to the users a common platform where they can share their achievements and recent activities, since users typically experience a sense of reward when their publications receive comments. It is possible to link the game to existing social networks like Facebook or Twitter, or create specialized forums like Steam does in its platform.

The second element is the story, i.e., the sequences of events that happen during the game and the events that occurred prior to the start of the game. It could be a linear story or it can change depending on the user actions; therefore, it is linked to the game mechanics that should enable the storytelling.

The third element is the aesthetics, i.e., how the game should look, the music and sound that go along with the story telling and the mechanics. Aesthetic aspects constitute the most direct relationship with the user experience and influence engagement and immersion.

Finally, the fourth element that links the previous three ones together is the supporting technology: "it is essentially the medium in which the aesthetics take place, in which the mechanics will occur, and through which the story will be told." [67]. Choosing one technology enables the game developers to do certain things but might make it very difficult to implement some others; choosing the most adequate technology is therefore a complex process.

7.3 AN OVERVIEW ON GAMIFIED WATER MANAGEMENT SYSTEMS

We now overview the most relevant ongoing projects and applications aimed at the design and implementation of water management systems and gamified applications in the context of resource efficiency and consumption awareness enhancement. Our aim is the identification of the key features and functionalities that such systems should offer to be adopted by water utility companies, as well as the design challenges and implementation approaches that are common to gamified applications.

7.3.1 URBAN WATER PROJECT

Urban Water [8] is a project funded by the European Union's Seventh Framework Program for research and technological development, and it was developed by 12 partners from 8 different countries. Among them there were three water utilities, one university, one video game development company, and some governmental institutions. The project was seeking to develop and test an intelligent water management platform for the efficient and integrated management of water resources in urban areas [79]. More in detail, the project aim was a 50% reduction of urban water waste over 5 years of operation. The platform provides tools to manage data coming from smart meters, process information to forecast water demand and water supply availability, detect leakages in the water supply network, support decision-making of water authorities and water distributors, and empower customers to reduce their water consumption. The project objectives also included a market analysis at European level to identify market and legal European requirements and new market opportunities to ensure commercialization of the platform; the development of a new multi-technology communication gateway prototype combining wired and wireless technology, ensuring maximum geographic coverage and privacy; and the deployment of robust and proven data management tools based on Azure cloud platform [3] guaranteeing efficient data management and customer confidentiality. The project identified 7 enabling features to increase efficiency in water usage [14]:

Gamified Approaches for Water Management Systems: An Overview

- Effectively estimate water demand in urban areas in order to efficiently manage water supply chains.
- Reduce waste of water and economic losses associated to leakages in the urban water distribution network.
- Smooth daily water demand daily peaks in order to allow distributors to save costs related to the management of urban water distribution networks.
- Guarantee efficient and secure computational data management on the basis of smart grids recent and upcoming deployments in Europe.
- Reduce operating and maintenance costs associated with water metering and billing in urban areas.
- Incentivize urban households to reduce their current consumption and soften the current European water demand peaks.
- Build effective partnerships and develop innovation synergies between equipment providers, ICT companies, and water distributors.

The platform consists of several components [22], the main one is the customer portal that offers consumption monitoring, consumption forecast, and billing access information. Besides these functionalities, the portal features a serious game, "Water Mansion". For the utility management, the platform provides a business dashboard that shows the consumption (real and forecasted), the billing per hour, and the notification of recent events; it provides access to other functionalities such as:

- the decision supporting tool that provides visual information about the water availability and demand predictions;
- the leakage detection system that allows night flow analysis and mass balance;
- hydraulic model building tools;
- demand influence tools that include the automatic and dynamic billing address, the dynamic consumption analysis and the configuration options of the Water Mansion game;
- the integration tools that include the notifications, alerts, management and monitoring.

The featured online serious game, Water Mansion [55], seeks to raise consumers' awareness about efficient water consumption and its relation with economic savings. In the game users should execute a series of tasks involving day-by-day actions, such as washing their hands, cleaning the dishes, or filling the swimming pool. Each of these actions increases the consumption and reduces the "gold" that the user owns. The objective is to learn how to reduce consumption in order to save more "gold." The users are awarded a certain amount of "green drops" if in a period of time the consumption is reduced with respect to previous periods.

7.3.2 WATERP PROJECT

WatERP [11] is another project funded by the European Union's Seventh Framework Program. It was developed and supported by the Staffordshire University, the technological center of Catalonia, and 7 other technological and governmental partners.

The project was focused on developing an "open management platform" supported by real-time knowledge on the water supply and demand that enables an integrated and customized monitoring of the entire water supply distribution system, with the purpose of reducing the gap between water supply and demand through this information interaction and processing. The project had two main objectives: an 8% water consumption reduction in water-scarce areas where water distribution is already efficient, and a 5% energy usage reduction in areas where the water is abundant and the distribution is efficient but the consumption is high. Additional goals were the increase of user awareness and the promotion of behavioral changes of water utilities and authorities (not end-consumers). From the technological point of view, the project aimed at identifying the key variables that must be monitored throughout the water supply distribution system to enable water supply and demand to be matched across the entire water supply network and while coping with water scarcity, drought, and vulnerability indicators; and at developing protocols for real-time data collection and storage ensuring data quality, reliability and consistency. The WatERP architecture consists of three main components:

- the *data warehouse*, which manages the processing and storing of consistent, continuous and usable water supply and usage data originating from heterogeneous sources (periodically or in real-time);
- the *decision support system* (DSS), which coordinates actions prioritizing water usage, improving distribution efficiency and reducing costs;
- the *demand management system* (DMS), which analyzes socioeconomic drivers and policies to improve the management of water demand.

Information from the whole supply chain (including, e.g., water sources and deposits, desalinization plants, distribution networks) are stored in the data warehouse, which makes it available to the users in real-time (with the possibility of personalizing data analytics) in order to support their decision making, policy making, water pricing, risk management and planning processes [27]. Demand forecasting is performed by leveraging advanced data analytics that calculate similarity indexes between the consumption pattern of the current day and historical consumption data available in the database, identify the most similar ones, calculate weights to be assigned to similar days, and finally output a prediction of the demand for the desired future date.

7.3.3 WATERNOMICS

Waternomics [10] is an ongoing project developed collaboratively by 10 institutions, including the national university of Ireland Galway, the UNESCO-Institute for Water Education, and the municipality of Thermi, in Greece. The goal of the project is to provide personalized and actionable information about water consumption and water availability to individual households, companies, and cities in an intuitive and effective manner, at a time-scale relevant for decision making. Access to this information will increase end-user awareness and improve the quality of the decisions for decision makers in companies and in governments [72]. The project aims to accomplish these objectives by combining water usage-related information from various sources and domains to offer water information services to end-users, supporting personalized interaction with water information services, conducting knowledge transfer from energy management systems to water management systems, enabling sharing of water information services across communities of users, and enabling open business models and flexible pricing mechanisms responsive to both demand and climate/environmental conditions. Additional project goals are the introduction of water footprints (i.e., demand response and accountability) in the water sector; the interactive engagement of the consumers to enable efficient water consumption and behavioral change; the enactment of ICT-enabled water management by providing relevant tools and methodologies for water-related issues to corporate decision makers and municipal area managers. The Waternomics platform consists of a four-layered architecture:

- The *hardware layer* includes the smart meters installed at different levels of the water distribution network, depending on the location site (e.g., at appliance level for households and at building level public buildings or offices).
- The *data layer* stores, processes, and analyzes the collected metering data to provide high level information reports (destined, e.g., to municipalities).
- The *support service layer* implements the data analytics, the monitoring and alerting systems, and responds to the information requests from the application layer [28].
- The *application layer* is designed as an application portal: it provides basic functionalities and can be extended by getting more applications from a marketplace. Users access the marketplace, select the application they are interested in based on their role and the activities they want to perform, and install it. This approach ensures great flexibility in terms of personalization [43] and makes the portal customizable for a wide range of user types and location sites: administrators can decide which application should be available for each role and existing applications can be enhanced by adding components. Third-party applications can also be made available through an API [43].

The applications are divided into 4 categories:

- *Monitoring*: Applications that allow users to monitor their consumption and have it visualized.
- *Learning*: Applications that provide educational and informative material to users.
- *Exploring*: Applications that allow users to explore the potentials of the dataspace in terms of analytics and related services.
- *Playing*: Gaming applications or applications with gamification elements that help users learn and educate themselves through playing or through interacting with each other in non-leisure contexts.

Additionally, the platform provides a set of independent components, also available through the marketplace, which enable the user to create custom applications by combining them (e.g., to create dashboards). Some of the base applications for the household users are:

- *Family Dashboard*: Every family is entitled to create its own dashboard using a set of available components and choosing those that fit their needs. The total consumption component provides a visualization through a graph bar, the documentation explains that showing only cubic meters measurement might not be representative for most of the users, so it provides alternative representations like the number of kilograms of apples that could be produced with the same amount of water, or the amount of CO_2 that was released in the environment to deliver that amount of water to the household, thus enhancing awareness of the family water consumption.
- *Consumption Timeline*: This component emulates the function of a social media timeline by presenting series of events in a chronological order. The points along the timeline represent disaggregate consumption events and provide comparison with previous week's events of the same type (e.g., washing-machine water consumption).
- *Notifications*: The user can set notifications preferences for each type of event, and receive alerts accordingly. For example, the users can set an alert when the total consumption for the month is 10% higher than the previous month.
- *Drought conditions monitor*: This application periodically receives data from the European Drought Observatory [EDO] and informs the users about drought periods, thus raising awareness of water scarcity.

For the administration side, the offered applications are:

- *Management Dashboard*: Managers are interested not only in monitoring the total consumption but also in controlling specific areas of the network. In the case of public buildings (e.g., an airport), it allows for monitoring specific building areas (e.g., the consumption of one of the terminals), as long as the metering infrastructure supports the collection of disaggregated data.
- *Technician Dashboard*: It provides overall information and notifications about critical detected events to support a timely on-field intervention from staff members.
- *Leak Detection*: The platform monitors the network and when a leakage is detected by the deployed sensors, the notification system sends alarms. Leakages can also be monitored through a graphical interface.
- *Fault Detection Diagnostics*: This application uses the baseline consumption and a series of rules to detect anomalies on the network flow, generating alerts about events that require validation or intervention.

For what concerns the game framework, the Waternomics platform [43] provides a leaderboard that shows the highest scores achieved by the users in the different game applications. The platform offers two trivia-like games. The first one is "Water Flavors": it asks the player to make educated guesses about how much water is necessary to produce certain products. In the case of a correct guess, the player is awarded with points. This way, the game aims at creating awareness by educating the player about the importance of water in the production of food and other products of common use. The information used in the game is retrieved from the Water Footprint Organization [9]. The second game is called "Water Saving Calculator": in the first stage it provides a collection of water saving tips; in the second phase the player is asked to answer a set of questions by performing calculations based on the information provided by the tips. Players get points for the right answers and extra information for the wrong ones. The objective of the game is to make the players aware of the amount of water they are consuming and the potential savings that they can reach by changing some behaviors. As mentioned before, the architecture allows for the inclusion of additional games and gamified applications: the point storage and leaderboard management is achieved by an API.

7.3.4 SMARTH2O

The SmartH2O project [7] is funded by the European Union's Seventh Framework Program for Research and Innovation. The project aims at creating a communication channel and a continuous feedback-loop between water users and utility companies, providing consumers with information on their consumption in near real-time while enabling water utilities to plan and implement strategies to reduce or reallocate water consumption. This can be achieved by exploiting collected information about how the consumers adapt their behavior as a reaction to different stimuli, such as awareness campaigns and changes in regulations or prices. To this aim, smart water meters are leveraged for collecting high frequency consumption data, which are used to provide high granularity information to water utilities on the state of the distribution network. At the same time, the collected information can be employed to stimulate a change in water consumption behavior. Accordingly, the SmartH2O system has been designed as a behavioral change support system (BCSS): "a socio-technical information system [...] designed to form, alter or reinforce attitudes, behaviours or an act of complying without using coercion or deception" [58]. The SmartH2O approach considers that a change in water consumption behavior can occur when underlying psychological determinants change through a combination of different incentive and persuasion strategies, acknowledging that both in the energy domain and in the water domain consumption data alone are not sufficient to induce a sustainable behavioral change [35] [53]. Rather than relying on smart metered consumption feedback alone, the systematic approach followed by SmartH2O is grounded in motivational theory and research on incentive models, employing different mechanisms to incentivize users to save water. This has resulted in an ICT platform for improving consumer awareness, available on the web and on mobile devices via a downloadable app. It incorporates smart metering, social computation, data visualization, big data analytics to model user behavior, and gamification to engage consumers in the process.

As inducing long-term behavioral changes is a difficult task that involves psychological, social, and cultural factors, the project incorporates a mix of multiple engagement strategies. All actions undertaken by the consumer, in all the different applications (web portal, mobile app, and even an educational game) are logged. Based on such action logs and on the metered consumption data, a gamified social game is performed in which consumers are encouraged to save water through a mechanism that assigns points, badges, and prizes based on the full spectrum of their actions; leaderboards and weekly/monthly competitions provide a social dimension to the game and increase engagement and motivation to participate by creating a sense of community. Moreover, through the game the users are encouraged to provide detailed profile information about their demographics and their household configuration: such data are greatly valued by utility companies, as their analysis can provide important insights on the factors that drive the demand trends.

Figure 7.2 shows the home page of the Consumer Portal developed in SmartH2O. The page displays the histogram of the consumption readings collected with the smart meter infrastructure.

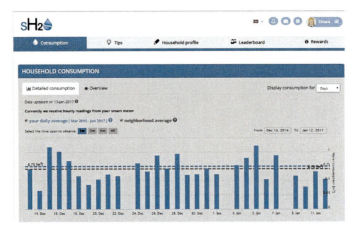

Figure 7.2: Homepage of the SmartH2O Consumer Portal.

To improve the awareness of the impact of consumption and start engaging the user in proactive water saving behavior, a complementary interface visualizes consumption data in an alternative way, shown in Figure 7.3.

In the overview visualization, the progress of the consumption in a certain period, e.g., the current week, is displayed as a tank filled with water. The tank metaphor is the basis for further incentive model elements, as discussed in more detail in Section 7.4. As a complement to the utilitarian interfaces provided by the consumer web portal and mobile app, SmartH2O has also designed a water board game, called Drop! (show in Figure 7.4), with a digital extension, called Drop!TheQuestion.

The Drop! board game is assigned as a reward to the users of the consumer web portal that achieve a minimum level of activity. It exploits a very popular home- and family-oriented entertainment scheme, called "push your luck." In this class of games, the players repeat an action (e.g., drawing a card) until they decide to stop, due to an increased risk of losing points in the next turn (e.g., drawing a negative card). The Drop! board game is designed around the basic idea that a game does not

Gamified Approaches for Water Management Systems: An Overview 185

Figure 7.3: Interface of the SmartH2O Consumer Portal showing an overview of the consumption progress.

Figure 7.4: The Drop! board game and its digital extension (left); scanning a card with Drop!TheQuestion (right).

necessarily need to be educational. The game metaphor is simple yet engaging: Lily is a young and clever girl who wants to save water. Lily's friend is a monster who does things in exactly the opposite way: the monster spills water (Figure 7.4). During the game, players do not need to answer questions or prove their knowledge to win, as winning is determined by luck. The cards showing Lily are good cards and let the player score positive points, while the monster cards give players negative points. At the end of the game, players can transform the monster cards with negative points into positive points, by scanning a QR code on each monster card and answering a question received through their mobile phone or tablet (as shown in the right part of Figure 7.4). By playing the game within a household, saving water becomes a topic of conversation. This stimulates a water saving culture within the household, which in turn is a strong predictor of water consumption [34]. Playing the board game and answering questions in the mobile app game increases knowledge. Users are incentivized to play the game in two distinct but related ways: the game design of the Drop! game itself, and the link with the gamified portal. Answering questions in the mobile game is awarded with points on the consumer portal.

The SmartH2O application is currently deployed in two real-world scenarios: the first in Switzerland, the second in Spain. A more detailed description of the SmartH2O application and the case studies will be provided in Section 7.4.

7.3.5 A BRIEF REVIEW OF SOME COMMERCIAL PRODUCTS

With the growing concern for water scarcity and governments of developed countries starting to change their regulations and policies about water management, the market offers important business opportunities for big technology companies that have already started to work on resource management projects. In the following, we focus on two commercial platforms, respectively offered by IBM and Oracle. IBM Intelligent Water [2] is part of the smart cities portfolio of the company: the platform offers situational awareness of operations at several levels and a range of advanced tools. It does not necessarily replace the legacy systems of the utilities, but works on top of those by providing visualizations and data analytics, raising alerts, starting workflows to respond to a given situation, and displaying information on dashboards that can be configured based on the user role and access level. Operational data are holistically leveraged to improve water management: the framework includes a water analytic component which aimed at reducing the non-revenue water through pressure optimization and pipe failure prediction; a monitoring component for smart meters to detect fraud; analytics components to improve irrigation, flood management and sewer overflows prevention. The framework has been successfully deployed in three study cases. The first is located in China, where the platform was used to manage the operation of a hydropower plant in collaboration with SimuTech solutions [6]. The platform supported the whole management cycle, from the planning phase of the hydropower operations to the monitoring of the state of the reservoirs and of energy production. A second study case is in the Netherlands, where the national Ministry of Water adopts the platform to counteract the effects of extreme weather conditions, predict floods and droughts, and monitor the quality of drinking water, by exploiting its capabilities of proactively proposing preventive actions to possible issues. Finally, the last study case is in Miami-Dade County, where the tool is mainly used to monitor and detect leaks: using geospatial maps, the operators get alerted about potential problems that can be visualized through dashboards displaying various analytics.

Oracle Utilities [4] offers a full suite of solutions for utilities, covering all aspects of management and planning. The suite is fully configurable to enable the utilities to adapt quickly to the changing regulations and ensures the full integration with legacy systems of the utilities. Some of the most interesting resource management applications of the suite are:

- The *metered data management* component, which deals with consumption data validation for consistency and storage;
- The *smart grid gateway*, which operates as a data hub to ensure two way communication between the smart meters and the utility enterprise applications;
- The *operational device management* component, which provides asset management of the smart grid and meters and allows for real-time configuration, automatic inventory, technical maintenance, and security verification;
- The *data raker*, which provides advance data analytics, data visualization and enrichment including meter malfunction detection, customer tampering detection, establishment of maintenance priorities, and billing exceptions detection;
- The *smart water network* component, which provides automated valve operations management, leakage detection, outage isolation, shutdown management, and geospatial visualization for segment isolation.

Another group of applications of the suite supports mission-critical utility operations and includes visualization applications that allows one to access, view, digitally annotate and collaborate on assets information and case-related documents (schematics, blueprints, 3D models, etc.). They also provide pre-built dashboards, extractors and schemas for the information of the other products. Finally, a set of complementary applications is designed for customer and billing support and dedicated to utilities serving residential, commercial, and industrial customers, with functions for managing payment and credit collection:

- The *customer relationship management* component supports sales and marketing by segmenting the users in groups with similar characteristics and providing insight into the group's specific needs, besides handling customer order management.
- The *rate management* component handles pricing design and analysis; using analytics, it evaluates the implications of the changes in prices offer to clients.

7.3.6 GAMIFIED APPLICATIONS FOR RESOURCE EFFICIENCY

Despite some successful studies, changing users' behavior via software is not an easy task. This is due to fundamental open questions such as how to design an effective application to achieve the desired results and how to maintain behavioral changes in the long-term. As an example, the response-relax effect is often observed: after the change induced by the initial exposure to the persuasive technology, the attention paid to the feedback reduces and users revert to their previous behaviors [60].

Computer games have recently been advocated as a promising tool for computer-mediated behavioral change. They create an immersive environment that can attract also kids and teenagers to serious topics and manifest a high persuasive potential. Moreover, games effectively exploit the procedural representation approach, i.e., a form of symbolic expression that uses processes rather than language to convey "how things work" [18]. In well-designed games, players can combine the processes embodied in the game and create new interactions beyond those considered by the game designers. This paradigm introduces new ways to persuade the player, which match well with the rhetoric concepts exposed by Fogg in [36] (triggers, motivation and ability) as the traditional means for persuasion through technology. Some recent contributions in environment and sustainability games focus on power conservation, environmental awareness, fossil energy use and water: Table 7.1 and Table 7.2 compare some persuasive games specific for water management along their main distinctive features:

- Mechanics describes the type of the game.
- Roles list the roles assumed by the players.
- Players define the number of players that engage in a game round.
- Feedback specifies whether the game can interact with players during a round.
- Issues summarize the main problems highlighted by the game.
- Focus refers to the core persuasive/educational objective of the game.
- Target identifies a specific population of players targeted by the game.
- Data defines the data collection method.
- Platform specifies the technological environment/device for which the game is designed.
- Technology characterizes the implementation languages and frameworks.

In summary, findings from existing literature suggest that gamification strategies and games with a purpose can be applied in resource efficiency contexts with a good user acceptance and successful results in savings (as supported by empirical studies such as [63]). However, the success depends on many factors, most of them related to the design of the incentive models and game mechanics motivating the usage of the systems and the specific system functionalities offered. The most important points and lessons from the literature include:

- Providing real-time and historical information of the consumption is important, but the data is useless if the consumer cannot associate a meaning to it, therefore informative measurements should go along with the real data (e.g., presenting cost in terms of money along with the consumption in cubic meters).
- Real-world rewards and prizes are a good strategy to keep user engagement, but it is important to select them according to the type of consumers. Having various prizes with different options seems to be a good solution when the player population is heterogeneous.

Table 7.1

Summary of water games features. Games analyzed: Water Wars [42], Atoll Game [33], Catchment Detox [69], FloodSim [61], and Aqua Republica [1].

	Water Wars	Atoll Game	The Basin Challenge/ Catchment Detox	FloodSim	Aqua Republica
Mechanics	Turn based	Rpg, computer assisted	Turn based	Turn based	Turn based
Roles	Stakeholders	Family	Policy makers	Flood policy, strategist	Mayor
Feedback	Message	N/A	Messages in game and leaderboard	Messages in game	Messages in game
Issues	Policies, variable weather conditions	Policies, variable weather conditions, scarcity	Policies, variable weather conditions, scarcity	Floods	Policies, variable weather conditions
Players	Multiplayer with chat	Up to 16 presential players	1-2 players	1 player	1 player
Focus	Interaction among inhabitants	Land/ water allocation conflicts	Manage a river catchment	Raising awareness on flooding policies	Conflicts and trade-offs in a river basin
Target	New Mexico residents	Tarawa atoll residents and policy makers	Teenage students	UK residents	Everyone
Data Collection	Interviews	Semi-automatic software	N/A	N/A	Numerical models (Mike Basin)
Platform	Web and mobile	PC supported board game	Web	Web	Web (portable)
Technology	N/A	Visual-Works and CORMAS platform	Flash	Flash	Unity

Table 7.2

Summary of water games features. Games analyzed: Irrigania [70], Run the River [5],WaaS [76], SeGWADE [52], and Drop! TheQuestion.

	Irrigania	Run the River	WaaS	SeGWADE	Drop! The Question
Mechanics	Turn-based strategy game	Model simulation	Model simulation	Model simulation	Trivia
Roles	Farmer	Decision makers	Decision makers	Water distribution systems managers	Water expert
Feedback	No feedback	In game alerts	Model-derived feedback	Continuous and instant feedback to players	No feedback
Issues	Governance and management of common resources	Balancing water use between various water consumers	River management	Water distribution systems design decisions	General water culture
Players	Single player	Single player	Teams	N/A	Single player
Focus	Water sharing policies, water scarcity	Water management policies	Water management policies under multiple perspectives	Drinking water distribution systems	Interesting facts on water and water consumption
Target	Students	Kids	Students, researchers, and professionals	Students	Family
Data Collection	Discussion in class	Based on actual and modelled historic data	Water system modelled with Integrated Assessment Meta Model (IAMM)	Feedback computed with hydraulic simulation engine based on EPANET	N/A
Platform	Web	Mobile phones and tablets	N/A	Windows, Linux, iOS and Android	Android mobile devices
Technology	VisualBasic, ASP.NET	N/A	Mixed (option card selection and model simulation)	HTML5 and WebGL	Java

- Games are good influencers when trying to make changes in consumer behaviors, but permanent changes are difficult to achieve. The game mechanics have to be powerful enough to keep the user mindset; a good recommendation is to strongly relate the actions and points in the game with actions in the real world.
- People are now aware of the problems related to resources and are willing to change their behavior, but constant, useful, and interesting information about what to change and how to change it should be always available on demand and offered by the game.
- Designing systems that can successfully stimulate behavioral change requires a systematic modelling of the behavioral change process and its implementation through different types of incentives (virtual, physical, social) adapted to the characteristics of specific target user groups [56].
- Social interaction and competition are powerful engagers and influencers of behavior; they should be exploited but in some cases moderation might be required and evaluation should be made about the effort required in relation to the benefits.

Further discussion and examples of serious games for the water management field can be found in [66].

7.4 SOME EXPERIENCES FROM THE DESIGN AND DEPLOYMENT OF THE SMARTH2O SYSTEM

In this section we consider some of the main experiences from the design and deployment of the SmartH2O gamified system for behavioral change. In particular, we show how design guidelines for gamified consumption visualizations have informed the design of the SmartH2O system, suggesting their suitability to guide the design of such classes of applications. We also summarize some of the first evaluation results, with respect to user feedback, system acceptance and participation dynamics, as well as first encouraging results of observed reduction in water consumption in a small scale pilot.

7.4.1 DESIGN GUIDELINES FOR VISUALIZATION AND GAMIFICATION IN WATER SAVING APPLICATIONS

Beyond leaning on findings from motivational and psychological theory in the design of the incentive model, the design of the SmartH2O system described in Section 7.3.4 also paid particular attention to effective use of visualization and its integration with gamification elements for supporting behavioral change. Based on the literature review of consumption visualization and use of gamified elements for water and energy conservation (as structurally similar domains [51]), initial guidelines were formulated and supplemented with findings from the SmartH2O requirements analysis workshops to yield a set of design guidelines for water and energy saving applications (see Table 7.3).

Accordingly, the SmartH2O system incorporated different types of visualization linked with gamified elements (such as saving goals) described in Section 7.3.4.

Table 7.3

Design guidelines for water and energy saving applications [51]

DG	Refined aspects from SmartH2O design cycle and user workshops		
a	Present interactive layered visualization (simplest by default)	c	Goals should be related to concrete actions users can perform
a	Use visual metaphors relating to users' consumption context	d	Feedback on consumption should be action-oriented and include saving tips embedded in the visualization
b	Present separate views for less vs. highly data-affine users	e	Real rewards should engage even more pragmatic users
b	Overview of consumption should trigger awareness and detailed information should point out concrete actions	e	Separate views for pragmatic and hedonic users should be considered
b	Units and analogies should illustrate consumption & savings	f	Common goals have the potential, to bring, e.g., neighbors closer
b	Showing consumption rather than smaller savings can raise awareness	f	Both are promising for different users embedded in the relevant social context: in-group, collaboration, intra-group competition

Most clearly, the application of the guidelines is reflected in the design of metaphor-based overview visualization. In a first version it employed a speedometer metaphor that showed daily and monthly consumption in comparison to one's average (Figure 7.5(a)), while in a second version, this was replaced with a water pipe metaphor (Figure 7.5(b)) to support more immediate cognitive mapping to the water topic and reinforce water awareness. The pipe metaphor formed the basis for integrating goal setting and gamified elements with the consumption visualization, as can be seen from the green box in Figure 7.5(b) (e.g., setting self-set goals and receiving double points when reaching them, visual alerts about current consumption with respect to set goals, using swiming pool analogy to illustrate the meaning of the amount of saved water, etc.). Similarly, while the bar chart overview visualization (see Figure 7.2) supported more pragmatically oriented users and needs, the water pipe visualization also supported hedonic aspects. Displaying the consumption average and the average of the neighborhood in the bar chart (Figure 7.2) supported social comparison. Linking achieved points to different types of collectable real rewards, addressed more pragmatic users for whom social comparison or symbolic awards (such as badges) were not enough to motivate action.

(a)

(b)

Figure 7.5: Metaphoric visualizations: a) speedometer-based display, b) water pipe metaphor.

7.4.2 EVALUATING USER RESPONSES AND FIRST EFFECTS ON WATER CONSUMPTION

The results of preliminary evaluation of the SmartH2O system in the first pilot in the municipality of Tegna (Switzerland), although reflecting a small initial test base confirmed the effectiveness of such a design, suggesting that the identified design guidelines are suitable for informing the design of such systems. There, the system has been deployed in two versions: as a basic portal implementing only the "pragmatic" elements, such as the interactive water consumption visualization and the water saving tips, and the gamified portal adding goal setting and the full spectrum of gamified incentives and rewards. After 4 months of basic portal trial with 40 users (households) user feedback collected with an online questionnaire (n= 15 / 37,5% response rate) revealed that most users found that the system made water conservation more interesting (11/15 users) with half of them finding it fun to use (7/15), indicating that even a non-gamified system incentivizes use (for more details see [56]). In particular, most respondents found the interactive water consumption chart easy to use (9 / 15 users) and almost all respondents (14 / 15 users) agreed that from the chart and the overview visualizations they could understand how much water their household consumed over time [56]. The perceived usefulness of the main design elements (the water consumption chart (8 / 15 users), consumption overview (7 / 15 users) and water saving tips (9 / 15 users)) was also relatively positively rated, and for all these three features most users also stated that they made them think about water conservation more often (11/15 users). Even though tentative in nature, and only based on a first simple version of the visualizations (bar chart view and speedometer metaphor), the results suggest that the designed visualizations were capable of incentivizing users to reflect on their consumption.

Following this initial assessment of user acceptance (based on the model by [77]) the evaluation of the gamified portal through the analysis of user activity was also performed. This system version also included the new consumption visualization using a pipe metaphor. To gain first insights into incentive dynamics on the gamified portal, user activity logs in the Swiss case were analyzed (Dec'15-Feb'16). After its launch in December 2015, little activity occurred on the basic portal compared to the gamified portal, which exhibited almost twice as many logins as the former [56]. This suggests that the gamification features prompted more interest, with activity peaking after a mail campaign to current users. An inspection of individual elements revealed that the consumption chart has been viewed and interacted with the most. Other key elements like the leaderboard, achievements, and actions panels were also accessed more than once by over half of the users and almost all users collected badges. These indicators suggest that the incentive model is capable of motivating users to access and use it, and thus to engage with water consumption information. The overall results suggest that even with a small user community and not a very long period for the incentive model to yield effects, the system was able to motivate different participation dynamics [46]. Even though these observed dynamics need to be assessed in more detail in the large scale pilot, such differences in responses to the incentive model highlight the importance of a holistic approach that comprises different motivational affordances to support the behavioral change process for all users [46].

Consumption effects were measured with respect to a historical yearly baseline for each user. We have observed that users with a low and medium consumption volume achieve the highest reductions (see [46] for details). After discounting the possible effect of the winter season on lower consumption (to an upper bound of 25-30%), a water consumption reduction of 3.4-8.4% that could be attributed to the use of the portal remained. This first encouraging result needs to be validated once detailed consumption logs are made available for the whole year, enabling us to make a meaningful comparison with the baseline.

Finally, some of the other lessons learned in the design, deployment, and evaluation of the SmartH2O system include:

- Gamification in combination with physical rewards can be a good way to strengthen the incentive induced by consumption feedback, as physical rewards have proven to be capable of engaging users who are not necessarily motivated by playful elements and associated virtual or social rewards. Business factors such as the available budget and the image the utility seeks to communicate to its customers shape the reward model, in terms of the type of rewards, the number of available rewards, and the user actions in the applications that make a user eligible for a reward.
- Whereas in traditional approaches the collection of user data requires substantial effort, in SmartH2O the gamified incentive model has proven to be capable of eliciting user-generated data about the household composition, water consuming appliances, presence of a garden or balcony and other factors that determine water consumption levels. These data allow utilities to model the water demand beyond what can be inferred from aggregated (smart metered) readings by clustering consumers based on socio-psychographic features and/or disaggregating consumption to the level of water consumption appliances [81], [48].
- Experiences from SmartH2O and other projects [35] [68] strongly suggest that the sustainability of effects on water consumption behavior cannot be taken for granted. Rather, utilities must be aware of the fact that relapse to "old" consumption behavior can and will occur. A significant share of water consumption behavior is habitual, which is notoriously difficult to change (e.g., [32]), requiring utilities to commit to a long-term implementation of behavioral change interventions and to allocate their required resources.

Such experiences from the SmartH2O project can support water utilities (or researchers) who intend to develop similar applications, as part of their water demand management strategy and as a way of developing new kinds of water consumption information services for their customers.

7.5 CONCLUDING REMARKS AND FUTURE DIRECTIONS

Gamification mechanisms appear to be promising and acceptable tools for incentivizing pro-environmental attitudes and promoting water conservation habits in urban contexts. Following this potential, several EU-funded projects focused on assessing the effectiveness of gamified tools to support water demand-side management, integrated with ICT solutions, as well as advanced metering infrastructure. In this chapter, we reviewed the most relevant experiences from state-of-the art projects, and presented the SmartH2O platform, a web-based gamified user portal that we developed as part of the EU-funded SmartH2O project and implemented in two real case studies, in Switzerland and Spain.

Our comparative analysis, as well as the preliminary results we obtained from real-world testing, suggest a few general take-home messages for the future exploitation of gamification mechanisms to design water demand management strategies. In particular, the number of experiences developing gamified approaches to water conservation and other environmental purposes in Europe and worldwide values such tools as alternatives/complements to more traditional top-down interventions (e.g., water use restrictions or mandatory retrofitting interventions). Besides, gamified applications do not purely constitute single-direction communication tools for giving behavioral suggestions to users. Rather, they can easily include interactive modules incentivizing users to provide data (e.g., demographics), which would be more difficult to collect and update with traditional surveys or una tantum data gathering campaigns. Therefore, their value to support user modelling and customized demand management activities is increased.

Yet, rigorous studies comparing exhaustively, with a complete set of quantitative and qualitative performance metrics, the state-of-the-art gamified methods to enhance awareness and behavioral change toward environmental issues do not exist. Therefore, findings and consideration on the effectiveness of such methods should be related to specific case studies and applications so far, and a generalization and integration effort is required.

Finally, the preliminary results of the SmartH2O project support the effectiveness of gamified approaches to pursue water savings, at least in the short-term, and we were able to draw some recommendations for the future development of such approaches (e.g., combining gamified mechanisms with real, physical rewards). Yet, their long-term effectiveness has not generally been assessed so far, as well as ways to keep it high and avoid the rebound effect. Therefore, future research and system implementations, involving researchers, utilities, and water consumers, should commit to a long-term design and testing of behavioral change interventions, possibly considering city-scale samples of heterogeneous users and designing alternative customized demand-side management actions.

7.6 REFERENCES

1. Acqua republica. [Online] http://aquarepublica.com/.
2. IBM Intelligent Water. [Online] http://www-03.ibm.com/software/products/en/intelligentwater.
3. Microsoft azure. [Online] https://azure.microsoft.com/en-au/.
4. Oracle Utilities. [Online] http://www.oracle.com/industries/utilities.

5. Run the River. [Online] https://play.google.com/store/apps/details?id=mdba.runtheriver.
6. SimuTech Group. [Online] http://www.simutechgroup.com/.
7. SmartH2O: An European Project on Water Sustainability. [Online] http://www.smarth2o-fp7.eu/.
8. Urban Water. [Online] http://urbanwater-ict.eu/.
9. Water Footprint Network. [Online] http://waterfootprint.org/.
10. Waternomics. [Online] http://waternomics.eu/.
11. Waterp. [Online] http://www.waterp-fp7.eu/.
12. Tactical technology collective, Visualising information for advocacy, 2013. [Online] https://visualisingadvocacy.org/.
13. L Atzori, A Iera, and G Morabito. The Internet of Things: A survey. *Computer Networks*, 54(15):2787–2805, 2010.
14. P Avila, A Cisneros, A Rodriguez, and L Tena. UrbanWater D6.2: UrbanWater platform architecture design. 2015.
15. R Bartle. Hearts, clubs, diamonds, spades: Players who suit muds. *Journal of MUD Research*, 1(1):19, 1996. [Online] http://mud.co.uk/richard/hcds.htm.
16. C D Beal, T R Gurung, and R A Stewart. Demand-side management for supply-side efficiency: Modelling tailored strategies for reducing peak residential water demand. *Sustainable Production and Consumption*, 6:1–11, 2016.
17. C Beal, R Stewart, D Giurco, and K Panuwatwanich. Intelligent metering for urban water planning and management. *Water Efficiency in Buildings: Theory and Practice*, pages 129–146, 2014.
18. I Bogost. *Persuasive games: The expressive power of videogames*. MIT Press, 2007.
19. D Börner, M Kalz, S Ternier, and M Specht. Pervasive interventions to increase pro-environmental awareness, consciousness, and learning at the workplace. In *European Conference on Technology Enhanced Learning*, pages 57–70. Springer, 2013.
20. T Boyle, D Giurco, P Mukheibir, A Liu, C Moy, S White, and R Stewart. Intelligent metering for urban water: A review. *Water*, 5(3):1052–1081, 2013.
21. T C Britton, R A Stewart, and K R O'Halloran. Smart metering: Enabler for rapid and effective post meter leakage identification and water loss management. *Journal of Cleaner Production*, 54:166–176, 2013.
22. A Broussel. Urbanwater D6.4: Urbanwater platform prototype. 2015.
23. R Cahill and J Lund. Residential water conservation in Australia and California. *Journal of Water Resources Planning and Management*, 139(1):117–121, 2012.
24. S K Card, J D Mackinlay, and B Shneiderman. *Readings in Information Visualization: Using Vision to Think*. Morgan Kaufmann, San Francisco, 1999.
25. R Cardell-Oliver. Water use signature patterns for analyzing household consumption using medium resolution meter data. *Water Resources Research*, 49(12):8589–8599, 2013.
26. D Charitos, I Theona, C Rizopoulos, K Diamantaki, and V Tsetsos. Enhancing citizens' environmental awareness through the use of a mobile and pervasive urban computing system supporting smart transportation. In *Interactive Mobile Communication Technologies and Learning (IMCL), 2014 International Conference on*, pages 353–358. IEEE, 2014.
27. C Chomat. WatERP D 1. 2: Generic functional model for water supply and usage data. 2013.
28. D Coakley, W Derguech, S Hasan, C Kouroupetroglou, Y Lu, J Mink, D Perfido, and

U Ul Hassan. Waternomics: D1. 3 System Architecture and KPIs. 2015.

29. A Cominola, M Giuliani, D Piga, A Castelletti, and A E Rizzoli. Benefits and challenges of using smart meters for advancing residential water demand modelling and management: A review. *Environmental Modelling and Software*, 72:198–214, 2015.

30. A Cominola, A Moro, L Riva, M Giuliani, and A Castelleti. Profiling residential water users routines by eigenbehavior modelling. In *8th International Congress on Environmental Modelling and Software*, Toulouse France, 2016.

31. A Cuzzocrea, I-Y Song, and K C Davis. Analytics over large-scale multidimensional data: the Big Data revolution! In *Proceedings of the ACM 14th International Workshop on Data Warehousing and OLAP*, pages 101–104, Glasgow, Scotland, UK, 2011.

32. U Dahlstrand and A Biel. Pro-environmental habits: Propensity levels in behavioral change. *Journal of Applied Social Psychology*, 27(7):588–601, 1997.

33. A Dray, P Perez, C Le Page, P D'Aquino, and I M White. Companion modelling approach: The AtollGame experience in Tarawa atoll (Republic of Kiribati). 2005.

34. K S Fielding, S Russell, A Spinks, and A Mankad. Determinants of household water conservation: The role of demographic, infrastructure, behavior, and psychosocial variables. *Water Resources Research*, 48(10), 2012.

35. K S Fielding, A Spinks, S Russell, R McCrea, R Stewart, and J Gardner. An experimental test of voluntary strategies to promote urban water demand management. *Journal of Environmental Management*, 114:343–351, 2013.

36. B J Fogg. *Persuasive technology: Using computers to change what we think and do.* Morgan Kaufmann Publishers Inc., San Francisco, CA, USA, 2003.

37. J Froehlich, L Findlater, M Ostergren, S Ramanathan, J Peterson, I Wragg, E Larson, F Fu, M Bai, S Patel, and J A Landay. The design and evaluation of prototype eco-feedback displays for fixture-level water usage data. In *Proceedings of the SIGCHI Conference on Human Factors in Computing Systems*, CHI '12, pages 2367–2376, New York, NY, USA, 2012. ACM.

38. P Gerland, A E Raftery, H Ševčíková, N Li, D Gu, T Spoorenberg, L Alkema, B K Fosdick, J Chunn, N Lalic, et al. World population stabilization unlikely this century. *Science*, 346(6206):234–237, 2014.

39. P H Gleick and M Palaniappan. Peak water limits to freshwater withdrawal and use. *Proceedings of the National Academy of Sciences*, 107(25):11155–11162, 2010.

40. F Groh. Gamification: State of the art definition and utilization. *Research Trends in Media Informatics*, pages 39–46, 2012.

41. J Hamari, J Koivisto, and H Sarsa. Does gamification work?–a literature review of empirical studies on gamification. In *System Sciences (HICSS), 2014 47th Hawaii International Conference on*, pages 3025–3034. IEEE, 2014.

42. T Hirsch. Water wars: designing a civic game about water scarcity. In *Proceedings of the 8th ACM Conference on Designing Interactive Systems*, pages 340–343. ACM, 2010.

43. C Kouroupetroglou, J Van Slooten, N Kasfikis, W Derguech, E Curry, S Hasan, S Smit, and D Perfido. Waternomics: WATERNOMICS Apps. 2016.

44. K Kyatric. Bartle's taxonomy of player types (and why it doesn't apply to everything), 2013. [Online] http://gamedevelopment.tutsplus.com.

45. K Ling, G Beenen, P Ludford, X Wang, K Chang, X Li, D Cosley, D Frankowski, L Terveen, A M Rashid, et al. Using social psychology to motivate contributions to online communities. *Journal of Computer-Mediated Communication*, 10(4), 2005.

46. I Micheel, J Novak, L Caldararu, C Rottondi, C Pasini, P Fraternali, R Marzano, C

Rouge, G J C N Bakkalian, M Velzquez M Bertocchi, M Melenhorst and A Rizzoli. Deliverable D7.2 WP7. The SmartH2O project, a European Project on Water Sustainability, 2016.

47. K Madani, T W Pierce, and A Mirchi. Serious games on environmental management. *Sustainable Cities and Society*, 29:1–11, 2017.

48. A A Makki, R A Stewart, C D Beal, and K Panuwatwanich. Novel bottom-up urban water demand forecasting model: Revealing the determinants, drivers and predictors of residential indoor end-use consumption. *Resources, Conservation and Recycling*, 95:15–37, 2015.

49. P W Mayer, W B DeOreo, E M Opitz, J C Kiefer, W Y Davis, B Dziegielewski, and J Olaf Nelson. Residential end uses of water. *PAWWA Research Foundation and American Water Works Association*, 1999.

50. R I McDonald, P Green, D Balk, B M Fekete, C Revenga, M Todd, and M Montgomery. Urban growth, climate change, and freshwater availability. *Proceedings of the National Academy of Sciences*, 108(15):6312–6317, 2011.

51. I Micheel, J Novak, P Fraternali, G Baroffio, A Castelletti, and A-E Rizzoli. Visualizing & gamifying water & energy consumption for behavior change. In *Workshop: Fostering Smart Energy Applications*, pages 555–564, 2015.

52. M S Morley, M Khoury, and D A Savić. Serious game approach to water distribution system design and rehabilitation problems. *Procedia Engineering*, 186:76–83, 2017.

53. M Nachreiner, B Mack, E Matthies, and K Tampe-Mai. An analysis of smart metering information systems: A psychological model of self-regulated behavioural change. *Energy Research & Social Science*, 9:85–97, 2015.

54. A Letiţia Negruşa, V Toader, A Sofică, M F Tutunea, and R V Rus. Exploring gamification techniques and applications for sustainable tourism. *Sustainability*, 7(8):11160–11189, 2015.

55. J Nielsen, A Rodriguez, A Broussel, S Flake, G Sinne, and P Avila UrbanWater D5.6: Game solution for customer empowerment using water consumption data. 2015.

56. J Novak, M Melenhorst, I Micheel, C Pasini, P Fraternali, and A E Rizzoli. Behaviour change and incentive modelling for water saving: first results from the SmartH2O project. In *Proceedings of the 8th International Congress on Environmental Modelling and Software*, Toulouse, France, 2016.

57. Houses of Parliament. Smart metering of energy and water: no. 417, The parliamentary Office of Science and Technology, London, U.K..

58. H Oinas-Kukkonen. A foundation for the study of behavior change support systems. *Personal and Ubiquitous Computing*, 17(6):1223–1235, 2013.

59. A V Pandey, A Manivannan, O Nov, M Satterthwaite, and E Bertini. The persuasive power of data visualization. *New York University Public Law and Legal Theory Working Papers*, 2014. paper n. 474.

60. G Peschiera, J E Taylor, and J A Siegel. Response–relapse patterns of building occupant electricity consumption following exposure to personal, contextualized and occupant peer network utilization data. *Energy and Buildings*, 42(8):1329–1336, 2010.

61. G Rebolledo-Mendez, K Avramides, S de Freitas, and K Memarzia. Societal impact of a serious game on raising public awareness: the case of FloodSim. In *Proceedings of the 2009 ACM SIGGRAPH Symposium on Video Games*, pages 15–22. ACM, 2009.

62. A Rizzoli, A Castelletti, A Cominola, P Fraternali, A D Dos Santos, B Storni, R Wissmann-Alves, M Bertocchi, J Novak, and I Micheel. The SmartH2O project and the role of social computing in promoting efficient residential water use: a first analy-

sis. In *Proceedings of the 7th International Congress on Environmental Modelling and Software*, 2014.

63. C Rottondi, A Facchini, and A Rizzoli. An agent based framework for residential water usage modelling under social stimuli. *8th International Congress on Environmental Modelling and Software*, Toulouse, France, 2016.

64. S Russell and K Fielding. Water demand management research: A psychological perspective. *Water Resources Research*, 46(5), 2010.

65. L E Nacke, S Deterding, R Khaled, and D Dixon. Gamification: Toward a definition. In *CHI 2011: Workshop Gamification: Using Game Design Elements in Non-Game Contexts*, Vancouver, BC, Canada, 2011.

66. D A Savic, M S Morley, and M Khoury. Serious gaming for water systems planning and management. *Water*, 8(10):456, 2016.

67. J Schell. *The art of game design: A book of lenses*. CRC Press, 2008.

68. P Wesley Schultz, A Messina, G Tronu, E F Limas, R Gupta, and M Estrada. Personalized normative feedback and the moderating role of personal norms: A field experiment to reduce residential water consumption. *Environment and Behavior*, 48(5):686–710, 2016.

69. A Science. Catchment detox, 2014. [Online] http://www.abc.net.au/science/catchmentdetox/.

70. J Seibert and M J P Vis. Irrigania–a web-based game about sharing water resources. *Hydrology and Earth System Sciences*, 16(8):2523–2530, 2012.

71. V Sempere-Payá, D Todolí-Ferrandis, and S Santonja-Climent. ICT as an enabler to smart water management. In *Smart Sensors for Real-Time Water Quality Monitoring*, London, UK., Mukhopadhyay, S. C. and Mason, A. (Eds.), pages 239–258. Springer, 2013.

72. S Smit. Waternomics: Key ideas, brochure & poster, 2015.

73. A L Sønderlund, J R Smith, C Hutton, and Z Kapelan. Using smart meters for household water consumption feedback: knowns and unknowns. *Procedia Engineering*, 89:990–997, 2014.

74. UNDESA. World Urbanization Prospects: The 2014 Revision, Highlights. Department of Economic and Social Affairs. 2014.

75. Oracle Utilities. Smart metering for water utilities: White Paper, 2009. [Online] https://www.oracle.com/industries/utilities/index.html.

76. P Valkering, R van der Brugge, A Offermans, M Haasnoot, and H Vreugdenhil. A perspective-based simulation game to explore future pathways of a water-society system under climate change. *Simulation & Gaming*, 44(2-3):366–390, 2013.

77. V Venkatesh, M G Morris, G B Davis, and F D Davis. User acceptance of information technology: Toward a unified view. *MIS quarterly*, vol. 27, no. 3, pages 425–478, 2003.

78. D Vine, L Buys, and P Morris. The effectiveness of energy feedback for conservation and peak demand: A literature review. *Open Journal of Energy Efficiency*, 2(1):7–15, 2013.

79. Urban Water. Document of work (DoW). 2015. [Online] http://urbanwater-ict.eu/wp-content/uploads/2015/12/DOW-UrbanWater-318602-2015-07-01.pdf.

80. R M Willis, R A Stewart, K Panuwatwanich, S Jones, and A Kyriakides. Alarming visual display monitors affecting shower end use water and energy conservation in Australian residential households. *Resources, Conservation and Recycling*, 54(12):1117–1127, 2010.

81. R M Willis, R A Stewart, D P Giurco, M R Talebpour, and A Mousavinejad. End use water consumption in households: impact of socio-demographic factors and efficient devices. *Journal of Cleaner Production*, 60:107–115, 2013.

82. A Zoha, A Gluhak, M A Imran, and S Rajasegarar. Non-intrusive load monitoring approaches for disaggregated energy sensing: A survey. *Sensors*, 12(12):16838–16866, 2012.

8 Integrated Brine Management: A Circular Economy Approach

DIMITRIS XEVGENOS, DESPINA BAKOGIANNI,
KATHERINE-JOANNE HARALAMBOUS, MARIA LOIZIDOU

CHAPTER HIGHLIGHTS

- Overview of desalination and brine management methods and challenges
- Seawater desalination market: case study of the Mediterranean
- Circular economy approach for seawater desalination brine: focus on industry desalination

---

8.1 INTRODUCTION AND MOTIVATION

According to the World Economic Forum, water crisis has been recognized as the first global risk in 2015 [31]. If we continue business as usual, global demand for water will exceed water supply by 40% by 2030, affecting half of Europe's river basins [10]. Desalination comprises a nonconventional water resource practice that is currently gaining importance internationally for filling the gap in the water balance. In theory, desalination can deliver unlimited amounts of freshwater, since seawater alone represents approximately 97% of the total of earth's water. Its capacity more than doubled between 2001 and 2011. As of June 2011, 15,988 desalination plants have been installed and operated in 150 countries producing a combined 66.5 million cubic meters of fresh water per day [28]. In practice, there are currently limitations in terms of high energy requirements and negative environmental impacts. According to UNEP-MAP, desalination brine effluent comprises one of the two major emerging threats to the Mediterranean Sea environment. UNEP-MAP acknowledges that in time desalination may be the only solution to secure water availability, but it stresses

---

D Xevgenos ✉

SEALEAU B.V. NL. E-mail: d.xevgenos@gmail.com

D Bakogianni • K-J Haralambous • M Loizidou

National Technical University of Athens, GR.

that sustainable desalination methods need to be developed. Desalination of seawater produces vast amounts of wastewater: as much as 2 liters for every liter of potable water produced. This copious wastewater poses significant management problems, as well as placing critical pressure on the sea environment (for seashore desalination plants) and/or the underground aquifers (for inland desalination plants). The first paper to note that the brine from desalination plants poses risks to the marine environment was in 1979. However, it took another 10 years before the scientific interest of impacts from desalination plants took hold, as reflected by the publication history [5]. Since the current desalination technologies are not capable of closing the wastewater loop ("brine loop"), a lot of energy goes wasted and valuable materials are lost. Mining metals and minerals from desalination brine was recognized by the World Economic Forum as one of the top 10 emerging technologies in 2014. These materials are valuable and if recovered can not only offset the treatment cost, but also create significant business opportunities. Global Water Intelligence (GWI) in its most recent report (2016), recognizes the "selective beneficial salt separation" as the most promising challenge of desalination in market terms that is estimated by GWI at a market value of $2 billion. This chapter aims to provide insights on integrated brine management, with a view to circular economy.

8.2 BRINE MANAGEMENT

8.2.1 SOURCES AND NATURE OF BRINES

According to [32], brines can be defined as a water stream with a high dissolved salts content (> 50 g/L), usually, but not exclusively, sodium chloride. These saline impaired water streams can be generated either:

(a) As a waste product from industrial processes; or
(b) As the concentrate stream from desalination processes.

These sources of brines are discussed briefly next.

Industrial Brines:

The process industry is the major source of chloride releases in Europe, with the chemical industry representing the vast majority, accounting for 11.5 million tons/year, the steel industry for 323,000, power sector 213,000, pulp and paper 58,000, and food and beverage for 37,000 tons/year. In Figure 8.1, the breakdown of NaCl users is provided for the case of Europe (2012 data). The salt consumed by these industries ends up as brine effluent. In Europe, 62.19% of salt consumption is attributed to chemical production (chlorine and caustic soda, sodium carbonate, sodium chlorate, etc), followed by food processing (7.64%), deicing (6.53%), water softening (3.72%), and other uses [24].

Moreover, industry is the largest non agricultural water consumer, accounting for 22% of the global water demand. According to Global Water Intelligence, other major industrial water users are the mining sector (13.2%), pulp and paper (12.2%),

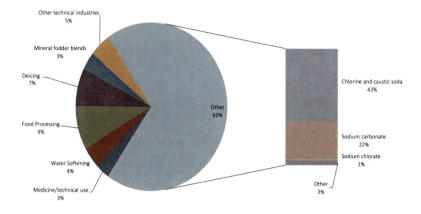

Figure 8.1: Salt consumption in Europe (2012). (Based on [24]).

power (9.5%), and food and beverage (4.2%). Industry can therefore play a significant role in tackling the water crisis. These releases are very complex effluents and represent today a big challenge for the companies both in terms of management and costs. There is a knowledge gap regarding the brine management and a lack of development of economic treatment solutions for brines. Brine discharges are recognized as one of the major threats to the aquatic environment, according to the United Nations Environment Programme. This is increasingly being translated into legislation in many parts of the world and Europe. Apart from an environmental challenge it will constitute an economic challenge, since nonconforming industries will have to pay fines.

Brines from Desalination Process:

Production of water from brackish or saline water from desalination plants can result in the production of large amounts of brines. Production of drinking water by desalination processes is a common practice in the Middle East and North African region (MENA) countries. In contrast, several industries around the world use water processed to a high quality by desalination techniques that result in a concentrate stream with a high salt content [32].

Worldwide, the total volume of desalinated water produced is 83,091,721 m^3/day derived from 14,220 plants [2]. The vast majority of desalination plants work in order to produce drinking water for municipalities. Figure 8.2 reveals that, another significant end-user of desalinated water is the industrial and power sector. As shown in Figure 8.3, the total capacity of desalinated water production is served mainly by seawater (salinity greater than 35,000 ppm) and brackish water (salinity between 1,000 and 35,000 ppm). Desalination of seawater produces vast amounts of wastewater: as much as 2 litres for every litre of potable water produced. This wastewater is called concentrate, retentate, or brine and it poses significant management problems.

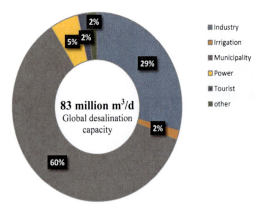

Figure 8.2: Global desalination capacity by end-user. (Based on [2]).

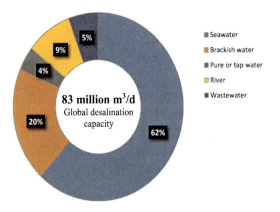

Figure 8.3: Global desalination capacity by feedwater type. (Based on [2]).

More information about desalination technologies is provided in Section 8.3. This contribution will focus on seawater desalination brine.

8.2.2 CURRENT CONCENTRATE MANAGEMENT OPTIONS

The desalination process produces a significant amount of wastewater: 2 L of brine for every liter of freshwater produced. The generated wastewater has to be managed properly, while the concentrate management options depend heavily on site characteristics and the wastewater volume. The cost for concentrate management depends heavily on site-specific and regulatory conditions, varying from a small fraction up to several times the cost of the main desalination system [21]. Historically, about 75% of the volume of concentrate generated in USA municipal membrane desalination plants is disposed to surface water [19].

These options typically include the following:

1. **Deep Well Injection (DWI)**. The brine is injected in underground porous geological formations. To avoid potential risks of aquifer pollution (through leaking), the brine is injected at depths below potable water aquifers. This technique has very good potential for economy of scale. However, DWI is not allowed in many locations. In the USA, nearly all plants applying the DWI method are located in Florida [15]. Wolthek, Raat et al. [30] provide results from a case study in Europe (Netherlands), indicating that the method is technically feasible for brackish water desalination [31].

2. **Evaporation Ponds**. This method is dependent on climate (ideal for dry climate with high evaporation rates), as well as land availability (at low cost). The cost of ponds is high for large volumes of concentrate, has little potential (if any) for economy of scale; however the operating costs are low. The costs are high mainly due to land acquisition costs and the need to apply impermeable liners. Enhanced evaporation ponds (such as WAIV) are a relatively new technology. Solids accumulating in evaporation ponds may need further management (typically disposed of in a landfill).

3. **Spray Irrigation**. This method is also called land application and it involves the beneficial use of brine effluent for parks, golf courses, or crop land. Land application is only suitable for low volume, low salinity streams, so it is not suitable for seawater desalination brine.

4. **Sewer Discharge**. This method involves the discharge of brine into an existing sewage collection system. This can be relatively low in cost, but it requires a permit from the local sewage authority, while in some cases the limits of Total Dissolved Solids (TDS) set by the authorities are so low (e.g., in the order of 1,000 ppm) that it is not possible to apply.

5. **Surface Water Dischage**. This comprises the simplest technique and it involves the discharge of the brine effluent into surface water bodies, namely rivers, lakes, and the sea. National Research Council [21] reports that 41% of the municipal desalination facilities recorded in the USA apply this technique (see also Figure 8.4). However, it can be said that nearly all desalination facilities of significant capacity use this technique worldwide. Environmental considerations associate with impacts on benthic communities (e.g., Posidonia Oceanica) [23], while some argue that these can be solved with proper design of the discharge system (diffusion, dilution, etc).

6. **Brine Concentrator/Zero Liquid Discharge** (ZLD). Brine concentrators (also called thermal evaporators) can increase the concentration ratio by a factor ranging between 1.67 to 90, but the typical applications report in the range of 10 to 50 [19]. The water recovery can then be in the range of 90-99%. Most of the brine concentrators employ energy efficiency techniques (mechanical vapor compression) and are of the vertical tube, falling-film type, with calcium sulfate seeding to prevent scaling. In order to reach ZLD, the remaining brine can be treated by evaporation ponds, crystallizers, or spray dryers as a final processing step [21]. The salt by-product needs further management, requiring disposing of it into a

landfill. This comprises an extra cost of this solution.

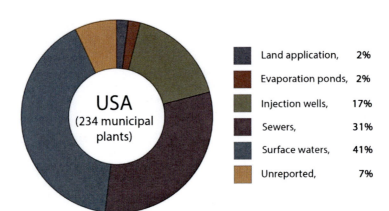

Figure 8.4: Use of brine management options in USA. (Based on [21]).

In the USA, the conventional disposal methods account for more than 98% of the municipal membrane desalination plants. The concentrate disposal methods have been well documented elsewhere [20]. Every concentrate management method has different costs, benefits, environmental impacts, and limitations. The costs are highly dependent on the concentrate volume.

Zero Liquid Discharge: The first ZLD systems were developed in the early '70s in the US, when increased salinity was observed in the Colorado River due to power plant discharges [25]. This technology normally comprises an evaporator and a crystallizer and results in high quality water recovery, while conforming to discharge regulations. However, these ZLD technologies are very capital intensive. ZLD systems also generate much solid by-product that has been disposed of in landfills. Thus ZLD systems have not been widespread because of burden of costs and waste generation. The average total spending on ZLD systems around the world is estimated to be between $100 and 200 million per year [12].

8.3 DESALINATION: TECHNOLOGIES, CHALLENGES, AND OPPORTUNITIES

Figure 8.6 represents the global market share of different desalination technologies, concerning online plants contracted since 2015. It reveals that the predominant technology is RO followed by MSF and MED processes.

8.3.1 DESALINATION TECHNOLOGIES

Desalination technologies are used to produce drinking water from brackish water and/or seawater. Industrial desalination technologies use either phase change or involve semipermeable membranes to separate the solvent or some solutes[1]. Thus, desalination techniques may be classified into the following categories (see also Figure 8.5): (1) Phase-change or thermal processes; (2) membrane or single-phase processes; and (3) hybrid processes. Below you can find a brief description of the main technologies per category.

[1] Phase-change or thermal processes

Multistage flash evaporation/distillation (MSF): MSF is a thermal distillation process that involves evaporation and condensation of saline water. The water is heated in a vessel where both temperature and pressure increase and then the heated water passes to another chamber at a lower pressure which causes vapor to be formed [6]. The sudden introduction of hot water into the chamber with lower pressure causes it to boil very quickly, almost exploding or "flashing" into steams. The vapor generated is condensed to pure water on tubes of heat exchangers that are cooled by the incoming feed water going to the first vessel, thus preheating that water and recovering part of the thermal energy used for evaporation in the first stage [3]. This process is repeated through a series of vessels in up to 40 stages.

Multiple-effect evaporation/distillation (MED): MED is a thermal process similar to MSF. The feed saline water is distributed in evaporators of different chambers (effects) placed in series. The evaporator tubes in the first effect are heated by steam and the resulting steam is condensed inside the evaporator tubes of the subsequent effect, where again vapor is produced. The surfaces of all the other effects are heated by the steam produced in each preceding effect [3]. In each effect the pressure is lower than the preceding one, allowing the water to be boiled without having to apply any additional heat [7]. This process is repeated within up to 16 effects [3]. The steam produced in the last effect is condensed in a separate heat exchanger, which is cooled by the incoming saline water, which is then used as preheated feed water for the desalination process.

Vapor Compression (VC): In this system, saline water is evaporated to be afterwards compressed in order to use the energy produced in the compression to evaporate new saline water [7]. VC thermal methods can be classified in two sub categories regarding the vapor compression method. In mechanical vapor compression (MVC) systems the pressure and temperature increase by a mechanical compressor. For this reason, its capacity is limited and these systems have between one and three stages, while most of them only have a single stage. This method finds application in small and medium plants. MVC units typically range in size up to about 3,000m^3/day. In thermal vapor compression (TVC) systems,

---

[1] S.A. Kalogirou, Seawater desalination using renewable energy sources, Progress in energy and combustion science 31(3) (2005) 242-281.

the vapor is compressed with a steam ejector. TVC units typically range in size up to about 36,000m^3/day [3].

Solar Distillation (SD): In SD systems, heat from the sun warms saline feed water in a glass-covered enclosure, in which the water rising vapor condenses on a glass cover [6, 17]. The droplets of distilled water that run down the glass are then collected in troughs along the lower edges of the glass, after which the distilled water is channeled to storage tanks [17]. This is a low-cost system that is usually used for small-scale, decentralized solar powered desalting technologies [3].

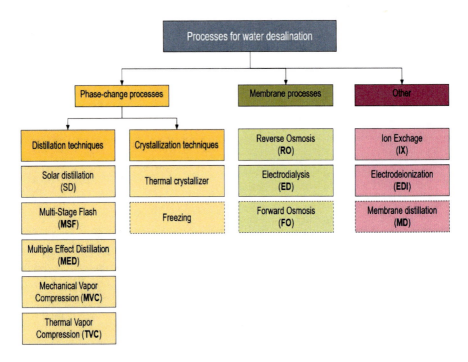

Figure 8.5: Processes for water desalination. Adapted from (Based on [13], [33]).

Eutectic Freeze Crystallization (EFC): In the EFC process, the feed is frozen continuously until the eutectic temperature is reached [17]. Further heat removal then generates both ice and salt crystals [27]. Specifically, ice is formed when the unsaturated solution is cooled to its freezing point, after which, the removal of additional heat induces the system to follow its freezing curve and the proportion of brine to ice decreases. At the eutectic point, ice and salt crystallize simultaneously [17].

Integrated Brine Management: A Circular Economy Approach

[2] Membrane or single-phase processes

Reverse Osmosis (RO): RO is a membrane separation process that recovers water from a saline solution pressurized to a point greater than the osmotic pressure of the solution [3]. Semipermeable membrane filters hold back the saline ions from the pressurized solution, allowing only the water to pass and leaving a concentrated brine solution on the feed-side of the membrane. The brine solution is typically between 10% to 50% of the feed water depending on the salinity and pressure of the feed water [6]. This process is continuous and separates the feed fluid into two streams without the need to regenerate the membrane [17]. RO has the ability to concentrate all dissolved and suspended solids, while producing a permeate that contains a very low concentration of dissolved solids [17]. RO membranes are sensitive to pH, oxidizers, a wide range of organics, algae, bacteria, depositions of particulates and fouling, therefore, pretreatment of the feed water is an important process step and can have a significant impact on the cost and energy consumption of RO [3].

Electrodialysis (ED): ED methods consist of an electrochemical separation process which uses a stack of ion-exchange membranes which are selective to positive and negative ions. Hundreds of cell pairs are included in an ED stack [34]. An electrical potential difference is used to separate ionic species from an aqueous solution and other uncharged components [6]. Because alternate membranes are permeable only to cations or anions, the water cells between the membranes are alternately depleted and enriched with salt ions [17]. The cation membranes allow only positively charged ions to diffuse through them [17]. Desalination occurs by the movement of anions and cations, without the water molecules, across the membrane [27]. ED methods can be categorized in electrodialysis reversal (EDR), electrodialysis metathesis (EDM), and electrodeionization (EDI).

[3] Hybrid processes

Hybrid desalination plants are mainly used in order to reduce energy cost and achieve optimized designs. Hybrid desalination systems combine both thermal and membrane desalination processes with power generation [14]. These systems are considered a good economic alternative to dual-purpose evaporation plants. Hybrid (membrane/thermal/power) configurations are characterized by flexibility in operation, less specific energy consumption, low construction cost, high plant availability, and better power and water matching [14]. By utilizing a combination of membrane, thermal, and/or resin technologies varying water quality can be achieved, depending on the technology used and the concentration of the feedwater.

As of June 2011, 15,988 desalination plants have been installed and operated in 150 countries producing a combined 66.5 million cubic meters of fresh water per day[2]. The desalinated water is used for different applications: municipal use (63%), industries (25.8%), power stations (5.8%), irrigation (1.9%), tourism (1.9%), and others. The sources of the water treated are: seawater (60%), brackish water (21.5%), river water (8.3%), wastewater (5.7%), pure water (4.3%), and brine (0.2%). Among the different desalination technologies, reverse osmosis (RO) predominates with a total share of 53%, followed by multistage flash distillation (MSF), multiple effect distillation (MED), and electrodialysis (ED). This pattern changes drastically across regions, with countries around the Arabian Gulf using mostly thermal techniques (MSF and MED), while almost all plants in Europe and the USA use RO (see also Figure 8.6).

Table 8.1

Main technical and operating characteristics by desalination technology (Source: [21], [9])

Desalination technology	RO	MSF	MED–TVC	MVC
Operating temperature (^oC)	< 45	< 120	< 70	< 70
Pretreatment needs	High	Low	Low	Very low
Energy	Electricity	Thermal	Thermal	Electricity
Heat consumption (kJ/kg)	-	250–330	145-390	-
Electrical consumption (kWh/m^3)	2.5 – 7	3 – 5	1.5 - 2.5	8 – 15
Module capacity	< 20,000	< 76,000	< 36,000	< 3,000
Produced water quality	200 - 500	< 10	< 10	< 10
Water recovery factor	35 - 50%	35 - 45%	35 - 45%	23 - 41%
Reliability	Moderate	Very high	Very high	High

The main challenges associated with the wider adoption of desalination technologies are the energy consumption and associated water treatment costs - and the management of the wastewater effluent, called brine. These challenges are discussed next.

8.3.2 ENERGY CONSIDERATIONS AND RENEWABLE DRIVEN DESALINATION

Desalination uses a significant amount of energy. Energy usage affects the technology's carbon footprint and hampers its wider deployment. Currently, RO is the most efficient water treatment technology, with its consumption ranging between 2 and 5 kWh/m^3 (according to the type of treated water, i.e., brackish or seawater). No thermal energy is required for driving the RO process (see also Table 8.1). However, even though RO is considered to be the most energy-efficient desalination technology, still it comprises the most energy consuming water treatment process.

---

[2]K.S. Tree, Water Industry Report Desalination, 2013. Available from: http://www.sdwtc.org/Resources/WEMI/Segment\%20Report/130226.Desalination.pdf.

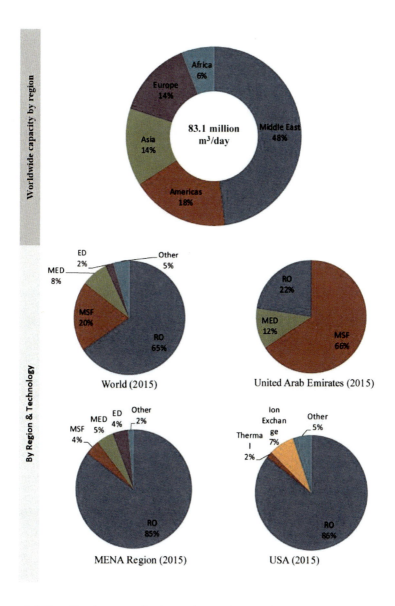

Figure 8.6: Desalination capacity by region and technology. (Data taken from [2]).

Renewable-Energy Desalination (**RES-D**): The main challenge of RES-D is how to make two different technologies work together. Even though both components comprise mature technologies, their combination is currently at the R&D stage. This is why RES-D plants are limited both in number and water production capacity. Kalogirou [16] reports that RES-D plants make up less than 1% of the total installed desalination capacity worldwide. Most of these plants are installed in arid areas, e.g., the MENA region, with capacities between a few m^3 up to 100 m^3.

RES-D systems have been the focal point of much research work worldwide. Many projects (such as AQUASOL, PRODES, ADIRA, AQUA-CSP, MEDESOL, and MEDINA) have examined and evaluated the barriers and challenges stemming from the coupling of these technologies (i.e., renewable source collector and desalination technology) with the aim to carry forward the RES-D techniques from the research to the commercial application development stage (see also Figure 8.6). One of the most often cited works on RES-D is the report titled "Roadmap for the development of desalination powered by renewable energy," which was developed in the framework of the European project PRODES [22]. The project PRODES has made significant contributions in collecting and reporting the available data for RES-D plants worldwide. Papapetrou, Wieghaus et al. [22] report that until 2009, 131 renewable powered desalination systems have been recorded across the globe.

It must be noted that the vast majority of the RES-D applications employ solar powered techniques with PV-RO being the dominant combination, amounting to a share of 31%. The different combinations of desalination processes coupled with renewable energy are provided in Figure 8.7. Solar energy techniques comprise proven, well-tested technologies, offering the potential of a reliable energy source for desalination practices. Solar powered desalination technologies can be divided into two broad categories: direct and indirect processes. The first category involves the solar still technology, while the second involves:

[1] solar collectors;
[2] solar ponds; and
[3] photovoltaic units.

Solar energy collector devices can drive both thermal desalination and membrane desalination systems.

The technical and economical potential of renewable energy (RE) resources vary significantly by country and region, across the globe. For instance, the MENA region has very abundant solar potential, with large area availability, meaning that the MENA region has a comparative advantage for solar applications and especially concentrated solar power (CSP) (FICHTNER 2011). In contrast, North European countries such as Denmark have a very good renewable potential for wind applications. The water production cost varies significantly according to the type of RES-D and the water type treated, as illustrated in Figure 8.8. Even though costs are prohibitive at the moment, cost reduction is anticipated due to technology improvements, learning effects, and scaling-up. Trieb [29] reports that RES-D cost is expected to fall to 0.9\$/m^3 by 2050.

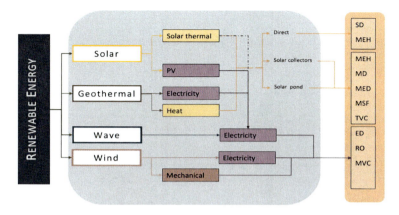

Figure 8.7: Possible combinations of desalination processes with renewable processes (Based on [13]; [33]).

8.3.3 VALORIZATION OF SEAWATER DESALINATION BRINE

This chapter focuses on the valorization of the brine effluent generated by seawater desalination plants. Seawater comprises a complex solution of water containing nearly all elements of the periodic system [11]. However it can be characterized sufficiently by 6 elements/compounds, since these represent more than 99% of the dissolved solids. These are chloride (Cl^-), sodium (Na^+), sulfate (SO_4^{2-}), magnesium (Mg^{2+}), calcium (Ca^{2+}), and potassium (K^+). The composition of seawater with an illustration of the mass quantities per major constituent for 1 kg of seawater is provided in Figure 8.8.

During the evaporation of seawater, salts start to precipitate gradually. Iron oxide (Fe_2O_3) and calcium carbonate ($CaCO_3$) start to crystallize first with the amount of Fe_2O_3 produced being negligible. Then calcium sulfate ($CaSO_4$) precipitates, followed by sodium chloride, and the so-called bitterns. The bitterns comprise a collection of magnesium, potassium, sulfate, and chloride salts [11]. The gradual deposition of these salts during evaporation of seawater is presented in Geertman [11] as a function of the density of the brine expressed as degrees Baume.

For the purposes of attributing a value of seawater desalination brine, the following methodology was used:

1. Calculation of mili-equivalents of ions
2. Calculation of of mili-equivalents of compounds (ionic balance)
3. Estimation of quantities of salts recovered by evaporation of 1 m^3 of brine

Following this methodology, we derive that for each m^3 of seawater desalination effluent, the following quantities of salts can be obtained:

- *NaCl*: 63.33 kg

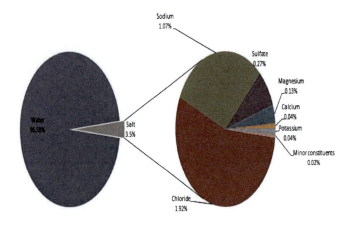

Figure 8.8: Chemical composition of seawater with salinity of 35%.

- $MgCl_2$: 7.51 kg
- $MgSO_4$: 3.38 kg
- $CaSO_4$: 2.97 kg
- KCl: 1.39 kg
- $CaCO_3$: 0.27 kg

According to current market prices for these salts and water, the economic value of the seawater desalination brine effluent was calculated. The breakdown by component is provided in Figure 8.9. It can be seen that water and sodium chloride represent most of the economic value. However, this depends also heavily on the purity achieved by the technologies applied, since there are certain materials that can be recovered, such as magnesium, that can have high market value in case of high purity.

8.4 INDUSTRIAL WATER USE AND DESALINATION

8.4.1 INDUSTRIAL WATER USE

According to [18], industrial water withdrawals is the largest nonagricultural water user, accounting for 20% of total withdrawals, followed by the domestic sector (10%). In western Europe and the USA, almost 50% of water withdrawals are for energy production (cooling water) and the majority of this water is discharged at higher temperature. However, these figures vary considerably across countries (see also Figure 8.10 for Europe).

Total industrial water withdrawal amounted to 846.7 billion m^3 in 2012, the vast majority out of which (78%) was used in once-through cooling systems. Water with-

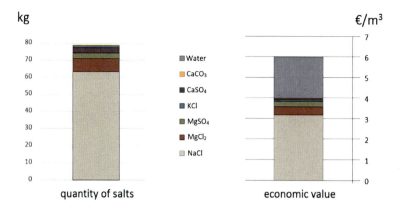

Figure 8.9: Quantity of salts and economic value per m^3 of seawater desalination effluent.

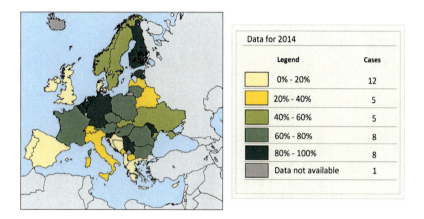

Figure 8.10: Industrial water withdrawal share, Europe, 2014.

drawal and water consumption is provided per major industrial sector at global scale in Figure 8.11.

As an example, 87.53% of the total water abstracted in Belgium is used for industrial purposes. The total water abstraction in Flanders (Belgium) in 2012 amounted to 3,127 million m^3 of which only 22.9% (or 716 million m^3) were consumed. The rest comprises mostly cooling water for industrial processes which is discharged back to the environment as wastewater (and often as waste heat as well).

These amounts of energy can be used to drive for example thermal based processes (such as ZLD systems) to make the business case more appealing. According to SPIRE [26] 20 – 50% of the energy used in industrial processes is estimated to be lost in the form of hot exhaust gases, cooling water, and heat losses from equipment and products.

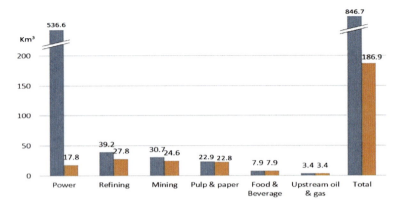

Figure 8.11: Industrial water withdrawal and water consumption in billion m^3/day by major industrial sector, worldwide, 2012 (source: [1]).

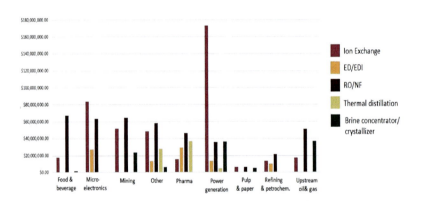

Figure 8.12: Total spending on dissolved solids technologies by industry sector and dissolved solids technology, worldwide, 2015.

Refineries

Petroleum refineries process crude oil into saleable oil products. A typical refinery will use up to two barrels of water for every barrel of oil produced [4]:

- Process water. This water is used for the following three main reasons: (a) various refinery processes (e.g., desalter, cracking of hydrocarbons); (b) tank and flare seal drains; and (c) cleaning operations. On average this quantity represents around 20% of the total water consumption
- Cooling tower make-up water. This water represents the largest amount of the water consumed, on average of around 51%. However, this varies greatly within the

sector, depending on the cooling system used (closed, open recirculating, once-through systems)

- Boiler feedwater. On average this water represents 32% of the total water consumed. Reverse osmosis and EDI/ion exchange (mixed bed) are used to meet the specifications.

Power industries

Power industries require considerable amounts of water, primarily for cooling purposes, while they account for the largest share of water withdrawals worldwide. The raw water can be taken from rivers, lakes, groundwater, or the sea. In a recirculating system, makeup water is required to compensate for water losses due to evaporation. Typically, the quality requirements of cooling water is lower than boiler feedwater and is limited to screening and filtering unit processes to remove suspended solids, followed by chemical addition (e.g., addition of sulfuric acid) to reduce scaling and corrosion. Further treatment may be required in order to meet contaminant levels in the blowdown stream. Water is mainly used at power plants for cooling purposes. Once-through systems typically consume 4% of the water withdrawn, while closed-loop systems consume 80% of the water withdrawn. Other water uses comprise process and technical water. For example, water for wet scrubbing (in the flue gas desulfurization units), as well as gas turbine compressors need to be cleaned with water and detergents periodically. Typically raw water is pretreated before being used: softening and demineralization. Softening can include flocculation, precipitation, sand filters, etc., while demineralization includes ultrafiltration, ion exchange, reverse osmosis, etc. The breakdown of the total spending on dissolved solids technologies is provided for the power sector, as well as other major sectors, as seen in Figure 8.12.

8.4.2 INDUSTRIAL USE OF DESALINATED WATER

Desalination is used mostly (60%) to produce an alternative water supply for utility systems, while industrial use of desalination is the second largest use of desalinated water (29%) (see also Figure 8.2). The use of desalination in industry is related to the production of ultrapure water for industrial processes. To do so, most of the industries use feedwater sources with low salinities such as tap water or river water. Most of the industrial desalination systems are installed in power plants, accounting for 35.6% of the total installed industrial capacity, for the production of boiler feedwater. The second largest industrial user of desalination is the oil and gas industry, including both the upstream oil and gas, as well as the refining sector.

Another application is also in mining operations, for the treatment of the discharge water stream. Finally, a smaller percentage of desalination systems is used for volume reduction purposes, related to zero liquid discharge systems. Most ZLD applications (as registered until 2000) refer to oil and gas (50%), followed by the power sector (23.7%), the pulp and paper sector (8.1%), and the refining sector (7.7%).

The largest share of desalination systems use reverse osmosis, followed by thermal desalination technologies (MSF and MED). However, some systems use also

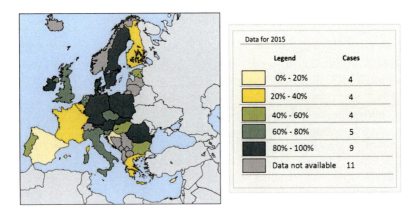

Figure 8.13: Share of desalinated water for industrial purposes in Europe. (Data taken from [2]).

ED/EDI and NF, but almost exclusively in industrial applications. Even though reverse osmosis can remove the vast majority of dissolved solids (in the order of 99.7%), the produced water is often unsuitable for industrial use. As a result, a further polishing step may be required such as ED/EDI or ion exchange. Nanofiltration is used mostly in offshore oil and gas industry. In Europe, the total volume of desalinated water produced is $8,830,979 m^3$/day derived by $2,235$ plants. Approximately 23% of the produced water is used for industrial purposes and 5% is used in power plants.

Spain holds first place in the production of desalinated water, followed by Italy. On the contrary, in terms of desalinated water for industrial purposes capacity, Italy holds first place, followed by United Kingdom. Additionally, in terms of the share of desalinated water for industrial purposes in the total volume of desalinated water in each European country, as shown in Figure 8.13 with dark green color, countries like Denmark, Sweden, and Germany hold first place.

8.5 INTEGRATED BRINE MANAGEMENT: THE MEDITERRANEAN CASE

In this section, the desalination market of selected Mediterranean countries is briefly presented. The data is retrieved from the most recent published report of Global Water Intelligence [2]. It was assumed that the desalination capacity is equal to the combined desalination of "online," "presumed online" plants, and plants "under construction." Moreover, the analysis considers the Mediterranean countries having a coastline directly on the Mediterranean Sea. These are: Spain, France, Italy, Greece, Turkey, Cyprus, Malta, Syria, Egypt, Libya, Tunisia, Algeria, Morocco, and including also Portugal and Jordan. Data was not available for the following Mediterranean countries: Bosnia & Herzegovina and Montenegro, while Slovenia, Croatia, Albania have minimal use of desalinated water, with an installed capacity of $4,861 m^3$/d, $432 m^3$/d, $1,672 m^3$/d, respectively. Thus, these six countries were not included in the analysis presented below.

8.5.1 MEDITERRANEAN DESALINATION MARKET

It must be mentioned that Mediterranean European countries (Cyprus, France, Greece, Italy, Malta, Spain, Portugal, Croatia) represent 81% of the total installed desalination capacity in Europe. The total volume of desalinated water produced in the MENA region is $15,613,932 m^3$/day derived by 2,910 plants. This number represents 18% of the world's operative installed desalting plants. Top producers are Spain (35.5%), Algeria (16.5%), Israel (14.2%), Egypt (7.9%), Libya (5.3%), and Italy (4.3%). The majority of the desalination plants use seawater as feedwater, (72%), followed by brackish water (16%), river water (7%), wastewater (3%), and tap water (1%) (see also Figure 8.14). The online desalination capacity by feedwater is given in Figure 8.15 for selected Mediterranean countries.

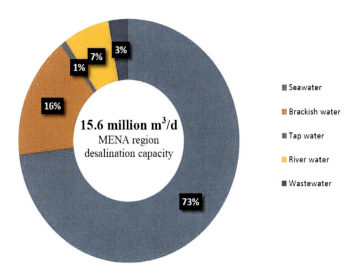

Figure 8.14: Online desalination capacity by feedwater type, MENA region. (Data taken from [2].)

Reverse osmosis is the most widespread technology in the Mediterranean region, representing 85% of the total desalination capacity in the region (see also Figure 8.16), followed by MED (4.7%), MSF (4.2%), ED (4.3%).

In most of the Mediterranean countries thermal desalination technologies comprise a small percentage of the total desalination capacity installed, with the exception of Libya, Lebanon, Italy, and Greece where the shares of thermal desalination technologies (MSF and MED) are 86.4%, 44.8%, 28.5%, and 13.9%, respectively.

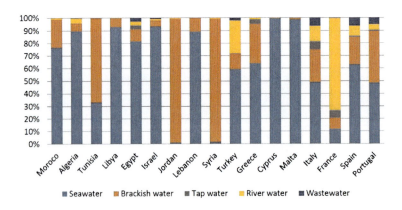

Figure 8.15: Online desalination capacity by feedwater type, breakdown by Mediterranean country. (Data taken from [2].)

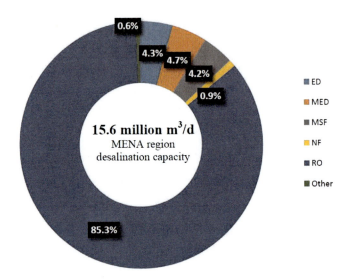

Figure 8.16: Online desalination capacity by technology, MENA region. (Data taken from [2].)

This larger share is attributed to the fact that in these countries, thermal desalination technologies are used for industrial use, most significantly in the power sector (see Figure 8.17 and also Section 8.5.2).

The same applies for ED/EDI and NF, which are not very common and are most often found in industrial applications, for the removal of specific compounds. For example, nanofiltration is mostly used in Latin America for sulfate removal for water injection in offshore oil and gas industry.

Integrated Brine Management: A Circular Economy Approach

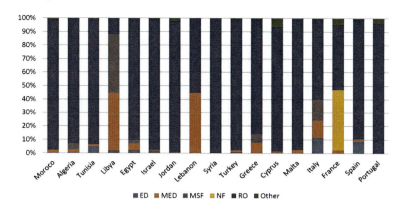

Figure 8.17: Online desalination capacity by technology, breakdown by Mediterranean country. (Data taken from [2].)

The high percentage of NF in France is attributed to the use of a NF plant for drinking water production out of surface water in northern France (Méry-sur-Oise). The total capacity of this plant is $140,000 m^3$/day [8], representing nearly half of the total desalination capacity in France.

In the Mediterranean, utility and public sectors comprise the main end-users of desalinated water, accounting for 75.1%, followed by industry (12%), tourism (4.6%), irrigation (4.6%), and power stations (2.8%) (see also Figure 8.18).

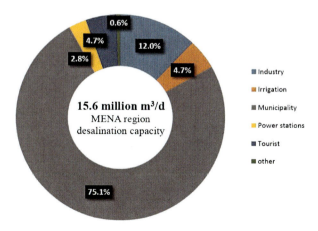

Figure 8.18: Online desalination capacity by end-user, MENA region. (Data taken from [2].)

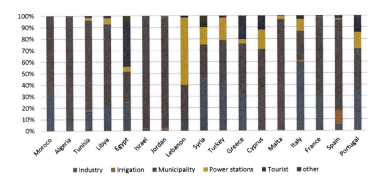

Figure 8.19: Online desalination capacity by end-user, breakdown by Mediterranean country. (Data taken from [2].)

The following section provides an in-depth insight on the industrial use of desalination in the Mediterranean with a focus on the power sector.

8.5.2 INDUSTRIAL USERS OF DESALINATION IN MEDITERRANEAN COUNTRIES

Most of the literature on desalination in the Mediterranean region is focused on drinking water desalination plants. This is attributed to the fact that the Mediterranean is a water scarce area and thus many countries resort to desalination to balance their water needs.

This chapter is focused on the desalination plants installed at industrial sites. As seen in Figure 8.20, the majority of the desalination plants in the Mediterranean countries are used for utility water production, with the exception of Turkey, Syria, Italy, and Portugal where the share of industrial desalination plants is large (see also Figure 8.19).

In the Mediterranean countries the amount of water produced for industrial purposes is lower than that of the other European countries. Approximately 15% of the produced water is used for industrial purposes and 3% is used in power plants. In Table 8.2 the breakdown of industrial desalination capacity is presented per Mediterranean country and technology.

Finally, in Figure 8.21, the online desalination capacity for the MENA region is presented by the industrial end-user, while in Figure 8.22, industrial desalination plants operating with the method of multi effect distillation (MED) are presented by Mediterranean country.

8.6 CONCLUDING REMARKS AND FUTURE DIRECTIONS

This chapter aims at providing a different approach in brine management, by viewing this problematic wastewater flow as a resource. After discussing the current brine

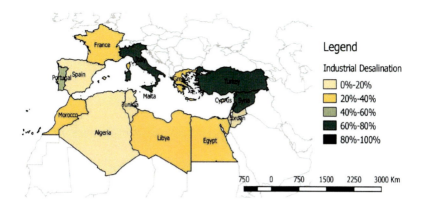

Figure 8.20: Share of desalinated water for industrial purposes in MENA region. (Data taken from [2].)

Table 8.2
Industrial desalination capacity (m^3/d) per Mediterranean country and technology (2015)

Country	MED	MSF	RO	EDI	ED	Other
Algeria	2,717	6,881	1,000			
Tunisia	860		3,713	817	240	
Libya	23,440	15,840	5,000			
Jordan	1,100					
Lebanon	14,670		4,500			
Syria			3,000			
Greece	4,600		1,740			
Malta	5,800					
Italy	3,511	28,954	30,544	1,362	3,560	1,152
France	1,600					
Israel			300			
Egypt	15,200	25,804	12,180			
Spain	7,520		26,674			7,642
Portugal			3,120			
Total	81,018	77,479	91,771	2,179	3,800	8,794

management techniques applied today worldwide, both for industrial and desalination brines, the authors estimate the economic value of seawater desalination brine. This is estimated at €6 per m^3 of brine generated by seawater desalination plants. The decision to valorize seawater desalination brine was made on the basis of the

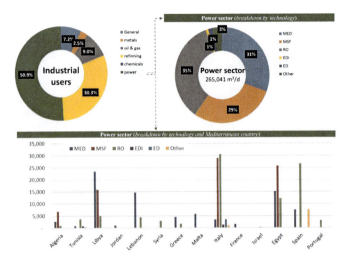

Figure 8.21: Online desalination capacity by industrial end-user, MENA region. (Data taken from [2].)

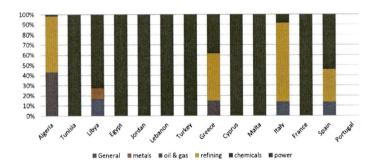

Figure 8.22: Online MED seawater desalination capacity by industrial end-user, breakdown by Mediterranean country. (Data taken from [2].)

application potential, meaning that the composition of seawater is approximately the same at global scale (contrary to brackish water which is characterized by large variations).

After the quantification of the economic value of brine, an analysis of the desalination sector in Europe is made and in particular in the Mediterranean region. In Section 8.1.2, desalination technologies are also described briefly. The analysis of the Mediterranean region was selected based on the fact that the analysis made is relevant for seawater, as well as due to the fact that this particular region is one of the largest desalination markets globally (representing 18% of the global installed desalination capacity) because of acute water scarcity problems.

In order to materialize this economic value, suitable technologies must be employed. The typical zero liquid discharge systems can be an option, but this results in end-products with high levels of impurities, which need actually further management, most often requiring disposal at landfills. Currently, there is no commercial application in the desalination sector that results in high purity products, while the GWI suggests that the "selective" beneficial salt separation is the top desalination challenge with a market value of $2billion. In any case, membranes are not able to reach to zero liquid discharge and thus a final step of (thermal) evaporation/crystallization is required. For this reason, this chapter is suggesting that the starting point of "circular desalination" would normally be the industrial sector, where plenty of waste heat is available. Typically, $20 - 50\%$ of the industrial heat is wasted. This heat could be used to drive brine treatment processes and recover valuable materials (mostly salts and high-quality water). Apart from the technological challenge, another potential barrier is also legislation. The Waste Framework Directive (Directive 2008/98/EC) sets out specific requirements for the establishment of end-of-waste criteria in order to characterize a recovered material from waste as secondary material.

In Section 8.5, the Mediterranean desalination market is discussed, focusing on industrial desalination. The analysis is made by Mediterranean country, technology, and industrial end-user. At the end of the section, further details are provided especially for the power sector, since this sector has extensive experience in running thermal desalination plants (accounting for 60% of total installed desalination capacity), as well as plenty of waste heat available (cooling water). Finally, the authors believe that the sector presents good possibilities as a niche market for "circular desalination" since the recovered products can be used locally. The high-quality distilled water can be used as boiler feedwater, while NaCl can be used to regenerate ion exchange resins. The power sector is the largest industrial user of ion exchange (see also Figure 8.12), due to the strict requirements to produce ultra pure water and avoid potential scaling problems in the process. According to GWI, the total installed desalination capacity in power plants, in the Mediterranean countries considered, amounts to $265,041 m^3$/d or 96.7 million m^3 per year.

Based on the analysis presented in this chapter, this brine effluent represents a potential economic value of €580M per year, which currently goes wasted through the loss of these materials, by discharging the brine effluent in surface waters.

8.7 REFERENCES

1. G Allen, H Brown, M Charamidi, I Elkins, C Gasson, and T Pankratz. (2015). Industrial water technology markets 2015. Oxford: Media Analytics Ltd.
2. F Alvarado-Revilla, H Brown, M Charamidi, I Elkins, E Filou, C Gasson, J Uzelac (2015). Desalination markets 2016. United Kingdom: Media Analytics Ltd.
3. AQUA-CSP (2007). WP 1: Review of CSP and Desalination Technology. Retrieved from: `http://www.dlr.de/tt/Portaldata/41/ Resources/dokumente/institut/system/projects/aqua-csp/ WP01_AQUA-CSP-Technologies-Final.pdf`.

4. P Barthe, M Chaugny, S Roudier, and L D Sancho (2015). Best available techniques (BAT) reference document for the refining of mineral oil and gas. Retrieved from: `http://eippcb.jrc.ec.europa.eu/reference/ref.html`.
5. T Bleninger, and G Jirka (2010). Environmental planning, prediction and management of brine discharges from desalination plants. Final report. Middle East Desalination Research Center. Retrieved from: `http://www.ifh.uni-karlsruhe.de/science/envflu/research/brinedis/brinedis-finalreport.pdf`.
6. R Clayton (2015). Desalination for water supply. Foundation for Water Research. Retrieved from: `http://www.fwr.org/desal.pdf`.
7. C J Cuenca (2012). Report on water desalination status in the Mediterranean countries. Retrieved from: `http://www.imida.es/docs/publicaciones/06_REPORT_ON_WATER\%20DESALINATION.pdf`.
8. B Cyna, G Chagneau, G Bablon, and N Tanghe. Two years of nanofiltration at the méry-sur-oise plant, france. *Desalination*, 147(1-3):69–75, 2002.
9. R Drew (2010). Desalination in Florida: Technology implementation and environmental Issues. Florida. Retrieved from: `http://www.dep.state.fl.us/water/docs/desalination-in-florida-report.pdf`.
10. European Commission. (2017). Water is too precious to waste. Retrieved from: `http://ec.europa.eu/environment/water/pdf/water_reuse_factsheet_en.pdf`.
11. M R Geertman(2000). Sodium chloride: Crystallization. Retrieved from: `http://worldtracker.org/media/library/Science/Chemistry/Crystalization,\%20Purification,\%20Separation/Encyclopedia\%20of\%20Separation\%20Science/Level\%20III\%20-\%20Practical\%20Applications/SODIUM\%20CHLORIDE\%20-\%20CRYSTALLIZATION.pdf`.
12. Global water intelligence. (2009). From zero to hero. The Rise of ZLD. Retrieved from: `http://www.epa.gov/region1/npdes/merrimackstation/pdfs/ar/AR895.pdf`.
13. V Gnaneswar Gude, N Nirmalakhandan, and S Deng. Renewable and sustainable approaches for desalination. *Renewable and Sustainable Energy Reviews*, 14(9):2641–2654, 2010.
14. O A Hamed. Overview of hybrid desalination systemscurrent status and future prospects. *Desalination*, 186(1-3):207–214, 2005.
15. J Jordahl (2006). Beneficial and nontraditional uses of concentrate. Retrieved from: `http://www.waterboards.ca.gov/water_issues/programs/grants_loans/water_recycling/research/02_006b.pdf`.
16. S A Kalogirou. Seawater desalination using renewable energy sources. *Progress in Energy and Combustion Science*, 31(3):242–281, 2005.
17. D H Kim. A review of desalting process techniques and economic analysis of the recovery of salts from retentates. *Desalination*, 270(1):1–8, 2011.
18. E Koncagl, R Connor, and M Tran (2014). Water and energy facts and figures. The United Nations World Water Development Report 2014. Retrieved from: `http://unesdoc.unesco.org/images/0022/002269/226961E.pdf`.
19. M Mickley (2008). Survey of high recovery and Zero Liquid Discharge Technologies for water utilities. Alexandria: Water Reuse. Retrieved from: `http://www.swrcb.ca.gov/water_issues/programs/grants_loans/water_recycling/research/02_006a_01.pdf`.

Integrated Brine Management: A Circular Economy Approach

20. C M Mickley (2001). Membrane concentrate disposal: Practices and regulation. Retrieved from: https://www.usbr.gov/research/dwpr/reportpdfs/report123.pdf.

21. National Research Council. (2008). Desalination: A national perspective. Committee on advancing desalination technology. Washington D.C.: National Research Council. Retrieved from: http://waterwebster.org/documents/NRCDesalinationreport_000.pdf.

22. M Papapetrou, M Wieghaus and C Biercamp (2010). Roadmap for the development of desalination powered by renewable energy. Stuttgart: Frauhnofer Verlag. Retrieved from: http://www.prodes-project.org/fileadmin/Files/ProDes_Road_map_on_line_version.pdf.

23. D A Roberts, E L Johnston, and N A Knott. Impacts of desalination plant discharges on the marine environment: A critical review of published studies. *Water Research*, 44(18):5117–5128, 2010.

24. Roskill. (2014). Salt: Global industry markets and outlook Fourteenth Edition. United Kingdom: Roskill Information Services Ltd.

25. A Seigworth, R Ludlum, and E Reahl. Case study: Integrating membrane processes with evaporation to achieve economical zero liquid discharge at the Doswell combined cycle facility. *Desalination*, 102(1):81–86, 1995.

26. SPIRE. (2013). SPIRE Roadmap. Retrieved from: https://www.spire2030.eu/uploads/Modules/Publications/spire-roadmap_december_2013_pbp.pdf.

27. A Subramani, and J G Jacangelo. Treatment technologies for reverse osmosis concentrate volume minimization: a review. *Separation and Purification Technology*, 122:472–489, 2014.

28. S K Tree (2013). Water industry report desalination. Retrieved from: http://www.sdwtc.org/Resources/WEMI/Segment\%20Report/130226.Desalination.pdf.

29. F Trieb (2005). Concentrating solar power for the Mediterranean region final report. Stuttgart, Germany. Retrieved from: http://www.dlr.de/Portaldata/1/Resources/portal_news/newsarchiv2008_1/algerien_med_csp.pdf.

30. N Wolthek, K Raat, J A de Ruijter, A Kemperman, and A Oosterhof. Desalination of brackish groundwater and concentrate disposal by deep well injection. *Desalination and Water Treatment*, 51(4-6):1131–1136, 2013.

31. World Economic Forum. (2015). Global risks 2015 10th Edition. Geneva. Retrieved from: http://www3.weforum.org/docs/WEF_Global_Risks_2015_Report15.pdf.

32. WssTP. (2012). Brines management. Research and technology development needs water in industry. Retrieved from: http://wsstp.eu/files/2013/11/ExS-Brines.pdf.

33. D Xevgenos, K Moustakas, D Malamis, and M Loizidou. An overview on desalination & sustainability: Renewable energy-driven desalination and brine management. *Desalination and Water Treatment*, 57(5):2304–2314, 2016.

34. Y Zhang, K Ghyselbrecht, R Vanherpe, B Meesschaert, L Pinoy, and B Van der Bruggen. RO concentrate minimization by electrodialysis: Techno-economic analysis and environmental concerns. *Journal of Environmental Management*, 107:28–36, 2012.

9 Model-Assisted, Real-Time Monitoring and Control of Water Resource Recovery Facilities

EVANGELIA BELIA, JOHN B. COPP

CHAPTER HIGHLIGHTS

- The development and implementation of model-assisted, real-time, plant-wide monitoring and control tools is expanding.
- Process models are invaluable tools for evaluating controller performance, including equipment and process constraints and reducing controller full-scale commissioning time.
- Operational constraints for reasons other than process performance can impact controller authority
- The tuning of controllers at full-scale is time consuming and must be carefully planned.

9.1 INTRODUCTION AND MOTIVATION

An integrated model-assisted approach to wastewater treatment plant operation and control can have a significant positive impact on the performance of a plant, and it is this premise that forms the basis for this chapter. Olsson et al. [8] gave an excellent overview of this concept and described the benefits of the model-assisted approach including:

- a reduction in the duration and frequency of water quality effluent excursions;
- a reduction in energy costs;
- the deferment of capital expenditures;

E Belia ✉

Primodal US Inc. E-mail: belia@primodal.com

J B Copp

Primodal Inc

- the optimal use of existing facilities;
- the improved resilience of the plant to cope with upsets.

These improvements are realized through the use of modern instrumentation and automatic control technology, combined with knowledge of the dynamic behavior of the wastewater treatment processes. This approach can overcome two major obstacles to optimization, encountered in current plants. The first is that treatment plants are typically designed based on steady-state conditions, yet they must cope with time-varying loads. The second is that existing control systems are often based on a series of isolated control loops that do not operate in an integrated manner. By using a dynamic model as the centerpiece of a comprehensive control system, integrated control becomes possible.

The tools and applications described in this chapter refer to integrated software tools that communicate automatically with supervisory control and data acquisition (SCADA) systems or local programmable logic controllers (PLCs), run dynamic mathematical process models (online or onside), perform data quality checks, diagnose the process state, and send control actions to plant equipment. For the purposes of brevity the authors refer to these tools as model-assisted, real-time, plant-wide monitoring and control (MA-RT-PW-MC) platforms.

The chapter includes a brief overview of the several MA-RT-PW-MC software tools already commercially available. Other products exist, but they have not been included because the authors are not aware of their application at full-scale. This brief discussion should provide a glimpse into the evolution of plant-wide control options.

The chapter concludes with a Michigan, USA, case study where one such tool is being used in a full-scale application. The case study demonstrates that improved sensor technology and new software solutions are making plant-wide real-time plant control possible and ready to be more widely accepted and integrated into current practice as we strive for energy optimization and resource recovery at these plants.

9.2 DRIVING FORCES IN REAL-TIME MONITORING AND CONTROL OF WRRF

Real-time monitoring and control of WRRF has had a long history, and dedicated MA-RT-PW-MC platforms started being developed in the early 1990s with the increased use of mathematical models in the wastewater industry and the increasing power of software. In recent years, the industry has seen increased activity in the development and application of these tools, in part, because the reliability of online instrumentation has improved significantly. The drivers for the continuous development and implementation of MA-RT-PW-MC platforms include [8]:

- the need to counteract plant disturbances caused by catchment and plant dynamics;
- the goal to reconcile multiple operational objectives which can be difficult;
- the aim to achieve compliance at least cost;

Model-Assisted, Real-Time Monitoring and Control of WRRFs

- increased plant complexity;
- improved actuators and sensors;
- the wider acceptance and installation of real time sensors;
- the increased acceptance of dynamic process models;
- the development of real-time data quality tools.

9.3 ESSENTIAL MODULES FOR AN INTEGRATED CONTROL SOFTWARE TOOL

MA-RT-PW-MC platforms can have a variety of functionalities. The complexity of the implemented tools and their functionality depends on the number of integrated control loops and whether they include automatic model identification/calibration and automated model-based control. The MA-RT-PW-MC platforms that exist today include various combinations of the following real-time modules:

- Low level signal tracking: to identify and report outliers, noisy or missing data, and excessive rates of change
- Data filters: to smooth noisy and/or biased data and identify trends contained within a time series
- Data quality analysis tools: to generate maintenance actions for online instruments
- Dynamic simulation model: running online for automatic process disturbance detection or offline as an analysis tool
- Dynamic parameter estimator: to automatically calculate changes in model parameters that are normally assumed to be constant but can vary in time, depending on the process conditions
- Advanced controller design: to create and test advanced control algorithms inside the dynamic modelling environment
- Auto-calibration: to ensure the dynamic model is continuously calibrated based on the most recent data
- Advanced fault detection for process diagnostics: to detect and identify process upsets as they occur
- Process variable optimization: to analyze the incoming data and calculate an optimized operational strategy
- Forecasting: to predict the process conditions in the future and identify potential problems or optimal operational setpoints
- Automatic controller tuning: to automate the determination of the controller parameters
- Uncertainty analysis: to estimate the uncertainty in the model / controller predictions
- Decision support tools: to help administrators allocate resources and make process decisions
- Operator interface: to simplify the interaction between those using the system and the underlying mathematical tools

The MA-RT-PW-MC platforms integrate the tools listed above into a cohesive software package that facilitates the free flow of information between the relevant modules, the SCADA system, and the local PLCs.

9.4 CHALLENGES OF INTEGRATED CONTROL SOFTWARE TOOLS

MA-RT-PW-MC platforms are complex software packages designed to run complex plants. The complexity increases with every running module and every integrated control loop running in real time. As a result, it is critical that any project scope be fully understood prior to starting any development. The key questions to be answered prior to embarking on such a project are:

- Should any control loops be linked together?
- Should the controllers be uni- or multivariate?
- Should the dynamic nonlinear process model be running online or onside?
- How much functionality should be running in real time?

The traditional consensus in the wastewater industry has been that most processes can be decoupled and successfully controlled with independent control loops making multivariable controllers unnecessary [8]. However, that traditional approach is now under pressure as WRRFs add resource recovery to their historic role of protecting human health and the environment. Resource recovery has created new challenges with the addition of new energy efficient, complex processes and it is unclear what types of control will be required in the future. However, even in conventional plant configurations, the coupling of certain control loops (e.g. aeration and SRT control) has demonstrated benefits [12] suggesting that the traditional consensus may need to be reconsidered.

The benefits of using dynamic process models for plant optimizations and control is obvious. As a system, WRRFs are stiff with large differences in the various process time constants, from seconds (air and liquid flow rates), to hours (concentration changes), to days or even months (shifts in microbial communities) [8]. Because most of the treatment processes are nonlinear, there are significant benefits to be realized if a dynamic process model is used for the design of control strategies and the optimization of process operations. However, the benefit of running the model online with parameter identification and autocalibration routines, for example, has not been sufficiently demonstrated.

The implementation of the different MA-RT-PW-MC platforms has faced major challenges since their inception. In the beginning, sensor reliability and computer processing capabilities placed limitations on the usefulness of such software. There were a few early adopters but the tools were either abandoned or were not able to achieve their intended potential and industry acceptance of this technology lagged behind the vision being promoted at the turn of the century. The adoption and further development of real-time data quality tools in the wastewater industry along with sensor and computer advancements has resulted in a resurgence in the development and implementation of MA-RT-PW-MC platforms more recently.

9.5 EXAMPLES OF MA-RT-PW-MC SOFTWARE PLATFORMS

9.5.1 THE INTEGRATED COMPUTER CONTROL SYSTEM (IC^2S)

The Integrated Computer Control System (IC^2S) based on technology developed by Hydromantis Environmental Software Solutions Inc. (www.hydromantis.com) was one example of an advanced model-based control system for WRRFs developed in the 1990s [1,15]. The ultimate objective of the IC^2S concept was controller design, process fault detection, and operator advice. Figure 9.1 shows the architecture of the IC^2S concept with links to the plant supervisory control and data acquisition (SCADA) system.

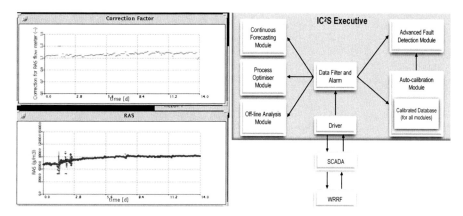

Figure 9.1: Design and implementation example of Hydormantis' Integrated Computer Control System (IC^2S) [1].

The IC^2S platform was implemented at a plant located in Sweden. The objective of the project was to check the mass balance around a final clarifier and calculate the sludge volume index (SVI) values from on-line flow and solids data using a calibrated model running on-line. The tool ran one calibrated model of the plant offline and a simpler model on-line. The simplified model was connected to the SCADA system, correcting for mass balance errors in the data and autocalibrating the complex offline model. More information on the specific application can be found in [15] and [1]. The IC^2S system had very limited success mainly because of immature sensor technology and the lack of acceptance of advanced control by the wastewater industry at the time.

9.5.2 STAR UTILITY SOLUTIONS™

During the same period, Krüger Inc. (www.kruger.dk), now a VEOLIA business unit, was developing the Superior Tuning and Reporting Utility Solutions™ (STAR) software package. The STAR system has enjoyed success in Europe with numerous applications since its initial launch in the early 1990s. The main functionality of the STAR system has been supervisory control. The system includes control modules for different unit processes as well as data quality and data reconciliation tools. Tools for cost and capacity optimization are also included and a web-based interface provides communication with operations staff (Figure 9.2). Examples of implementations of the STAR system can be found [7] and [9].

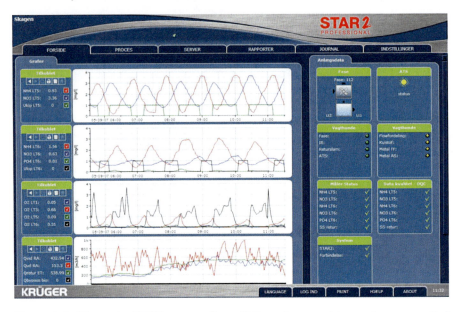

Figure 9.2: The early STAR system from Krüger (http://www.kruger.dk/krugeras/ressources/files/1/8176,STAR_control-UK.pdf).

9.5.3 IBM ADVANCED SIMULATION, FORECASTING, AND OPTIMIZATION PLATFORM

The IBM Research team based in Haifa, Israel (www.research.ibm.com/labs/haifa/), embarked on the development of a prototype simulation, forecasting, and optimization platform. Their Internet of Things pilot included real-time data quality, a plant state estimator, statistical forecasting and optimization algorithms to improve plant efficiency (Figure 9.3). The tool estimates the plant's state in real time and generates a plan for the operations staff to follow. The prototype was implemented in November 2014 at a treatment plant in Lleida, Spain, as part of a joint project between IBM and the plant operator, FCC Aqualia Spain ([3, 14]).

9.5.4 INCTRL SOLUTIONS MONITORING AND CONTROL PLATFORM

More recently, inCTRL Solutions Inc., a Canadian company (www.inctrl.ca), and ifak Magdeburg eV (www.ifak.eu/en), a German research institute, have teamed up to create a new software platform. The inCTRL Monitoring and Control Platform (M&CP) focuses on sensor maintenance, real-time data quality, and

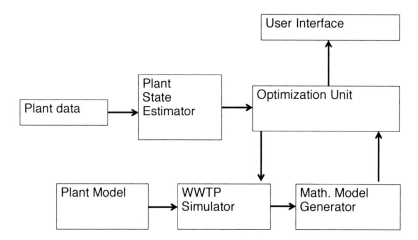

Figure 9.3: IBM advanced simulation, forecasting, and optimization platform [2].

advanced control design. The tool uses a dynamic simulator (SIMBA#, ifak Magdeburg, Germany) to design and test control solutions. The designed control solutions can be implemented in the plant PLCs or used as implemented in the simulation environment and run through the Monitoring & Control Platform. The first full-scale application of the inCTRL M&CP is at the City of Grand Rapids WRRF in Michigan, USA, and its implementation is described in the case study below (Figure 9.4).

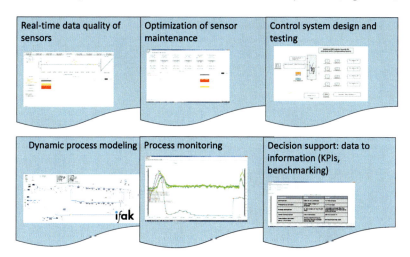

Figure 9.4: inCTRL monitoring and control platform [5].

9.6 CASE STUDY: REAL-TIME CONTROL AND OPTIMIZATION OF THE GRAND RAPIDS WRRF TARGETING ENERGY REDUCTION

9.6.1 THE PLANT

The Grand Rapids WRRF has a design flow of 230,000 m^3/d (61 MGD) and a peak wet weather flow of 340,000 m^3/d (90MGD). The plant currently treats on average 144,000 m^3/d (38 MGD). It serves a population of approximately 270,000 people in 11 communities. The plant also receives a significant amount of industrial wastewater from the local breweries, landfills, soap and detergent manufacturing industries, pharmaceutical manufacturing industries, and other food and beverage industries. The plant has screening, aerated grit removal, rectangular primary clarification, activated sludge tanks, and circular final clarifiers. The effluent is disinfected by UV disinfection prior to discharge. The treatment processes are divided into two parallel "plants"; namely, the North and the South plants. The Grand Rapids WRRF process flow diagram as depicted in the SCADA overview screen is shown in Figure 9.5.

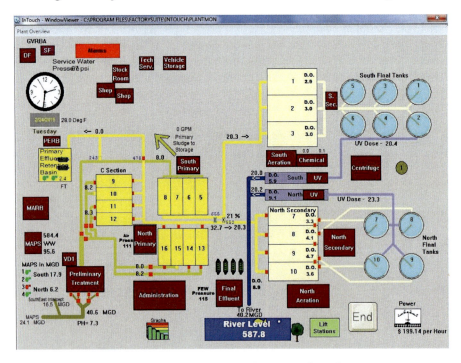

Figure 9.5: Grand Rapids WRRF process flow diagram.

The plant has to meet year-round total suspended solids (TSS), total phosphorus, and fecal coliform targets and seasonal carbonaceous biological oxygen demand (cBOD) and ammonia targets as shown in Table 9.1.

The dynamics of the plant influent are impacted by the large, flat catchment as

Table 9.1
GR WRRF current effluent permit.

Parameter	Monthly	7-Day	Daily	Units	Monitoring Frequency	Sample Type
Carbonaceous biological oxygen demand (cBOD$_5$)						
June 1 – Sept. 30	16		21	mg/l	Daily	24-Hr Comp.
Oct. 1 – May 31	23		38	mg/l	Daily	24-Hr Comp.
Total Suspended Solids (TSS)	29		44	mg/l	Daily	24-Hr Comp.
Ammonia (as N)						
June 1 - Sept. 30			8.5	mg/l	Daily	24-Hr Comp.
Oct. 1 – May 31			18	mg/l	Daily	24-Hr Comp.
Total Phosphorus (as P)	1.0			mg/l	Daily	24-Hr Comp.
Fecal Coliforms	200	400		ct/100 ml		Daily Grab

well as the concentrated landfill waste dumped at the head of the works. Figure 9.6 shows the impact of the trucks discharging landfill leachate at the head of the works (spikes between 6 am and 4 pm). During weekends, the plant does not receive leachate. Figure 9.7 shows the timing of the arrival of the peak load to the head of the aeration tank as well as the delay in the peak exiting the aeration tank.

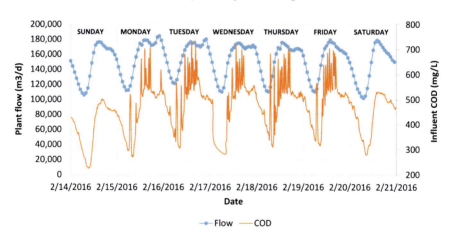

Figure 9.6: Grand Rapids WRRF influent flowrate and COD concentration.

9.6.2 GRAND RAPIDS WRRF JOURNEY TO REAL-TIME MONITORING AND CONTROL

While interest in real-time control was maturing and developments in sensor technology were advancing, the City of Grand Rapids, started investing in real-time measur-

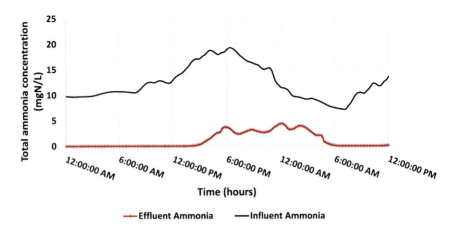

Figure 9.7: Grand Rapids WRRF influent and effluent ammonia nitrogen concentration.

ing instruments. In 2014 the City installed six LiquID units by ZAPS Technologies that measure multiple parameters including: BOD, COD, solids, nitrate, and turbidity. The plant also invested in a Hach®Ion Selective Electrode (ISE) Ammonium probe and Hach®Amtax sc Ammonium Analysers. These instruments brought to light the true dynamics of the influent and the response of the process. As a direct result of these observations, the plant decided to venture into advanced process control. They started by using online signals from the ammonia probes and analyzers to implement ammonia-based dissolved oxygen control in one half of the plant (South) and they also started using LiquID signals to modulate the UV lamp dose.

Following these initial efforts, the City decided to embark on an ambitious project to combine real-time data and dynamic simulation for integrated plant control. Primodal US Inc. (www.primodal.com) was hired to put a team together to execute the project and the inCTRL Monitoring and Control Platform was selected as the software for the project.

9.6.3 PROJECT OBJECTIVE AND TASKS

The Grand Rapids WRRF has no difficulty meeting its effluent permit, however it has set a goal to reduce energy consumption and operational costs. Therefore the key objective of the project was to achieve energy savings by extending the plant automation and making full use of the installed on-line instrumentation. The execution of the project required a plan consisting of several coordinated tasks as shown in Figure 9.8.

The project established an energy use benchmark to be used as a quantitative measure of success and several key parameter indicators. This was followed by identifying the available control handles for each of the control objectives [4].

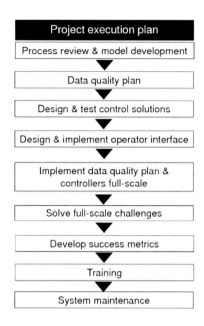

Figure 9.8: Grand Rapids project execution plan [6].

9.6.4 PROCESS MODEL DEVELOPMENT

A dynamic process model of the plant was developed and calibrated. The model captured plant dynamics resulting from the influent, intermittent pumping of sludge and the various recycles. In addition to the biochemical processes, the simulated plant layout included detailed models of the aeration, and control systems. With the help of the process model, control scenarios were developed for: a) enhanced biological phosphorus removal, b) aeration, c) sludge retention time (SRT), d) disinfection (UV), e) wet weather operation, and f) grit aeration. In addition, the model was used to investigate integrated control strategies for processes that interact with each other (Figure 9.9). The process interactions investigated were: a) ammonia-based aeration control and sludge retention time control, b) carbon flow path control for load equalization, digestion, and biological phosphorus removal, and c) wet weather grit aeration control. In addition, the process model was used to support and develop control options for future plant modifications including the planned concentrated waste (CW) storage tank that will receive industrial loads, blend them with selected side streams, and then use that mixture for load balancing.

The model evaluations indicated that the most significant cost savings could be achieved by implementing ammonia-based aeration control (ABAC) integrated with SRT control [12]. The following sections will discuss the key project steps involved in the development and implementation of inCTRL's patent-pending ABAC-SRT controller tested at the Grand Rapids WRRF.

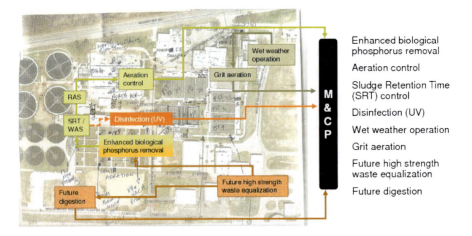

Figure 9.9: Grand Rapids WRRF, Michigan, integrated modelling and control scenarios.

9.6.5 DATA QUALITY OF ONLINE SENSORS

The most significant obstacle encountered in real-time monitoring and control is the quality of the data collected by the on-line instrumentation. Before any control options were implemented, the quality of the existing instrumentation was verified by collecting 20 reference measurements and comparing them to the on-line values. The sensor slope and offset were adjusted if the data quality algorithms running in the M&CP indicated a problem (Figure 9.10). This initial evaluation established sensor accuracy and repeatability. For certain sensors, correcting their installation and verifying the signal transmission to the SCADA proved necessary. Figure 9.11 shows the installation of a turbidity/TSS sensor in the mixed liquor of the Grand Rapids plant before and after the installation was corrected and the sensor was recalibrated.

Following this initialization period, maintenance measurements were collected for each on-line instrument at set intervals. The M&CP platform continuously checks the data quality of the on-line signals in real-time using a unique set of algorithms. When a potential error is detected, the system goes into Warning mode and corrective actions are triggered. If the system validates an error, it switches into Alarm mode and the sensor signal is excluded from the control logic, and "safe mode" is activated for the specific control loop.

The objective of the QA/QC step is twofold. Firstly, it is crucial to secure high data quality from the on-line signals so confidence in the data being fed to the control routines is assured and secondly the QA/QC step is necessary to optimize the maintenance of on-line equipment which is expensive and labor-intensive.

Model-Assisted, Real-Time Monitoring and Control of WRRFs

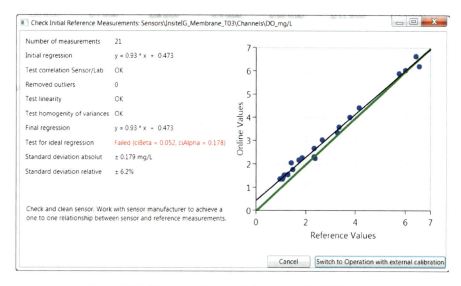

Figure 9.10: Data quality check for an on-line DO probe.

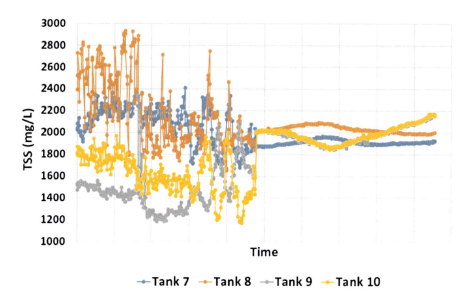

Figure 9.11: TSS concentrations before and after the correction of the sensor installation and its recalibration.

> KEY MESSAGE: The operation of the sensors and actuators needs to be verified prior to the implementation of any control solution. For continuous maintenance, optimized workflows need to be created to avoid overburdening plant staff.

9.6.6 THE ABAC-SRT CONTROLLER

The ABAC-SRT controller consists of two cascades linked by a supervisory controller as shown in Figure 9.12. The first cascade consists of the ammonia-based aeration control and the second cascade consists of the SRT control. The supervisory controller links the two by calculating an optimal SRT that maintains a target average DO concentration.

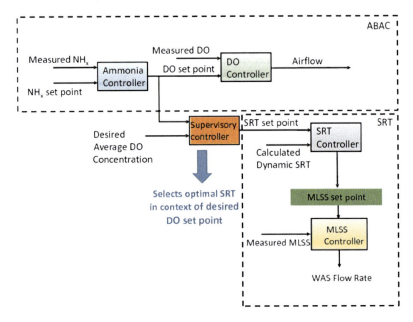

Figure 9.12: Ammonia-based aeration control coupled with SRT control [12] (© 2017 inCTRL Solutions Inc. Used with permission. The technology illustrated in the figure is patent pending. All rights reserved).

9.6.7 CONTROLLER TESTING IN THE PROCESS MODEL: ABAC-SRT CONTROLLER

Following the verification of all the real-time instruments, the process model was used to test and tune the controller. Prior to this project, the North plant was operating with a DO setpoint aeration control as shown in Figure 9.13A. The first modification

Model-Assisted, Real-Time Monitoring and Control of WRRFs 245

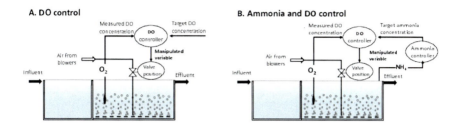

Figure 9.13: DO and ammonia based feedback aeration control [10].

was to test an ammonia-based aeration control strategy (ABAC) as shown in Figure 9.13B.

All the equipment constraints as well as the operational limitations imposed by the installed equipment and plant operations were included in the model and this procedure allowed the project team to visualize the controller performance as predicted by the model prior to its implementation. Figure 9.14 shows the model predictions for an effluent ammonia target of $2mgN/L$ and two different minimum bounds in the DO setpoint. In this case, the minimum DO bound was $1mg/L$ in Figure 9.14A and $0.5mg/L$ in Figure 9.14B. The upper bound was $3mg/L$ for both. This exercise demonstrated that even with this type of control, the ammonia at the analyzer location was unlikely to approach the setpoint because the lower bound on the DO controller was too high. In addition, the minimum airflow imposed by operations to ensure adequate mixing in the aeration tanks meant that the airflow was always greater than required to maintain the ammonia at the setpoint. The results showed (Figure 9.14) that the controller has restricted control authority and that the DO and airflow would be consistently on their lower bounds meaning that the plant would be consistently over-aerated in this scenario. In addition, the model was used to estimate the DO concentrations down the aeration basin based on the diffuser taper and air distribution. This was necessary in this case because each North train has only one DO probe.

In addition to tuning and setpoint evaluations, the model was used to compare the predicted airflow and energy costs to the base case scenario. Figure 9.15 shows the type of analysis that can be done with a calibrated model. The figure shows a comparison of aeration requirements between the historic control strategy (no ABAC) shown in the light blue bar and two ABAC solutions with different DO setpoint bounds. The historic control has minimum and maximum bounds on the DO setpoint of 1 and $5mg/L$, respectively. The ABAC scenario results shown with the dark blue bar had the same bounds on the DO setpoint. The red bar shows a more aggressive DO control strategy with the ABAC controller on and minimum and maximum bounds on the DO setpoint of 0.5 and $3mg/L$, respectively. The figure clearly shows that this more aggressive control strategy achieves even higher aeration savings.

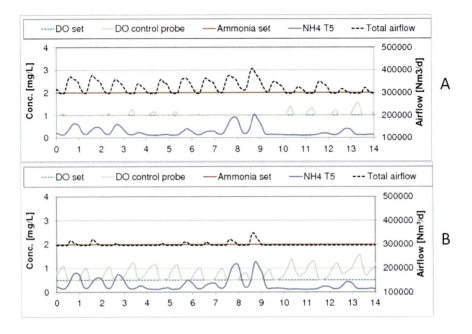

Figure 9.14: Model predicted controller performance during testing for a two-week period with minimum DO setpoints of $1 mgN/L$ (A) and $0.5 mgN/L$ (B).

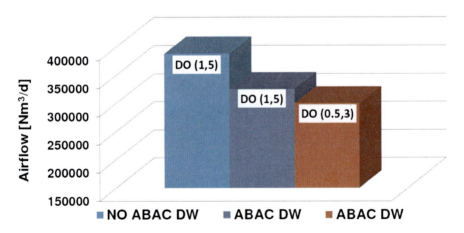

Figure 9.15: Cost comparison of two control strategies.

Following the ABAC testing in the model for the North plant, the implementation and testing of a new SRT controller began. Originally, the plant was using a spreadsheet to estimate target SRTs based on plant performance and season. To more appropriately control the SRT at the plant, inCTRLs ABAC-SRT controller [12] was implemented in the model and tested.

This controller includes several cascaded control loops that can work independently (Figure 9.12) or in combination. Figure 9.16 shows the model-based evaluation of two loops of the cascade, the ABAC and the TSS controller, working independently.

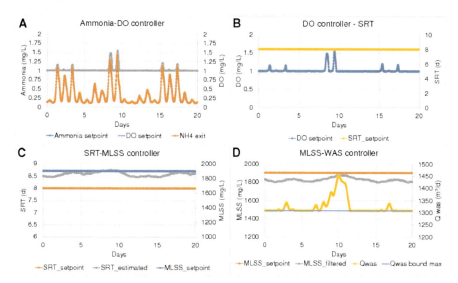

Figure 9.16: Independent ABAC and mixed liquor suspended solids (MLSS) control.

Figure 9.16A shows the ammonia-based aeration control calculating the DO setpoint based on the ammonia concentration exiting the activated sludge tank. The ammonia setpoint was set to $1 mgN/L$ and the lower bound of the DO setpoint was set to $1 mg/L$ as well. In this simulation the SRT controller is decoupled from the ammonia-based aeration controller. In addition, only one controller of the SRT cascade — the MLSS-WAS part — is operational. Figure 9.16C shows the MLSS-WAS controller estimating the WAS flow based on a target MLSS setpoint. The controller cannot reach the setpoint in this simulation because the WAS flow rate has an imposed lower bound to avoid running the pumps at their lower capacity setting. Figures 9.16B and 9.16C show the SRT setpoint at a constant value as this part of the cascade is not activated.

Figure 9.17 shows a simulation of the coupled ABAC-SRT controller. Figure 9.17A shows the ammonia-based aeration controllers calculation of the DO setpoint. Figure 9.17B shows the setpoint sent to the SRT controller which changes the SRT as required and the SRT controller varies the MLSS setpoint (Figure 9.17A). Figure 9.17D shows the changes in WAS flow rate required to achieve the MLSS setpoint.

> **KEY MESSAGE:** Operational bounds for reasons other than process performance can impact controller authority.

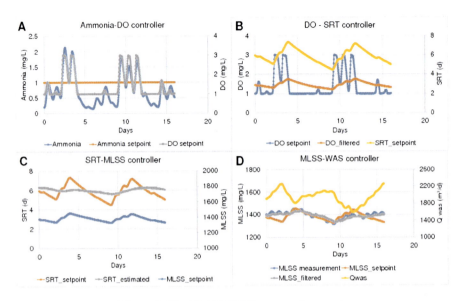

Figure 9.17: ABAC-SRT integrated control.

> **KEY MESSAGE:** The model is an invaluable tool for evaluating controller performance, including equipment and process constraints and identifying issues in areas of the plant that are not monitored.

9.6.8 CONTROLLER IMPLEMENTATION AND FULL-SCALE TESTING: ABAC-SRT CONTROLLER

The modelling work provided sufficient evidence to support the full-scale implementation of the ABAC-SRT control strategy. The optimal controller settings identified by the model for the ABAC controller were implemented in the M&CP platform and activated. As with any control system, the controller tuning constants had to be fine-tuned in the full-scale implementation because only at full-scale are all the system delays and the true process time constants evident. Figure 9.18 shows the initial controller tuning as suggested by the model and Figure 9.19 shows the refined tuning following several months of operation. The figures show that the controller still has limited control authority because of the constraints imposed. In this case, the most important constraints impacting the controller performance are the minimum airflow imposed to ensure mixing and the minimum allowable DO setpoint set by the plant operators. The result is periods of over aeration during low load conditions.

The next step following the ABAC tuning was the activation (one at a time, in reverse order) of the three cascaded controllers linking the ABAC to the Supervisory DO, SRT, and wastage rate control (Figure 9.20).

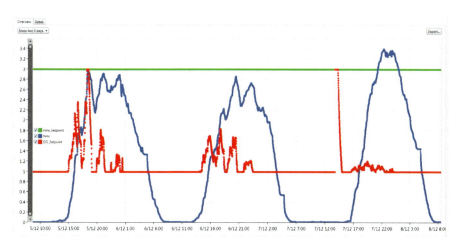

Figure 9.18: Initial full-scale implementation of ABAC. Green: Ammonia target, Blue: ammonia measurement at end of aeration lane, Red: controller defined DO setpoint manipulated by the ABAC controller.

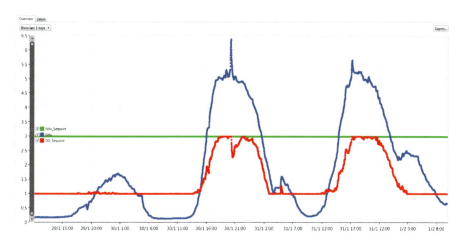

Figure 9.19: ABAC controller performance after re-tuning. Green: Ammonia target, Blue: ammonia measurement at end of aeration lane, Red: controller defined DO setpoint manipulated by the ABAC controller.

The first controller to be activated was the MLSS-WAS (mixed liquor suspended solids-waste activated sludge flow) controller. This part of the cascade attempts to maintain a user defined MLSS target concentration by manipulating the sludge wastage rate (Figure 9.21). The on-line MLSS measurements are heavily filtered to avoid large variations in wastage rate, which would have a negative impact on the thickeners downstream.

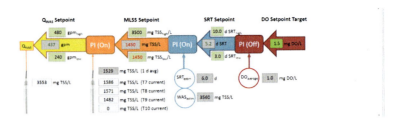

Figure 9.20: Supervisory DO-SRT controller cascade.

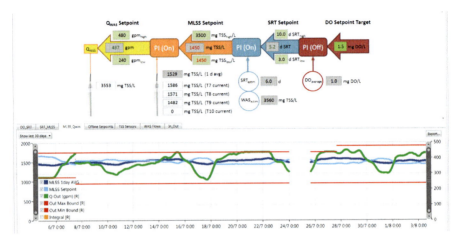

Figure 9.21: TSS-WAS controller commissioning. Green: waste pump flow rate, Dark blue: MLSS measurement, Light blue: MLSS setpoint, Red: waste pump setpoint bounds.

The next steps to be implemented include the commissioning and full-scale tuning of the SRT-MLSS controller which will attempt to maintain a user defined target SRT by calculating an MLSS setpoint which will subsequently be used in the MLSS-WAS controller. The final step will be the commissioning of the Supervisory controller which calculates the SRT setpoint depending on the desired average DO target. The plant is continuing to test all the control loops at full-scale.

> KEY MESSAGE: it takes time to implement and tune controllers at full-scale and it should be done so that lower-level controllers are activated and tested before their setpoint controllers in the cascade are activated.

9.7 CONCLUDING REMARKS AND FUTURE DIRECTIONS

Recent years have seen an increase in the development of real-time plant-wide monitoring and control and it is envisioned that the development and implementation of MA-RT-PW-MC tools will expand as WRRFs collect more data and savings are demonstrated. Implementing such a control solution is challenging due, in part, to real-world issues. As a result, the successful implementation of MA-RT-PW-MC requires an inclusive incentivized approach from the outset [11]:

- Include all disciplines in the development and knowledge transfer
- Create incentives for all disciplines involved in the project
- Incorporate existing operations in the new control scheme
- Incorporate new controls into existing operator procedures and operator work schedule

Implementation requires a sophisticated software tool with diverse functionality. In this chapter several such tools have been described and the application of one, the inCTRL monitoring and control platform, was discussed in detail. Integrated tools such as the ones described in this chapter provide a common platform for all the disciplines involved and encourage collaboration and integration of knowledge. Converting the data to information for decision support can only be accomplished through advanced tools but, the operation of WRRFs and especially the design of successful control systems requires expertise from a number of fields. And, it is the integration of that expert knowledge that leads to the successful implementation of these advanced control tools.

ACKNOWLEDGMENTS

The authors would like to acknowledge the support received from Michael Lunn, the Manager of the Environmental Services Department of the City of Grand Rapids, Michigan, as well as all the staff at the Grand Rapids WRRF.

9.8 REFERENCES

1. E Belia and I Takacs. Integrated on-line systems in wastewater treatment. In *2002 Borchard Conference, Ann Arbor, Michigan, USA*, 2002.
2. E Belia Optimization of energy, operations, and CIP knowledge development forum on intelligent water systems. 2016.
3. IBM. How we used IoT data to optimize wastewater treatment. Retrieved from `https://www.ibm.com/blogs/research/2016/08/used-iot-data-optimize-wastewater-treatment/`.
4. M Lunn and E Belia Simulation and control of the city of Grand Rapids WRRF. In *MWEA 91st Annual Conference, June 19th - 22nd*, Boyne Mountain Resort, Michigan, 2016.
5. M Lunn and E Belia Fifteen years of progress in model assisted real-time monitoring and control. In *Borchardt Conference, Feb 21-22*, Ann Arbor, Michigan, 2017.
6. M Lunn and E Belia Real-time control targeting process stability and energy reduction at the grand rapids WRRF. In *Intensification of resource recovery forum, August 10-12, 2017*, Manhattan College, New York, 2017.

7. M K Nielsen and A Lynggaard-Jensen. Superior tuning and reporting (star)-a new concept for on-line process control of wastewater treatment plants. *Water Science & Technology*, 28(11-12):561–575, 1993.

8. G Olsson, M Nielsen, Z Yuan, A Lynggaard-Jensen, and J P Steyer. *Instrumentation, control and automation in wastewater systems*. IWA Publishing, 2005.

9. C Printemps, G Manic, C Lemoine, M Payraudeau, T Munk-Nielsen, T Onnerth, and C Ame. Development of a model based tuning tool for simulation of star control at Blois WWTP. In *7th International Conference on Hydroinformatics HIC, 2006*, Nice, France, 2006.

10. L Rieger, R M Jones, P L Dold, and C B Bott. Ammonia-based feedforward and feedback aeration control in activated sludge processes. *Water Environment Research*, 86(1):63–73, 2014.

11. L Rieger and G Olsson. Why many control systems fail. *Water Environment and Technology*, 24(6):42–45, 2012.

12. O Schraa, L Rieger, J Alex, and I Miletic. Ammonia-based aeration control with optimal SRT control: improved performance and lower energy consumption. In *12th IWA Specialized Conference on Instrumentation, Control & Automation*. Quebec City, Quebec, Canada 11-14 June, 2017.

13. Veolia. Water Treatment Technologies. Retrieved from `http://technomaps.veoliawatertechnologies.com/star-utility-solutions/en/references.htm`.

14. S Wasserkrug, E Israeli, F E Maneiro, E Belia, J Giribelt Argiles Jaume, A Zadorojniy, and S Zeltyn. Value based wastewater treatment plant operation through advanced simulation, forecasting and optimization. In *87th Annual WEF Technical Exhibition and Conference (WEFTEC2014)*, 2014.

15. A Westlund and I Takcs. Extracting SVI from on-line flow and solids data. In *8th Conference of the International Association on Water Quality (IAWQ) on Design, Operation and Economics of Large Wastewater Treatment Plants (LWWTP's)*, Budapest, 1999.

Part III

Applications and Emerging Topics: Water Natural Resources

10 Methods for Estimating the Optical Flow on Fluids and Deformable River Streams: A Critical Survey

KONSTANTINOS BACHARIDIS, KONSTANTIA MOIROGIORGOU,
GEORGE LIVANOS, ANDREAS SAVAKIS, MICHALIS ZERVAKIS

CHAPTER HIGHLIGHTS

This chapter:

- Presents a comparative study of the optical flow methods for fluid flow estimation, a topic of significant importance for unattended cyber-physical environmental monitoring, emphasizing the methodological advantages and limitations of each methodological class;
- Goes past the description of the state-of-the-art by providing experimental validation and ranking of the corresponding methods;
- Presents a comparison of their performance using ground truth data of a real-world river monitoring application.

10.1 INTRODUCTION AND MOTIVATION

Climate change appears to be one of the most significant challenges that the scientific community will have to cope with in the next few decades. The weather and climate shifts present a continuous trend that potentially will affect not only the environment, but also human lives and activities. In particular, the efficient management of water resources becomes a necessity, following the global climate change trend. Many places around the world already experience changes in rainfall density with

---

K Bacharidis • K Moirogiorgou ✉ • G Livanos • M Zervakis

Technical University of Crete, School of ECE, GR. E-mail: dina@display.tuc.gr

A Savakis

Rochester Institute of Technology - Department of Computer Engineering, US.

255

frequent occurrences of extreme flood events [25]. The United Nations World Water Development Report 2015 [19] emphasizes the significant potential of climate change on the distribution and availability processes of water resources. It is clear that we need effective ways to ensure water sustainability in order to avoid water crisis phenomena in the next decades.

One crucial aspect of water sustainability is the preservation of its quality. This can be achieved through continuous monitoring of water delivery sites, like springs, rivers and any site that is located in coastal and inland hydraulic basins. Nowadays, water quality and distribution management are performed by installing sensor networks that can track water flow and measure water contents. There are many commercial solutions to equip river monitoring stations, mainly using sensors that record physical (flow, suspended sediment, temperature) and chemical (conductivity, dissolved oxygen, nitrate, pH, heavy metals) water quality parameters. The main disadvantage of this kind of monitoring is the inability of operation during extreme weather events. In the case of extreme flood events, the water flow exhibits drastic increase in terms of level and mean velocity. Furthermore, the strong force of moving water affects the surroundings (human structures and environment) in many ways [4]. In the case of temporary river watersheds, which are ephemeral and episodic streams, the temporal variability of flow is a key feature for the prediction of catastrophic events. Temporary rivers constitute 30% in the Mediterranean region and at least 42% of the Greek territory. Based on recent trends, the number of temporary rivers will likely increase in the near future, due to the impacts of human activity that cause increased water abstraction. The prediction of floods and the implementation of the Water Framework Directive (WFD-2000/60/EC) states, in general, that all water bodies in Europe have to be managed at the watershed scale [73]. To relate this directive with river management solutions, the design of sustainable solutions for monitoring the water resources becomes a demanding task of great significance. One efficient way to automatically record the river flow during "flash" flood events is using optical sensors, since they can record the stream surface flow and use analytic methods to calculate the flow discharge during the event without the need of human presence. Alternatively, one could use commercial flow meters that can provide discharge information using accelerometers or ultrasound technology, but either a human operator is still needed in order to insert and operate the sensors inside the water stream or the total cost of the equipment is quite high. In any of these cases, the use of optical sensors sounds quite challenging, as long as the accuracy of the calculated water flow parameters remains high.

An optical monitoring system can be part of an integrated water flow monitoring system, such as a new concept of cyber-physical systems (CPS), targeted exclusively to streams monitoring [51]. In that respect, a new alternative on intelligent management of the water life cycle can be designed and used to address the impact of climate change on water natural sources. Moreover, it can provide useful information about water discharge to the smart grid structures already deployed on such springs or rivers, for better management of the available water resources or even as an alert in case of a "flash" flood event.

Our study explores the state of the art on optical fluid-flow estimation schemes and attempts both quantitative and qualitative comparisons of these techniques. In general, fluid flow estimation is a valuable tool for decision making in many real-world applications. In environmental sciences, temporary river monitoring may provide information about "flash" flood events (e.g. [14], [24], [72]), while sea monitoring may provide information about tsunami events (e.g. [61], [62]). The field range of applications engaging fluid flow extends to monitoring many different types of processes in medical, mechanical, or chemical engineering [3]. The use of video and imaging data provides a continuous, non-intrusive and low-cost method for monitoring and estimating motion characteristics. However, the motion estimation procedure can be quite difficult, due to the dynamic nature of the fluid dynamics. Non-rigid motion differs from the motion of rigid bodies, mainly due to forces altering the motion of the fluid locally, depending on the field of application (wind and gravity for the environmental cases, chemical accumulation for washout procedures, etc.). This dynamic nature of fluid motion necessitates the use of mathematical models combined with physical constraints generated by the properties and concentrations of fluid contents, in order to model the multidirectional and multiscale formulations encountered in such motion patterns.

Finally, the chapter is organized as follows: Section 10.2 presents a categorized theoretical study of the optical flow methodologies, providing an analysis of the key concepts and variations of each methodology class. In Section 10.3, we present a comparison of each methodology class and we formulate an experimental comparison scheme in order to investigate the estimation accuracy of the different methodology classes. Finally, Section 10.4 refers to the way optical flow can provide helpful results and conclusions when we consider real-world applications.

10.2 METHODOLOGICAL ANALYSIS OF ALGORITHMS

10.2.1 PARTICLE-BASED METHODS

Particle-based methods are mainly simple block matching techniques combined with correlation matching and feature tracking methodologies ([14], [46], [26]). The main idea is that the velocity field of particles that exist in the flow can be associated with the fluid velocity field under the assumption that these particles follow the same motion pattern as the fluid irrespective of their material properties. As particle can be considered an element or a group of elements (pattern), natural or artificial, which can be used as a tracer. The particle-based approaches are performed either under laboratory conditions or under exterior (outdoor) conditions in the case of insitu processing. As illustrated in Figure 10.1(a), 10.1(b), in each case, the fluid is seeded with particles which are either physically illuminated ([77], [38], [58], [49]) or with the help of an appropriate light source, such as lasers in laboratory tests.

In the case of optical flow estimation through tracing particles, two main methodologies have been developed, the particle image velocimetry (PIV) method and the particle tracking velocimetry (PTV) method, which both use correlation schemes and feature selection image processing techniques. The main difference between these

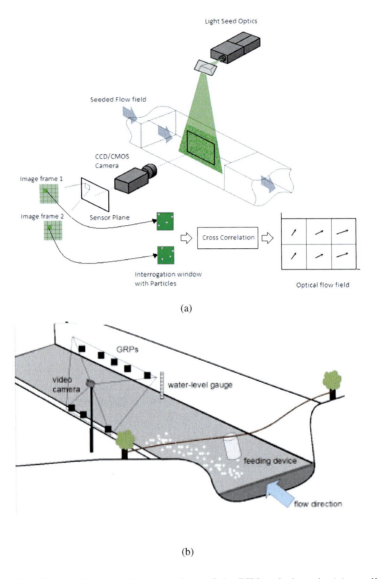

Figure 10.1: The implementation procedure of the PIV technique in (a) small-scale flows such as the measurement of fluid flows using laser beams as tracer pointers, (b) river high flow conditions [39].

two approaches is that PTV methods track a single particle along the series of the video frames, so it is appropriate for "low" seeding density flows. PIV approach, on the other hand, is more appropriate for "medium" or "high" seeding density flows, since it is capable of tracking groups of particles. Furthermore, PIV allows the estimation of the Eulerian velocity, that is focused on specific space locations through which the fluid flows, observing the way the flow passes through a selected location. PIV approaches yield an average particle motion in the space based on a group of velocity vectors. PTV however, is used to describe the Lagrangian motion, the motion of an individual fluid parcel in time and space. The parcel's/particle's motion vector is derived by tracking the displacement path of this individual particle in a sequence of images [16]. Essentially, the Eulerian velocity in PIV approaches is estimated by adopting a Lagrangian motion estimation approach but for a very short duration.

In any particular approach, particle-based methodologies assume a certain type and density of the particles beyond the methodology for the estimation of the optical flow field itself. In the case of artificial tracers, it is important to use particles with characteristics that allow them to follow the fluid motion. The shape and density characteristics, as well as the fluid density and its dynamic viscosity are some of the factors that affect the tracking success of the particle. In order to select the appropriate particle characteristics, an acceptable relation among the particle motion $\mathbf{u}_{particle}$, the flow motion $\mathbf{u}_{flow}$ and the real instantaneous relative velocity $\mathbf{V}_{inst}$ between the fluid and the particle must be defined. As an example, in case of spherical-shaped particles, the Basset–Boussinesq–Oseen (BBO) method, presented by Basset in 1888 [7], associates forces, pressure gradients, and fluid density in a way that provides an estimation of the particle's desirable characteristics. The estimation is based on the condition that the particle's Reynolds number Re_p and instantaneous relative velocity $\mathbf{V}_{inst}$ satisfy Stokes' drag law [50], where low Reynolds numbers reflect smooth and constant flows with dominant viscous forces (laminant flows), while high Reynolds numbers represent turbulent flow characterized by instabilities, such as vortices or eddies. A more recent review on the influence of seeding particles parameters in PIV and PTV approaches, for the case of a turbulent fluid flow is presented in Hadad and Gurka [27].

Another important task that has to be performed before the particle-based methodologies provide fluid motion estimation is the particle detection in the fluid flow. Usually, the particle detection is done through local and dynamic thresholding procedures ([65], [20], [15]), particle-mask-correlation operators [67], or feature selection strategies combined with correlation metrics on the estimated optical flow field ([64], [52]).

In order to perform particle-based methodologies, the image has to be divided in interrogation regions and the accuracy of the estimation of the displacement vector within a given interrogation window affects the accuracy of the total results. The accuracy of the estimation is affected by a number of factors, ranging from the particle characteristics (shape, size, density, etc.) to the seeding density of the particles within the flow.

Concerning the latter, the particle density N_I affects the amount of in-plane displacement F_I and the amount of out-plane displacement, F_O, of the particle which all in accordance with the size of the interrogation window determines the accuracy of the motion field estimation [40]. The out-of-plane motion reflects the motion of an object across the z-axis of the 3D coordinate system, in the form of translation or rotation. In the image coordinate system, this kind of motion is not reflected as object motion but rather as a change in the dimensions of a static object (e.g. magnification) (Figure 10.2). Instead, the motion of this particle in the x or y axis is correctly displayed and identified as motion in the image coordinate system, usually referred to as in-plane motion. In terms of image displacement, the out-of-plane motion is manifested in the form of gradients, which in high-valued cases can affect and degrade the estimation of in-plane motion.

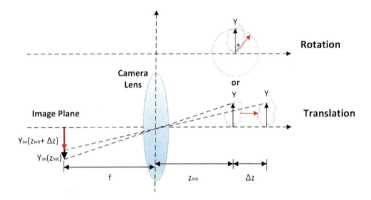

Figure 10.2: The manifestation and effects of the out-of-plane motion in the object's representation in the image plane.

Regarding the particle density, Rafael et al. [60] has shown that larger particle densities reduce the measurement uncertainty given a specified particle shift. The combination of the particle density and the in- and out-plane displacements within the interrogation window form a condition for the accuracy of displacement estimation expressed as the minimization of the product $N_I \cdot F_I \cdot F_O$ ([40], [60], [78]). Keane and Adrian (1992) [40] argue that for the case of PIV measurements in high density flows the tracer particle image-pair density should be greater than 7, i.e. $N_I \cdot F_I \cdot F_O > 7$.

Particle Image Velocimetry (PIV)

Particle image velocimetry (PIV) provides an estimation of the displacement of a single pattern/object within a template window by searching the location of the most similar pattern in next time stamp (see Fig. 10.3(a), 10.3(b)). PIV focuses on the image properties and intensity changes over space and time, whereas PTV considers the object(s) and the preservation of object properties over time and space.

The velocity is calculated from the displacement of the image pattern along the image pair divided by the corresponding time interval. The approaches that have been developed for the PIV model can be grouped into three basic categories based on the formulation of the search window as well as the hardware used for particle tracking. More specifically, in the present study we examine approaches that are based on (a) the window-size and image deformation, (b) the phase cross correlation and (c) the stereoscopic camera concatenations.

(a) Image and window deformation approaches: In window-based approaches, the main factors that affect the estimation are the size and shape of the interrogation window. The simplest window-based motion estimation approach is to use a window with fixed size and shape and just apply a correlation coefficient as a matching criterion to extract the motion [14]. Such approaches, however, are constrained by the assumption that the motion is restricted within the window. This assumption makes them susceptible to out-of-plane deviations of the estimated motion vectors. A solution to this problem is to adjust the window's size and shape according to the previously estimated motion field. Approaches in this concept involve the use of multistage iterative evaluation methods that enable the adjustment of the position and the shape of the window ([46], [37]), as illustrated in Fig. 10.3(a).

In image deformation techniques, the interrogation window in the second image is deformed, in terms of rotation, shape, and size, using bilinear interpolation or a weighting function in order to maximize the correlation result ([38], [56], [5]). This modification can address decision conflicts due to large interrogation windows or truncation effects of particles in small window sizes. In such approaches, the particle region described by a reduced number of grid nodes is spatially deformed according to a displacement model (defined usually by a region correlation criterion) and matched to the frame $t + 1$ within an augmented region, as illustrated in Figure 10.3(a), 10.3(b). The motion estimate is given either by the mean displacement of the pixels belonging to the particle in the examined region or the displacements of co-gravities of the two (different shape) regions.

(b) Phase cross-correlation approaches: Phase filtering and phase correlation procedures, which incorporate a series of optimized filters to the generalized cross-correlation scheme between the interrogation area and the reference pattern are mostly used to this category of methods. The aim of these techniques is to increase the accuracy of the correlation procedure by enhancing the standard cross-correlation implemented in the Fourier domain through a series of filtering procedures. The main disadvantages of this implementation relate to the spectral leakage due to under-sampling and window-bound discontinuity (finite window sizes) and the aliasing effects produced by the periodic boundary constraint in the discrete Fourier case. In such cases, Fourier based cross-correlation produces large estimation deviation, especially in cases where the window shifts are larger. Through the incorporation of filtering procedures, such as spatial masking, phase filtering, and Gaussian transforms of the phase correlation, these phenomena can be minimized. For example, the addition of a pair of adaptive smoothing filters prior to the Fourier transformation as prefiltering operations can reduce the effect of noise in the estimation of correlation. This scheme is known as the generalized cross correlation (GCC) and results in increasing the estimation accuracy due to the restriction of the background noise effect. Around this concept there have been numerous approaches developed, such as the use of phase filters and Gaussian kernels prior and posterior to the FFT corre-

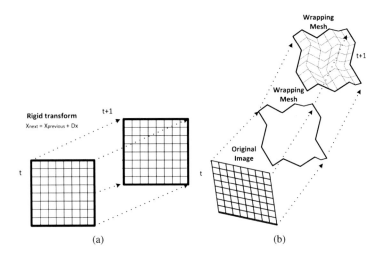

Figure 10.3: An example of a particle pattern deformation. (a) Deformation using a rigid transformation on the search block, preserving the overall shape of the pattern, (b) Deformation on the whole grid points of the pattern.

lation operator ([22], [21], [55], [69]) or weighted filter approaches [75] to detect the existence of an object of the current scene in the reference scene.

The incorporation of such filters aims to the enhancement of a phase term in the DFT domain reflecting the expected displacement; when the reference and candidate regions match an appropriate displacement, then the correlation of regions results in phase cancellation in the spatial domain and the generation of a linear phase term in the DFT domain, reflecting a delta-like function at the correlation peak. The particle's displacement within the interrogation window is then estimated based on the position of the peak (see Fig. 10.4).

The filter $W(u,v)$ can be formed as a matched spatial filter to implement cross correlation in the frequency domain, or aid in removing unnecessary information in the correlation structure [75]. The latter form eliminates magnitude information, producing a robust correlation estimate unaffected by scaling, shape, or size factors of the particle(s), which can be defined, for example, in the form of the phase-only-filter (POF):

$$\mathbf{W}_{POF}(u,v) = \frac{1}{|\mathbf{F}_2(u,v)|} \qquad (10.1)$$

with $\mathbf{F}_2$ being the candidate region in the frequency domain.

The correlation efficiency and the accuracy of displacement estimation depend on the sharpness of the peak and the signal-to-noise ratio (SNR).

The main advantage of implementing phase correlation in the DFT domain instead of traditional correlation in the spatial domain is that the resulted peak is sharper and at higher SNR [75]. However, the main drawbacks of phase correlation approaches relate to their conditioning on image characteristics, i.e. the filters used are heavily

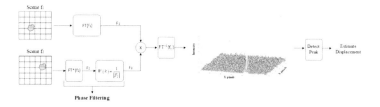

Figure 10.4: Generalized cross-correlation scheme, where the filter $W(u,v)$ (POF filter) is generated from the magnitudes of the two scenes in the frequency domain.

dependent on the signal-to-noise ratio ([55], [75]).

(c) Stereoscopic PIV approaches (SPIV): The incorporation of stereoscopy in the PIV models aims to reduce the deviations in the estimation of in-plane displacements due to the perspective error introduced by larger out-of-plane displacements. In conventional single camera PIV systems, the out-of-plane displacements are due to the fact that the location of each particle does not correspond to the camera axis and, thus, the depth perception results in a deviation between the apparent and the true in-plane displacements. By using a stereoscopic system, one can eliminate this deviation by acquiring a pair of simultaneous images of the particle from different directions, so as to reproduce the depth perception (essentially the out-of-plane motion) and improve the computation of the true displacement.

The stereoscopic PIV approaches based on the alignment of the camera system define two classes: (1) the translational and (2) the rotational systems. The overall identification and tracking of the particle is performed using the previously presented window-based or image-deformation techniques, while more details on the SPIV approaches can be found in Raffel et al. [60] and Prasad [57].

In the translational formulation (see Fig. 10.5(a)), the axes of the camera pair are parallel and the cameras lay symmetric to the systems axis. This simplifies the computation of the matching scene elements between the two views, since only the x-axis position and translation information needs to be matched as well as providing well-focused images, due to the perpendicularity of the optical axes of each camera system to the illuminated area. Various SPIV approaches have been studied for the estimation for liquid flows using the translational camera alignment model ([58], [45], [47]).

In the rotational formulation (see Fig. 10.5(b)) on the other hand, the optical centers of the camera pair are no longer parallel, i.e. the cameras are rotated over a viewing angle. This rotation results in the increase of accuracy for the out-of-plane motion estimation due to the reduction of the perspective error. However, this layout results in a non-uniform magnification effect for the image domain. To cope with this issue, the image plane is further rotated under an angle bounded by the Scheimpflug condition, which ensures colinearity between the image plane, the lens plane and the particles, increasing also the focusing efficiency. The rotational model has been extensively applied on SPIV systems due to the accuracy increase in the out-of-plane

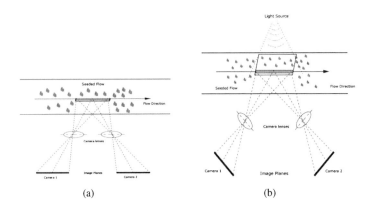

Figure 10.5: (a) The camera configuration for the translational stereoscopic PIV, (b) The camera configuration for the rotational stereoscopic PIV, using an angled lens arrangement.

motion ([76], [33]). However, it is susceptible to errors introduced by the calibration and reconstruction procedures (i.e. the reconstruction of the 3D velocity field), where the accuracy depends on the viewing angles of the cameras [2].

Particle Tracking Velocimetry (PTV)

Particle tracking velocimetry (PTV) uses only a single particle as a tracking pattern. PTV approaches require low seeding density in order to be able to identify and track the particle. The steps of particle identification and tracking are crucial for the success of this method. The difference between these two approaches lies in the density of the flow. In particular, multiply-exposed single image techniques are implemented in "low" seeding density flows where the particles are sparse, avoiding in such a way particle occlusion and overlapping in single frames. On the other hand, singly-exposed multiple image PTV techniques can be implemented in a relatively "higher" density flow since they track a given particle in sequential frames. The overall success of a PTV approach relates to several aspects including (1) the particle density, (2) the identification and tracking procedure, and (3) the robustness to noise. With respect to motion field estimation, PTV approaches can be grouped into two categories, (a) block matching correlation-based and (b) differential-based estimation. In both classes, the displacement of the particle is estimated either by using the initial optical flow field estimates for the entire image domain to guide particle matching or by first isolating the particle and then tracking it within a predefined region in subsequent frames. In PTV approaches, the feature-based description of the particle is more crucial for the overall accuracy compared to the PIV methods since a single particle must be isolated and tracked from image to image. Moreover, using stereoscopic imaging we can enable photogrammetric mapping to be used to track the three-dimensional coordinates of a particle, leading to volumetric reconstruction

Methods for Estimating Optical Flow on Fluids and Deformable River Streams   **265**

of the particle's motion. As presented in the previous section, stereoscopic layouts allow us to constrain the trade-off between in-plane and out-of-plane motions that in high-density particle flows (where a PIV approach is used) can reduce the estimation accuracy, a phenomenon that does not occur in PTV approaches since a single particle is tracked in low-seeding density flow cases.

(a) Block matching approaches: This scheme employs block matching motion estimation techniques to extract initial displacement estimates. Correlation metrics such as the sum of squared differences (SSD) or the mean squared error (MSE) are used in order to find the best candidate block corresponding to the reference block.

$$
\begin{aligned}
SSD &= \sum_{i=0}^{N-1} \sum_{j=0}^{N-1} (f(i,j,t) - f(i+u,j+v,t+1))^2 \\
MSE &= \frac{1}{N^2} \cdot \sum_{i=0}^{N-1} \sum_{j=0}^{N-1} (f(i,j,t) - f(i+u,j+v,t+1))^2
\end{aligned}
\tag{10.2}
$$

where N corresponds to the block size, $f(i,j,k)$ is the intensity of pixel (i,j) of the k-th frame and (u,v) the pixel displacements at the subsequent frame.

After the initial flow field estimation, particle identification is performed by isolating the particle region and then associating and updating the optical flow field associated to the particle. Along these grounds, Stitou and Riethmuller [65] presented a PTV algorithm where particle extraction is based on intensity-level thresholding determined by the local intensity distribution. As a first step, velocity distributions are computed using a cross-correlation scheme within blocks viewed as grid points followed by particle matching. In other words, the procedure begins with the splitting of the image into a number of blocks, then it formulates the grid and the estimation of the displacement at each of the centered pixels of each block, known as grid points, is performed. The associated velocity of a particle linked to multiple grid points is derived by interpolating the velocity estimates of these grid points in the neighborhood of the particle, as illustrated in Fig. 10.6.

(b) Differential methods: Differential methods incorporate variational optical flow approaches that rely upon the preservation of intensity assumption, such as the method proposed by Lucas–Kanade ([48], [10]) or the one proposed by Horn–Schunck [35] to estimate the optical flow field of a region. Differential methods are based on the optical flow constraint equation (OFC) derived from the assumption that intensity structures of local, time-varying image regions are approximately constant under motion for at least a short duration:

$$
\nabla \mathbf{I} \cdot \mathbf{u} + I_t = 0 \tag{10.3}
$$

where I defines intensity, $\nabla \mathbf{I} = (I_x, I_y)$ the spatial gradient, I_t the temporal gradient, $u = (u,v)$ the optical flow vector.

In such methods, the estimation of optical flow field is based on the computation of spatial and temporal image derivatives that allow the estimation of the optical

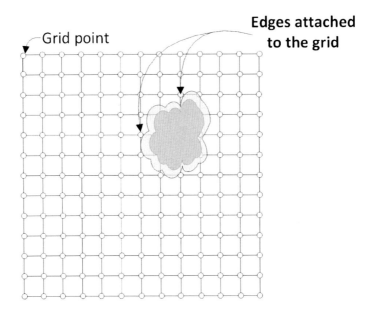

Figure 10.6: Grid points involved in particle velocity estimates. The object's motion is associated with the velocity estimates computed at the grid points lying on the particle edges.

flow field $u = (u,v)$ through the minimization of an energy based cost function. However, when relying on image derivatives for the estimation of the optical flow field we end up with an ill-posed problem due to the existence of the so-called aperture problem [13] which allows the computation of the optic flow component normal to image edges. In differential optical flow methods, this ill-posedness can be overcome through the use of smoothing techniques and assumptions. Based on these, two philosophies have been developed: (a) Local approaches that use spatial constancy assumptions on the optic flow field (e.g., Lucas–Kanade method), and (b) Global approaches that add an initial assumption on the solution of the optic flow field that constrains the estimate with a regularizing smoothness term.

In order to incorporate these methods in the particle tracking concept, existing approaches utilize the image-gradients to define particle descriptors that allow the particle to be isolated and tracked through the frame series with the use of cross-correlation metrics ([64], [34], [12]). In such approaches, the particle's motion field is defined by the corresponding motion vectors extracted via the differential methodology that was used to estimate the optical flow.

As a representative example, Shindler et al. [64] presented an image gradient driven particle tracking method that utilizes a differential optical flow estimation method to estimate the particle's optical flow field. Specifically, in order to estimate the optical flow field of the examined region the differential method (a modified version of Lucas–Kanade's method) solves the optical flow equation using sum-of-

Methods for Estimating Optical Flow on Fluids and Deformable River Streams **267**

squared-difference (SSD) of the pixel intensity levels between interrogation windows in subsequent frames. The velocity estimates are used to define a particle search region in the subsequent frame and moreover, the second derivatives of the pixels in both regions are used to extract corner features that will describe the particles and allow their identification.

The cost functional SSD over the windowed region W, between the reference image I_1 at time t and the candidate image I_2 at time $t + s$, given that the two regions are related with a motion field $\mathbf{u} = (u, v)$ is defined as:

$$SSD(u) = \int \int_W (\mathbf{I}_2 - \mathbf{I}_1)^2 \, dx$$

Minimizing the SSD metric with respect to $\mathbf{u} = (u, v)$ and following the Taylor-series based analysis presented in Lucas–Kanade optical flow estimation method we end up with a least squares estimation scheme:

$$\begin{bmatrix} \int \int_W I_x^2 \, dxdy & \int \int_W I_x I_y \, dxdy \\ \int \int_W I_x I_y \, dxdy & \int \int_W I_y^2 \, dxdy \end{bmatrix} \cdot \mathbf{u} + \begin{bmatrix} \int \int_W I_x I_t \, dxdy \\ \int \int_W I_y I_t \, dxdy \end{bmatrix} = 0 \qquad (10.4)$$

This formulation written in the form of matrices leads to the estimation of the optical flow field by solving a least squares problem for the motion vector u:

$$\Leftrightarrow \mathbf{G} \cdot \mathbf{u} + b = 0$$
$$\Leftrightarrow \mathbf{u} = -\mathbf{G}^{-1} \cdot b \qquad (10.5)$$

where G is a Harris matrix and b is the mismatch vector. The system of linear equations can be solved provided that the matrix G is invertible and both eigenvalues of the matrix are real and positive (requirement for good particles with definite corner points).

In order to isolate and track the particle in subsequent frames Shindler et al. [10] utilize the local displacement to define a search region. The particle is defined and isolated through the extraction of its corner points, based on the region's Harris matrix. Finally, to deal with overlapping particle cases Shindler defined a minimum distance between the extracted features and their corresponding barycenters' driven by knowledge of seeding density characteristics.

This class of particle-based methods combines the advantages of differential method with the particle rigidity to improve the estimation of the optical flow field. The solution of the optical flow equation through the differential approaches enables the calculation of the local velocity vector that can be used as a predictor for the successive step of particle pairing through frames. This reduces the search region resulting in significant improvement of the computational cost.

10.2.2 PROBABILISTIC METHODS

Another approach for estimating the optical flow field of fluids relies on the stochastic modelling of the intensity displacement field within the optical view, often using a Bayesian framework. The majority of the inference techniques consider Gaussian models describing the prior and likelihood information given the observations ([17], [30], [31], [43], [44], [9]). The main idea of these techniques is that the motion vector of a pixel at a certain time is considered as a random variable with a probability distribution function, so that we can use a conditional model to connect the image intensity, the unknown velocity field, the prior motion assumptions and the motion likelihood models.

The estimation of the motion field $\hat{u}$ essentially calls for the estimation of posterior probability of the motion field $p(\mathbf{u}|\mathbf{I})$ given a function $I(\cdot)$ describing the image data. According to the Bayesian framework the posterior probability is determined by two factors, the likelihood or conditional probability $p(\mathbf{u}|\mathbf{I})$ and the prior probability $p(\mathbf{u})$.

Conditional Probability Models

The conditional probability in the case of motion estimation describes the observed data given the actual realization of the underlying motion field. The crucial step in this formulation is the selection of the appropriate function ϕ for the observed data representation. One way to define ϕ follows a stochastic formulation, by assigning a probability of selection for each destination position of the candidate region in the next frame. Chang et al. [17], for example, sets as a discrete probability density function that describes the probability of a pixel (x_s, y_s) actually moving to the pixel $(x_s + \Delta x_i, y_s + \Delta y_i)$ in the next frame. This data relation between sequential frames leads to the spatio-temporal autoregressive (STAR) model:

$$\mathbf{I}(x_s, y_s, t) = \sum_{i=1}^{D_s} A_i \cdot I(x_s + \Delta x_i, y_s + \Delta y_i, t_s + \Delta t_i) \tag{10.6}$$

where A_i reflects the probability of displacement as a coefficient indicating the correlation degree between the pixel (x_s, y_s) at time t and the pixel $(x_s + \Delta x_i, y_s + \Delta y_i)$ at the destination neighborhood at time $t + \Delta t_i$.

In this form, the function ϕ describing the probability of displacement based on the observed data becomes:

$$\phi(\Delta x_i, \Delta y_i) \approx \frac{A_i}{\displaystyle\sum_{j=1}^{D_s} A_j} \tag{10.7}$$

Using this representation, the likelihood function relating the data given a motion model estimate can be defined by associating the discrete probability function ϕ with a continuous function representing an assumption of the motion model. For example, Chang et al. [17] models the conditional probability in the form of:

$$p(\mathbf{I}|\mathbf{u}) = \prod_{x_s y_s}^{N_s} (\sum_i^{D} (\phi(\Delta x_i, \Delta y_i) \cdot g(\Delta x_s - \Delta x_i, \Delta y_s - \Delta y_i))) \tag{10.8}$$

Methods for Estimating Optical Flow on Fluids and Deformable River Streams **269**

with g being a differential function serving as a description of the data given the motion estimate, ϕ is the function describing the data in a probabilistic formulation, D is the candidate region of displacement for the pixel and N_s is the neighborhood of the pixel.

This formulation imposes a smoothness constraint on the description of the observed data and provides a form of prior weighting on the selection probability of each candidate pixel. Furthermore, the estimation accuracy is highly dependent on the appropriate selection of the differential function in order to avoid diminishing the importance of the displacement probabilities.

Another way to define the function ϕ is to associate motion directly with the intensity conservation assumption. In a more direct representation scheme, Héas et al. [32] defines ϕ based on the optical flow constraint equation (OFC, Eq. 10.3), formulating the data representation as a linear combination of the image temporal discrepancies and an observation operator on a pixel grid:

$$\phi(\Delta x_i, \Delta y_i) = \frac{1}{2} \cdot \sum_{x,y \in \Omega} (I_t(x,y) + \nabla \mathbf{I}(x,y,t+1) \cdot \mathbf{u}(u,v))^2 \tag{10.9}$$

where I defines the intensity, $\nabla \mathbf{I} = (I_x, I_y)$ is the spatial gradient, I_t the temporal gradient, and $\mathbf{u}(u,v)$ is the optical flow vector.

In this approach, the conditional probability (likelihood) is defined in the form of a Gibbs distribution with the function ϕ defining the energy of the observed data:

$$p(\mathbf{I}|\mathbf{u}) = \frac{1}{G_\phi} \cdot e^{-\beta_d \cdot \phi(\Delta x_i, \Delta y_i)} \tag{10.10}$$

where G_ϕ is the normalization constant and β_d represents a free model parameter.

This representation leads to continuous probabilistic representation, similar to Chang's with the difference that the data model now involves an energy configuration scheme instead of a probabilistic-based formulation.

Prior Probability Models

The second term of the Bayesian estimation is the prior. The prior probability reflects the initial knowledge about the motion of the optical flow field. The most common distribution used as such an initial estimate is that the random variable vector $\mathbf{u}$, describing the optical flow field, follows a Gibbs distribution with a probability distribution function as:

$$p(u) = \frac{1}{G_{prior}} \cdot e^{-\lambda_{prior} U} \tag{10.11}$$

with G_{prior} being a normalization constant and U being the energy term describing the motion (usually a linear relation between the partial derivatives of the global optical flow field $\mathbf{u}$).

The appropriate selection of the prior is important, since it acts as a smoothness factor preventing the velocity flow of having abrupt variations and discontinuities.

Optical Flow Field Estimation

The estimation of flow field $\mathbf{u}$ is, as previously mentioned, the estimation of the posterior distribution of motion $p(\mathbf{u}|\mathbf{I})$. In terms of Bayesian formulation, the posterior probability is estimated using the maximum a posteriori rule, which for the case of the optical flow field is formulated as follows:

$$\hat{\mathbf{u}} \propto \mathrm{argmax}_{\hat{u}} p(\mathbf{I}|\mathbf{u}) \cdot p(\mathbf{u}) \qquad (10.12)$$

where $p(\mathbf{I}|\mathbf{u})$ is the conditional probability describing the observed data given the actual realization of the underlying global flow field and $p(\mathbf{u})$ is a probability describing a prior knowledge for the motion.

Under the assumption that the prior motion model estimate follows an exponential distribution we end up with:

$$\begin{aligned}
\hat{\mathbf{u}} &\propto \mathrm{argmax}_{\hat{u}} p(\mathbf{I}|\mathbf{u}) \cdot p(\mathbf{u}) \\
&\equiv \mathrm{argmax}_{\hat{u}} ln(p(\mathbf{I}|\mathbf{u}) \cdot e^{-\lambda U}) \\
&\equiv \mathrm{argmin}_{\hat{u}} - ln(p(\mathbf{I}|\mathbf{u})) + \lambda U \\
&\equiv \mathrm{argmin}_{\hat{u}} L\{\phi(\Delta x_i, \Delta y_i)\} + \lambda U \\
&\equiv \mathrm{argmin}_{\hat{u}} f_{data} + \lambda f_{smooth}
\end{aligned} \qquad (10.13)$$

with $L\{\}$ denoting a functional term based on the conditional distribution and U being an energy based smoothness factor.

The maximization of the MAP rule for the case of exponential motion distributions as in Eq. 10.10 is proportional to minimizing a cost function consisting of a smoothness function term and a data function term depending on ϕ. Overall, the minimization criterion can be written in the form:

$$C = f_{data} + \lambda f_{smooth} \qquad (10.14)$$

where f_{smooth} is the smoothness term following the prior motion distribution and f_{data} is the data term describing or penalizing the observed data with λ being the Lagrange multiplier.

The data term f_{data} is essentially a function describing the relation between the observed data based on the underlying flow field. In terms of the classical optical-flow formulation, this term describes errors in the rate of change in image brightness given the estimated motion field. The smoothness term f_{smooth} describes the prior knowledge about the motion of the optical flow field and imposes a smoothness constraint on the flow. Following the two approaches presented for the formulation of the conditional probability function, we can observe that both result in the minimization of a cost function reflecting different forms of MAP estimation.

The probabilistic formulation leads to a global motion-field estimate since the final MAP formulation relates the likelihood models of all image pixels and leads to a highly dense motion field. However, Chang's scheme may also be manipulated to derive local motion estimates for pixel blocks, by using the displacement probabilities A_i in Eq. 10.10 for each neighborhood. In this form, the local motion estimate is associated with the pixel position in the candidate neighborhood of the subsequent frame with the highest probability of displacement A_i. The difference between global and local schemes is that the former provides smoother and more robust flow estimates, while the latter scheme results in a loss of overall detail restricting the estimates only to an average block level. On the other hand, by restricting to nearby object boundaries the local scheme provides more accurate results for occluded cases and boundary motion, making it ideal for particle tracking.

Models Increasing Estimation Accuracy

The Bayesian probabilistic scheme relies on the appropriate selection and formulation of the prior and likelihood models. Thus, to increase the accuracy of estimation it is essential to search for means of improving the estimation of hyper-parameters and/or the likelihood describing the conditional model. Towards this direction, various approaches have been proposed that involve the use of hierarchical schemes that derive initial estimates for the motion field that are then fed to a hyper-parameter minimization scheme that consists of motion models inspired by statistical and physical properties of the fluid motion (e.g., assign parabolic families of Gaussian distribution to each candidate pixel) and finally, estimates the hyper-parameters by describing the deviations produced by observation outliers and motion discontinuities ([30–32]). The addition of fluid property inspired motion models forms a link to the next section that introduces the fluid's properties in the motion estimation model.

10.2.3 DIFFERENTIAL METHODS GUIDED BY FLUID PROPERTIES

In differential optical flow estimation techniques, the image domain is assumed to be continuous and the optical flow is computed using the spatiotemporal derivatives of the pixel intensity. The partial derivatives of the intensity, with respect to the spatial and temporal coordinates, are used to express the energy channeling between the frames. In association with probabilistic techniques, the model of intensity dynamics is derived from the conservation of mass or the optical density flow in space and time. Differentiation over space and time derives a term in the form:

$$f_{data} = \nabla \mathbf{I} \cdot \mathbf{u} + I_t \tag{10.15}$$

which should be bound to zero. Associated optimization methods derive dense optical flow fields by minimizing a cost function derived by the optical flow constraint equation f_{data} and a global constrain function f_{smooth}:

$$\min \int \int_{x,y} f_{data}^2 + \lambda^2 \cdot f_{smooth}^2 \, dxdy \tag{10.16}$$

A further strength of differential techniques is the potential for adding more constraints into the optimization framework, which associate to physical or dynamic phenomena. Considering the case of fluid flows, the basic idea is to consider that the density of features in the flow is altered by the motion in a local level. This allows us to incorporate the properties of fluid mechanics, such as conservation of mass, or fluid models describing fluid phenomena, such as wave generation models, in order to model and constrain the image brightness variation:

$$\nabla \mathbf{I} \cdot \mathbf{u} + I_t = \text{Fluid Property or Model}$$

The optical flow constraint equation is now modified, containing an additional term which describes the divergence of the optical flow based on the affine transformation model. We can again estimate the optical flow field, in a global way, by minimizing the same cost function in which now the optical flow constraint equation satisfies the constraints denoted by the fluid properties.

Based on whether the brightness variation constraint is based on fluid properties or in fluid motion phenomena we can discretize the methodologies into two categories: (1) methods based on properties of fluid mechanics and (2) methods based on the physical properties of waves. The first class, as is obvious, uses properties such as the mass conservation constraint, whereas the latter uses wave generation theory to model the changes in the image brightness.

Methods Based on Fluid Mechanics

The physical properties of fluid mechanics, such as mass or brightness conservation assumptions can be used as constraints for the fluid motion ([18], [79]). In terms of image intensity, one of the most used fluid mechanics driven properties is the conservation of mass assumption with significant work presented by Schunck [63], Wildes et al. [79] and Nakajima et al. [54].

The basic idea behind this class is based on the fact that the density of the fluid $p(x,y,z,t)$ can be associated to the 3D velocity field $\mathbf{V}(x,y,z,t) = (U(x,y,z,t), N(x,y,z,t), W(x,y,z,t))$ using the conservation of mass assumption [72]:

$$\nabla(p \cdot \mathbf{V}) + \frac{\partial p}{\partial t} = 0 \tag{10.17}$$

The image intensity values are associated with the density of their corresponding object as:

$$I(x,y,t) = \int_{z_1(x,y)}^{z_2(x,y)} p(x,y,z,t)\,dz \tag{10.18}$$

where z_1 and z_2 are the surface boundaries of the object being imaged. By imposing the surface boundaries on Eq. 10.17, based on the work presented by Fitzpatrick [23], we can express it in terms of the image intensity as:

$$\nabla_{x,y}\mathbf{I} \cdot \mathbf{u} + \frac{\partial}{\partial t}I = -[p \cdot n \cdot \mathbf{V}_{z_1}^{z_2}] \tag{10.19}$$

in which V is the 3D velocity estimate, p is the density and u is the 2D optical flow field derived from the weighted average, of the initial 3D velocity field V with the

Methods for Estimating Optical Flow on Fluids and Deformable River Streams **273**

density p as:

$$\mathbf{u} \equiv \frac{\int_{z_1}^{z_2} p\mathbf{V}_{x,y}\,dz}{\int_{z_1}^{z_2} p\,dz} \qquad (10.20)$$

Equation 10.19 is known as the continuity equation denoting the association of the 3D fluid flow with the 2D flow derived from the image under the assumption that the conservation of mass law is satisfied.

Applying the conservation of mass flow continuity equation to a temporally varying image yields:

$$I_x u + I_y v + I u_x + I v_y + I_t = 0 \Leftrightarrow \nabla_{x,y} \mathbf{I} \cdot \mathbf{u} + I_t = 0 \qquad (10.21)$$

where $\mathbf{u}(u,v)$ is the optical flow field.

In fact, the continuity equation is enhanced with additional physical constraints along with a smoothness velocity constraint, to further restrict the velocity motion field and ameliorate the effects of noise ([79], [54]). For example, Wildes et al. [79] imposes an additional smoothness constraint c_s to the continuity equation c_c and then follows the methodology of Horn–Schunck [35] to minimize the combined constraint equation with respect to the flow parameters (u,v):

$$\min \int\int (k \cdot c_c + c_s)\,dxdy \qquad (10.22)$$

Methods Based on the Physical Properties of Waves

The second approach to incorporate physical properties is by attributing the brightness alteration of the fluid flow to wave phenomena ([8, 53, 61, 62]). Along these lines, Schunck [63] presented a model for indoor wave generation combining a sinusoidal wave equation with pattern matching to estimate a regular wave change. Saikano [61], on the other hand, presented a wave-based optical flow framework for fluid observation in which a wave generation equation is used to model the image brightness changes in a multidirectionality irregularity (MI) model:

$$I_w = \sum_{m=1}^{M} a_m^* \cos(k_m^{x^*} x \cos \theta_m^* + k_m^{y^*} y \sin \theta_m^* - 2\pi f_m^* t + \varepsilon_m) \qquad (10.23)$$

where I_w is the image intensity at the pixel (x,y) as defined from the multidirectional model, a_m^* is the amplitude, $(k_m^x, k_m^y)^*$ are the wave number components, f_m^* is the frequency, θ_m^* is the orientation, ε_m^* is the noise, and M is the number of the cosine functions used to describe the wave.

The multidirectional irregularity (MI) model is incorporated into the optical flow equation to explain deviations in the estimated brightness changes through the temporal derivative of image intensity. Under this assumption the spatiotemporal variation of optical intensity must now follow the variation of the expected intensity following the wave properties ([29]):

$$\nabla \mathbf{I} \cdot \mathbf{u} + I_t = \frac{d(I_{wave}(x,y,t))}{dt} \qquad (10.24)$$

The estimation of the wave-related parameters and the optical flow components is performed through the minimization of an objective function consisting of three terms, one referring to the observed data and two smoothness constraints, one for the wave model and one for the optical observations ([61], [62]):

$$E(\mathbf{u}, \text{wave parameters}) = f_{data} + f_{smoothImage} + f_{smoothWave} \qquad (10.25)$$

The characteristic difference and advantage of this method is that discontinuous motion in images with inhomogeneous brightness is estimated in a visually plausible way reflecting these expected discontinuous motion patterns, whereas the methodologies based on brightness conservation can estimate rather smooth and uniform motion.

10.3 COMPARISON AND EXPERIMENTAL RESULTS

10.3.1 METHODOLOGY DESCRIPTION AND EVALUATION

Comparing different algorithmic schemes in the case of fluid motion poses a challenging task, since there are several issues to consider regarding the different forms of motion, as well as the various optical or mechanical assumptions of the algorithms. Some approaches rely on the existence of particles in the flow whereas others introduce knowledge on the properties of the fluid examined. In this section, we attempt a comparison of algorithms applied on similar problem formulations. Since the focus of this study is on optical surveillance, we restrict comparison to algorithms dictated by the optical surface information rather than the physical properties of the medium. The latter could be used as an addition property influencing the optical flow constraint towards improving the estimation results in a similar form for all optical-based algorithms considered. In order to perform a fair comparison, we need to define a set of features that will determine the distinction among classes and illustrate the advantages or disadvantages of each method. Towards this direction, we select the following features to characterize the algorithmic performance: (a) the need of particles or tracking features, (b) whether the resulted estimate optical flow field is dense or sparse, (c) whether the method operates on a global or local basis, and (d) the extent of displacement that can be identified and how it is dependent on the interrogation window. A qualitative comparison based on these features is presented with quantitative comparisons of selected algorithms on the same realistic framework pertaining to the estimation of fluid flows to follow.

The first class we have defined considers particle-based methods. Such approaches clearly rely on the existence of particles or other natural features present in the flow, such as foam formations. The motion field is found by tracking these features/particles through the frame series. Since the majority of methods in this class employ block matching, the resulted flow field is sparse. Even for the case of hybrid methods, such as the one of Shindler et al. [64] that uses dense motion models, the particle dependence leads again to sparse flow fields. The block-matching methodologies used in the PIV class are heavily dependent on the window size, implying severe restrictions to the amount of displacement that can be identified. This limits the accuracy of the particle dependent methodologies for flow cases expressing large displacement vectors. Furthermore, the window size depends on the particle size and the seeding density, which in turn depends on the flow characteristics (flow speed, the flow scene, wind phenomena, etc.). Finally, the fundamental assumption of such methods requires that the particle's motion pattern follow the motion of the fluid.

The second algorithmic class relies on intensity-defined probability density functions (pdfs) describing the image data. This allows the derivation of denser optical fields compared to the previous class. The probabilistic methods are also bound to windowed regions for the extraction of the pdfs. In some methods, such as Bacharidis et al. [9], the displacement identification range is affected by the window size. A solution to this problem is the use of a series of frames instead of a frame pair in the pdf estimation step [17]. This further enhances the displacement range allowing the estimation of the optical flow field for both small and large displacements. Moreover, despite the pdf estimation on a local window for each pixel, the estimation process proceeds in a global way utilizing information from all available regions and extending a global optimization framework to the entire image. This global nature allows the estimation of a unique velocity vector everywhere, as a characteristic of the fluid flow. In essence, this class approaches the OFE from a stochastic perspective with the aim of modelling its parts in relation to the prior knowledge of the motion field and the conditional adjustments given the observations. The solution can be performed in a local scale by conditioning the motion within the neighborhood or at a more global one by letting the conditioning model extend to the entire motion field. Thus, the solution schemes can resemble the local approach of Lucas–Kanade [10] or the more global scheme of Horn–Schunck [35].

The third class is also particle independent and relates the differential optical flow estimation with fluid properties. It is apparent that the restriction here is the knowledge of material parameters, but the overall estimate is consistent with the actual fluid motion. The flow field derived is dense and global, since the relation between frames is achieved through the optical flow constraint bounded from the fluid properties of the entire medium. If we compare this with the aforementioned classes we can readily infer that, if applied correctly with the proper assumptions, it surpasses the previous two in terms of estimation coherence. However, it is case-dependent implying strong dependence on the prior knowledge of fluid characteristics. Furthermore, the global mechanics relation to the optical flow constraint leads to complicated minimization schemes.

Overall, the particle-based methods are much simpler to implement but the estimate is constrained by the interrogation region considered, a factor that enforces locality on the estimate instead of a global interconnected flow field. The probabilistic methods treat the OFE in a stochastic way that may effectively adjust the data-dependent and regularization metrics. This class forms a good compromise between the restrictive requirements of particle-based schemes and the increased fluid mechanics conditions of the differential fluid-mechanics methods. A comprehensive overview of the advantages and disadvantages of each class can be found in the following Table 10.1.

Table 10.1

Method class theoretical basis comparison

Method Class	Description	Advantages	Disadvantages
Particle image based	Flow field is estimated by tracking particles/features floating in the fluid through the frame series. The displacement of the particle corresponds to the motion field of the fluid.	*Simple and quite accurate. *Is easily combined with other motion estimation method classes boosting its accuracy.	*Need of particles. *Particles' features affect the accuracy. *Accuracy is, also, window- size dependent, i.e., reduced accuracy for large displacements. *Local method, sparse flow field.
Probabilistic based	Flow field is estimated using a Bayesian minimization scheme based on probability density functions derived from the image data.	*No need of particles in the flow. *Global and local coherence of the motion estimates. *Can identify small and large displacements.	*Accuracy depends on a priori knowledge and the minimization success of the cost function. *Increased computational cost, for the cost function minimization task.
Differential physical property based	Flow field is estimated by minimizing a differential scheme derived from the bounding of the optical flow constraint equation from the physical properties of the fluid. The physical properties are used to describe the variation of the image intensity.	*No need of particles *Highly accurate, relates the fluid properties with the estimated motion field. *Global method, motion vectors are inner-related.	*Accuracy depends on the a priori knowledge of the fluid properties. *Increased computational cost due to the minimization of a multiparameter model.

10.3.2 EXPERIMENTAL RESULTS

We proceed now with the actual testing of realistic problems. We focus on estimation schemes relying on the optical formulation of motion, including methods from the first two categories. In particular, we consider three basic types of fluid motion that can be encountered in both laboratory and real-world experiments. The motion models used to formulate the experimental dataset consist of a simple linear motion, a vortex pair rotational motion pattern and a membrane-like motion arrangement. These motion patterns are formulated synthetically by applying the appropriate motion function to a set of moving particles. The generation process follows the synthetic motion generation scheme based on Kalman filter, which is implemented in the PIVLab tool ([71], [70]). The flow fields are seeded with particles of 3-pixel diameters, with a size variation of ± 0.5 pixels, having a displacement range from 0 to 4 pixels. Synthetic images are used as "ground-truth" patterns, since the flow motion pattern and the particle seeding are well defined. In particular, ground truth motion fields are derived from the motion equations used to create the synthetic images.

The algorithms compared are (a) a Gaussian weighted Lucas–Kanade method ([48], [10]), (b) the classic Horn–Schunck method [35], (c) a probabilistic approach [9] using normal pdfs, and (d) a local particle image velocimetry tool (OpenPIV) ([68], [1]) that utilizes a cross-correlation algorithm with shifting windows to accommodate large displacements and reduce the effect of out-of-plane motion (see Section 10.2.1). For the probabilistic approach, we have used two implementation schemes. The first is a local based approach in which the motion field is estimated through the probability of displacement A_i, as derived from the local density function ϕ (Eq. 10.7), by simply setting as the displaced position the position associated with the highest probability. The second is a global scheme that estimates the global motion field through the minimization of a cost function derived from a MAP formulation (Eq. 10.14).

In every experimental case, we have used a macro block formulation of 16 x 16 pixels, leading to a sparser and easier to comprehend flow field. This was performed so as to first match the first 2 methods with the block based formulated methods (c) and (d). The aim behind the selection of these methods is to illustrate the difference between the fluid directed methods from the classical optical flow approaches as well as the difference between local and global philosophies in the estimated optical flow field.

The evaluation metrics used to assess the accuracy of each method are the angular error (AE) and the endpoint error (EE) [11]. The angular error is determined by the dot product of the motion vectors divided by the product of the motion-vector lengths, followed by the inverse cosine transformation; it depicts accuracy of the directional information on the estimated motion field. On the other hand, the endpoint error is defined as the absolute squared error between the estimated and the ground truth motion vectors; it primarily depicts the magnitude accuracy for the estimated motion field. We evaluate each method's accuracy based on the average and standard deviation (STD) of these error metrics over the entire image, which are presented in the following tables.

Results

The test dataset considers fundamental motion patterns encountered in the fluid motion. The motion patterns considered simulate simple linear displacements, a Rankine vortex formulation containing a pair of vortices and an image pattern representing a membrane fluidity case. Figures 10.7(a), 10.7(b), 10.7(c) show an image example of the dataset, along with the ground truth flow fields implemented in the dataset.

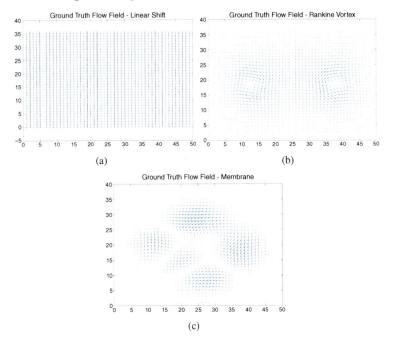

Figure 10.7: The ground truth motion vectors for first dataset as derived based on the motion pattern generation equations used in the PIVLab tool ([71], [70]). (a) Ground truth flow field for the linear shift case, (b) Ground truth for the Rankine vortex pair case, and (c) Ground truth for the membrane case.

The following figures present the estimated flow vector fields for each algorithm tested. Each figure is followed by a numerical table presenting the average and standard deviation AEs and EEs achieved by each method, respectively.

The linear case motion implies small deviations in amplitude in the range of $[0,3]$ pixels. In order to allow for estimating such small deviations, the smoothing factors in all methods are not as severe as they should be in uniform motion. From the results, we observe that the local probabilistic as well as the local particle-based method are superior in estimating the velocity field both in direction and magnitude compared to the differential approaches. More specifically, we observe in Table 10.2 that the average angular and endpoint errors for particle-based and probabilistic methods range from 2 to 5 degrees and from 0.17 to 0.4 pixels, respectively, implying good estimation efficiency. The local versions of probabilistic (Figure 10.8(a)) and particle-based (Figure 10.8(e)) methods preserve the discriminant information of the region and thus result in high fidelity estimates. The global probabilistic approach shows a varying motion field (Figure 10.8(b)) resulting in increased error rates compared to its local version. This is primarily due to convergence in local minima created from

Methods for Estimating Optical Flow on Fluids and Deformable River Streams 279

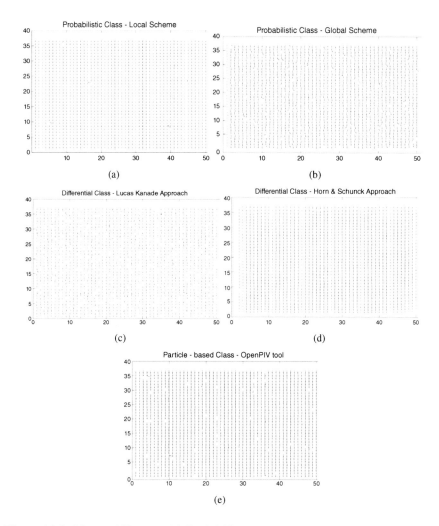

Figure 10.8: Linear shift case; (a) Probabilistic approach with a local implementation, (b) Probabilistic approach with a global implementation, (c) Weighted Lucas–Kanade, (d) Horn–Schunck, and (e) OpenPIV tool.

the randomly varying nature of the motion field. This is a characteristic problem associated with global optimization techniques [59].

Differential approaches do not perform well in recovering magnitude information and result in significant underestimation of the motion effects. The weighted Lucas–Kanade method shows the worst performance in both amplitude and direction estimation, where the error is increased compared to either the global Horn–Schunck method or any other examined algorithmic class. This is due to the locality of the implementation scheme with the use of small window size or small extent of the

Table 10.2
Evaluation metric results for the linear shift motion case

Linear shift	Average AE (degrees)	Average EE (pixels)	Std AE	Std EE
Probabilistic - Local	1.5339	0.1732	3.8183	0.1693
Probabilistic - Global	4.8251	0.3719	4.1595	0.1613
Differential- Weighted Lucas–Kanade (local)	8.2863	0.9624	4.6570	0.1090
Differential- Horn–Schunck (global)	2.8214	0.5508	0.8821	0.1052
Particle-based OpenPIV tool (local)	1.8335	0.2338	3.8095	0.2470

Gaussian weighting function, which forms a trade-off between uniform and locally varying estimate. Thus, it appears that the small size of the estimation window and the weighting (smoothing) function has beneficial effects only in abrupt and varying motion fields with large deviations, where it effectively imposes a constraint factor and increases the estimation accuracy. The global form of Horn–Schunck's method allows better preservation of the directional information of this (slightly varying) flow field, yielding low angular errors with a small deviation factor. Notice that this non-iterative global scheme preserves more consistent angular information than its iterative global counterpart in Figure 10.9(b). However, from Table 10.2 we observe that Horn–Schunck's method is inferior to the recovery of magnitude information compared to the probabilistic and particle-based methods, due to the smoothness of the flow field which overweights the motion amplitude at each region.

In summary, the local probabilistic approach attains the best performance, showing a relative angular error below 10% and an average endpoint error of 0.17 pixels. Moving a bit further, we observe that local approaches generally provide better estimates than global ones, which is mainly due to preservation of the local region characteristics and the low noise presence in our test cases. However, one should be aware that local schemes may result in reduced estimation accuracy when the selection of the neighborhood size fails to capture the range of motion and noise is present. In non-ideal cases where the noise effect is large, the local methods will lose accuracy and will perform worse compared to the global ones.

For the second test of a rotational motion pattern with two vortexes, the estimated motion field varies smoothly with a diminishing effect in magnitude while approaching the vortex's center. The particle-based methods (Figure 10.9(e)) attain more accurate estimates compared to the other techniques, succeeding in depicting both the direction variation as well as the magnitude change of the vortex pattern. The average and endpoint errors remain low showing a highly accurate estimated field, with

Methods for Estimating Optical Flow on Fluids and Deformable River Streams   281

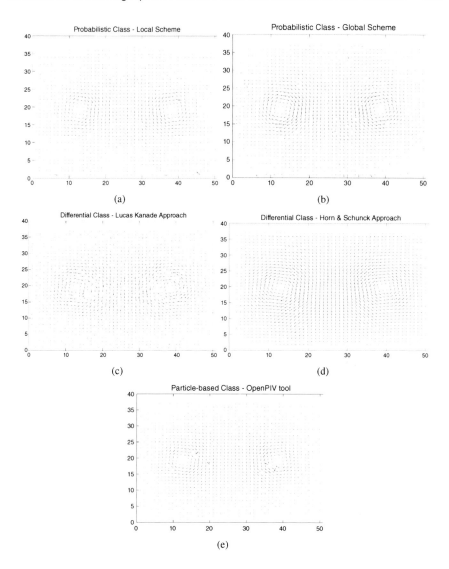

Figure 10.9: Rankine vortex pair case; (a) Probabilistic approach with a local implementation, (b) Probabilistic approach with a global implementation, (c) Weighted Lucas–Kanade, (d) Horn–Schunck, and (e) OpenPIV tool.

the error attributed to a small number of incorrect estimations which, however, show large deviations. The local probabilistic approach (Figure 10.9(a)) follows second in performance taking into account both angular and endpoint errors, but with almost twice the error compared to particle-based schemes. Nevertheless, the standard deviation of probabilistic schemes is smaller implying that the estimates are more

Table 10.3

Evaluation metric results for the vortex pair of the rotational motion case.

Linear shift	Average AE (degrees)	Average EE (pixels)	Std AE	Std EE
Probabilistic - Local	10.3751	0.4530	4.0843	0.1297
Probabilistic - Global	12.8997	0.3997	4.2051	0.1320
Differential-Weighted Lucas–Kanade (local)	13.0720	0.4999	6.9035	0.3148
Differential-Horn–Schunck (global)	10.0056	0.5989	3.2550	0.3860
Particle-based OpenPIV tool (local)	5.4748	0.2172	5.0570	0.8076

compound. This is a useful algorithmic attribute in more random cases where the estimated displacement is expected to show large deviations affected by noise. However, when expanding to global estimates with the probabilistic approach, we observe a smoother flow field (Figure 10.9(b)) but with increased angular and endpoint errors, possibly due to non-optimal selection of the global minimization scheme parameters. This is a clear demonstration of the importance of minimization parameters to the convergence process. Nevertheless, we notice that the majority of the motion vectors are estimated accurately, preserving the pattern morphology in terms of magnitude and direction compared to the corresponding ground-truth motion vectors. The differential approaches show increased estimation accuracy compared to the previous cases. Horn–Schunck's method (Figure 10.9(d)) manages to retain the directional information of the flow, ranking second in the angular error estimation, however, it fails to accurately estimate the magnitude of the motion vectors yielding the highest endpoint error. The reason behind this deviation can be attributed to the regularization term, which provides a smoother and coherent flow field but fails to retain the motion detail of each region. The locally weighted Lucas–Kanade scheme on the other hand, manages to counterbalance the smoothness loss with an accurate magnitude estimate. The directional information loss shows an angular error of 13 degrees and the magnitude information loss an endpoint error of 0.5 pixels. The weighting factor and the local relation between neighboring pixels utilized by the Lucas–Kanade approach manages to recover the overall motion in magnitude and directional aspects (Figure 10.9(c)). The increased error of the weighted Lucas–Kanade method compared to the local probabilistic and particle-based methods can be attributed to pixels near the vortex center, which are characterized by more abrupt velocity changes compared to their neighboring pixels which has a negative impact on the motion estimate, reducing its accuracy.

Methods for Estimating Optical Flow on Fluids and Deformable River Streams **283**

Overall, the winner of this test case is the particle-based method which by exploiting the increased local spatiotemporal consistency in the neighborhood of a "clear" particle attains good performance with a relative error of almost 10% and an average endpoint error of 0.2 pixels. However, in cases where the region's characteristics are not discriminant, e.g. noise presence, we can expect the particle-based method and the local based methods to be less accurate compared to the global schemes since they are more susceptible to noise.

Table 10.4

Evaluation metrics results for the case of a membrane motion pattern

Membrane	Average AE (degrees)	Average EE (pixels)	Std AE	Std EE
Probabilistic - Local	10.7231	0.2874	2.0170	0.1648
Probabilistic - Global	13.0046	0.2915	4.7230	0.1371
Differential- Weighted Lucas– Kanade (local)	12.5725	0.2675	1.8950	0.1097
Differential- Horn–Schunck (global)	20.9866	0.5029	5.0894	0.2584
Particle-based OpenPIV tool (local)	5.7395	0.1428	2.1470	0.2308

The final motion pattern, i.e. the membrane motion, is similar to a wave-like motion exhibiting large motion deviations within each region. For this motion case, the highest accuracy was achieved by the particle-based method, with the probabilistic approaches performing second and the differential methods following in evaluation. The particle-based approaches show an angular error of about 6 degrees and an endpoint error of 0.14 pixels, which as before are attributed to specific wrong particle estimates that reach high error values affecting both the error metrics and the visual result of the motion field (Figure 10.10(e)). On the other hand, the local probabilistic methods show a more compound motion field (Figure 10.10(a)) in terms of magnitude deviation, but have about twice as high of an angular error when compared to the particle-based approach. Thus, we can say that the wrong estimates for the local probabilistic case are both in magnitude and in direction, but with their values do not show large deviation from the ground truth motion. The global probabilistic scheme (Figure 10.10(b)) reflects the same problem as before, associated with the convergence parameters. Its smoothness term favors smoother flow field showing a coherent flow pattern similar to the ground-truth, but with a small deviation in estimation accuracy reflected by angular and endpoint measures. Nevertheless, the motion field is compact with a large percentage of accurate estimates both in direction and magnitude, as compared to differential methods. Differential approaches perform last in this test case with the Lucas–Kanade local scheme being slightly better (Figure

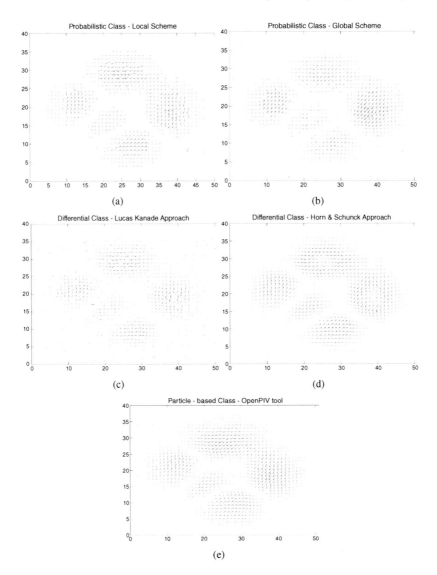

Figure 10.10: Membrane fluidity case: (a) Probabilistic approach with a local implementation, (b) Probabilistic approach with a global implementation, (c) Weighted Lucas–Kanade, (d) Horn–Schunck and (e) OpenPIV tool.

10.10(c)), but with errors similar to the global probabilistic method. However, we observe that the resulting motion field is denser and the deviation in the angular error is smaller compared to the global probabilistic scheme. Horn–Schuck's scheme (Figure 10.10(d)) provides the worst estimate due to the fact that its global scheme along with the smoothness constraint enforce strict coherence between regions in the form of smoothness, leading to a continuously diminishing motion estimate towards the center of each dense membrane.

Summarizing the result section, the experimental results indicate that problem-oriented methods (OpenPIV tool and statistical methods) provide more accurate and stable results for the tested motion patterns compared to classical optical flow methods. However, the PIV techniques require the existence of either natural or artificial particles to be tracked over successive frames. The existence of illumination consistency over frames is also essential in traditional optical flow methods, such as the Lucas–Kanade and Horn–Shunck methods. The parameters in these methods can control the accuracy of either angular or magnitude estimation, but in a conflicting fashion; improving magnitude deteriorates angular estimation. In more general environments, the statistical methods present advantages in modelling the motion field in terms of both angular and magnitude accuracy. In particular, the local implementation is more flexible, with the global one depending heavily on the convergence parameters. In general terms concerning local or global scheme selection, all three test problems signify that the former outperform the latter implementations. The locality allows different regions to maintain illumination information, an important aspect for the estimation of non-rigid and multidirectional motion patterns. In addition, if we want to take advantage of existing tracers in the flow, especially for particle based methods, we must efficiently address the problem of occluding objects in the flow. In this case, a local approach can lead to more accurate estimates exploiting the fact that the objects have different orthogonal motion components allowing their discrimination. On the other hand, global differential methods often fail to distinguish occluding objects because they use the spatial gradients of the objects, which are similar for each one of the occluding objects.

Despite the above properties, as mentioned earlier, the local-based approaches are more susceptible to noise compared to global ones. Indeed, global methods are more accurate, especially on the overall smooth-shaped motion, due to the smoothness constraint and the global nature of estimation that allows the motion information to spread over the image domain and cover all the homogeneous regions. Global schemes also derive dense motion fields, in contrast to the sparsely computed motion vectors in local schemes.

In terms of physical properties, a liquid flow creates a unique velocity vector for its main motion everywhere, implying strong relation in the motion of adjacent pixels. This justifies the imposition of a global relation among the motion vectors of pixels; however, the solution of the global model is often obtained under restrictive assumptions that deviate from reality. In our experiments, the global version of the statistical approach imposes such a global relation, yet it attains less accurate results compared to its local version. This inefficiency of the global scheme is due to the existence of overlapping particles in the test scene, as well as due to the non-optimal selection of the parameters used in the minimization of the global cost function. Nevertheless, the global statistical model attains more accurate results compared to both Lucas–Kanade (local) and Horn–Schunck's (global) methods that have been developed based on object rigidity. Table 10.5 provides a summary of the qualitative remarks on the overall performance of the presented methodology classes.

10.4 CHALLENGES IN REAL-WORLD APPLICATIONS

The main motivation for this study was the fact that many of the presented approaches have been tested on real-world applications (and so, in real-world conditions), which reveals the direct impact of the optical flow techniques application to well-addressed monitoring cases. In most cases, preprocessing and post-processing steps are necessary in order to produce accurate estimates, which can be a really challenging procedure since it depends on the specific application of real-world conditions. In this section, we will focus on a real-world application example on river monitoring and optical flow estimation in order to list the main challenges that we have to phase when this application is considered.

In river flow estimation, many parameters affect the correct flow estimation. The type, characteristics, and number of camera sensors to be used, the position and angle of view of the camera(s) at the monitoring scene, and the type of the motion estimation algorithm that is more appropriate for the specific application are only some of the issues that someone has to deal with in order to estimate the river flow in an accurate and effective way. Beyond these issues, usually pre- and post-processing steps are necessary since the monitoring is not performed in a controlled lab space but under exterior conditions.

Figure 10.11 presents an example of a river field setup at Clear Creek near Oxford in Iowa. Bradley et al. [14] performed particle image velocimetry (PIV) methodology for the river flow estimation, a technique that uses simple correlation metrics between the reference frame and each candidate region in order to determine a similarity rate. Similar approaches are found in the studies of Fujita and Komura [24], Tsubaki et al. [72], Hauet et al. [28], etc.

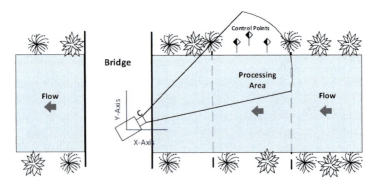

Figure 10.11: Bradley's system layout at Clear Creek near Oxford, Iowa. The selected area is videotaped by a camera laid on top of a bridge [14].

The estimation of the 3D velocities is derived from the particle tracking process

Table 10.5
Summary of the qualitative remarks on the overall performance

Method Class	Linear Motion Patterns		Rotational Motion Patterns		Membrane-Like Fluidity		Remarks on Overall Performance
	Direction ality	Magnit ude	Direction ality	Magnit ude	Direction ality	Magnit ude	
Probabi listic methods	*Local: ~2 degrees AE	*Local: ~0.2 pixels EE	*Local: ~10 degrees AE	*Local: ~0.45 pixels EE	*Local: ~10 degrees AE	*Local: ~0.29 pixels EE	**Local approach** *Preserves region's discriminant information. *Locality can be a benefit for small velocity flows even when abrupt directional changes in motion occur.
	*Global: ~5 degrees AE	*Global: ~0.4 pixels EE	*Global: ~12.9 degrees AE	*Global: ~0.4 pixels EE	*Global: ~13 degrees AE	*Global: ~0.29 pixels EE	**Global approach** *Susceptible to optimization related issues (convergence). *Smoothness factor affects directional and amplitude estimation accuracy. Acts beneficially for higher velocity flow cases leading to smoother motion estimates but with an impact on amplitude estimation.
Particle -based methods	*Local: ~2 degrees AE	*Local: ~0.25 pixels EE	*Local: ~5.5 degrees AE	*Local: ~0.22 pixels EE	*Local: ~5.8 degrees AE	*Local: ~0.14 pixels EE	**Local approach** *Preserves region's discriminant information. *Particle characteristics and occlusion cases affect overall accuracy. *Locality can be a benefit for small velocity flows even when abrupt directional changes in motion occur.
Differe ntial methods	*Local: ~8.5 degrees AE	*Local: ~0.95 pixels EE	*Local: ~13 degrees AE	*Local: ~0.5 pixels EE	*Local: ~13 degrees AE	*Local: ~0.27 pixels EE	**Local approach** *Preserves region's discriminant information. * Adding a weighted smoothness factor in accordance with locality of the estimate will reduce accuracy in small range, smoothed flow cases but can increase accuracy in abrupt cases.
	*Global: ~3 degrees AE	*Global: ~0.55 pixels EE	*Global: ~10 degrees AE	*Global: ~0.6 pixels EE	*Global: ~21 degrees AE	*Global: ~0.5 pixels EE	**Global approach** * Denser and more compound flow estimates with small deviations between the estimated vectors *Smoothness factor affects directional and amplitude estimation accuracy. Acts beneficially for higher velocity flow cases leading to smoother motion estimates but with an impact on amplitude estimation.

(Continued)

Table 10.5

Summary of the qualitative remarks on the overall performance (continued)

Method Class	Linear Motion Patterns		Rotational Motion Patterns		Membrane-Like Fluidity		Remarks on Overall Performance
	Directionality	*Magnitude*	*Directionality*	*Magnitude*	*Directionality*	*Magnitude*	
Motion case and method suitability	**Local methods** show greater accuracy in the estimation of a small to medium velocity range linear motion cases since they tend to show high robustness under noise despite being sparser.

Among the examined local-based approaches the choice lies between the use of a probabilistic and a particle-based approach. The first, although having higher complexity compared to the latter, removes the need of particle-specific motion estimation. | | **Both methods** tend to show weaknesses in rotational flows, due to the abrupt directional changes. The smoothness factor in Global methods manages to preserve amplitude information but has an impact on estimated flow directionality. In local methods directionality is better estimated but with a loss in amplitude accuracy for high velocity flow cases.

Small /medium velocity flows Local approach (increased accuracy when particles present), **but** sparser flow estimates.

High velocity flows Global approach with appropriate selection of the contribution of smoothness factor. | | | | **Local methods** Wave-like motions show large deviations in flow directionality within each region. Local methods manage to preserve both directional and amplitude information in small/medium velocity scaled flows making the ideal for such flow cases.

Small /medium velocity flows Local approach (increased accuracy when particles present), **but** sparser flow estimates.

High velocity flows Global approach will lead to smoother and denser estimates, with lower amplitude estimation loss, **but** higher loss for flow directionality estimation. |

through a projective transformation using control points with known real-world co-ordinates, placed at river banks. These control points are used for the estimation of the model parameters and, so, the estimation of the 3D velocities.

The monitoring conditions of the specific setup reveal the need for pre- and post-processing steps and will be used as a case-study application in order to exploit the most common issues that need to be addressed when we perform optical flow estimation in real-world applications.

One of the issues that we have to deal with is the illumination invariance. Non-uniform field illumination and its variations over the daylight generate background noise that results in weak contrast and dark spots on the images. Without the necessary preprocessing, motion estimation may result in a number of false estimates, while particle identification and tracking will incur erroneous estimation. Both probabilistic and particle-based approaches are prone to uneven illumination.

The simplest approach for the correction of uneven illumination is to perform global histogram equalization [41]. However, in image regions determined by low frequencies, histogram equalization will result in contrast loss. This effect is reduced through local equalization, e.g. adaptive histogram equalization ([41], [42]). A differ-

Methods for Estimating Optical Flow on Fluids and Deformable River Streams **289**

ent approach to deal with uneven illumination is to fit the illumination variation into a mathematical model. The most common assumption is that uneven illumination forms an additive signal, usually modeled as a Gaussian random variable:

$$I(x,y) = I_{ideal}(x,y) + B(x,y) \qquad (10.26)$$

where I is the image under uneven illumination, I_{ideal} is the ideal image with uniform illumination and B is the image background. The background term forms a low frequency signal and essentially denotes the additive signal producing uneven illumination ([74], [36], [66]). However, this approach often results in high variations of contrast among brighter and darker regions. A solution to this is to divide the background image with the original image, resulting in a more even contrast distribution [66].

Another common approach is to use the object motion as a means of illumination invariant metric in fluid flow estimation. This approach does not correct the image for illumination effects but it is valid to have false motion vectors due to the uneven illumination. This is based on the fact that non-uniform illumination may create false motion effects. The methodology manages to separate the pixels that appear to have motion due to instant illumination variation from those that follow an actual motion scheme [6]. The kurtosis metric is used for this task that equals zero for normally distributed data, while high values are assessed to pixels with actual motion.

Finally, a common issue that we have to deal with when optical flow estimation is performed in real-world applications and that is worth-mentioning is the existence of erroneous motion estimates, that often occur and need to be eliminated. In the case of river flow, erroneous motion vectors tend to appear close to the river banks and around rock formations within the river flow. These vectors act as outliers to the main magnitude and/or direction flow trend and appear due to scattering effects as the water encounters the banks/rocks. Applying a clustering task to the estimated optical flow field based on the direction of motion, using the main motion trend as an indicator, can be effective [9].

10.5 CONCLUDING REMARKS AND FUTURE DIRECTIONS

This survey presents a critical overview and analysis of existing approaches for the estimation of fluid flow. We examine the theoretical basis of each approach and reveal advantages and disadvantages of each methodological class. In particular, our analysis focuses on the special assumptions and constraints posed, in order to efficiently model the non-rigid nature of fluid motion. We attempt a qualitative comparison based on these properties and we also design a framework for quantitative comparisons using realistic models of fluid motion.

The problem of motion estimation inferred on fluids and, in general, on non-rigid objects poses specific challenges pertaining to the inherent deformations of the scene under consideration. There exist three forms of algorithmic approaches that attempt to address these issues. The first one focuses on artificial or naturally formed particles considered as rigid bodies, whereas the second type models the optical formation of the scene by means of probability density estimation under the assumption of short-time optical consistency. Finally, the third class enforces the kinematic properties of fluid matter in the optical flow estimation, in order to adjust to heavy deviations from the original model. Alongside the algorithmic developments, we also address requirements of real-world applications. The surveillance of natural scenes, such as rivers, water basins, etc., requires rather abstract estimation of the regular motion at a global level over the entire viewing area. For instance, the prediction of floods or other abrupt phenomena is based on the spatiotemporal average flow, whereas the estimation of water supply is based on a rather crude profile of surface speed. For such reasons, particle-based techniques have found extensive applications, even in commercial products. However, the localization of particles is not always possible, as, e.g., in surveillance of larger phenomena (water elevation or tsunami) on the seashore or in the deep sea. We should also note that particle-based techniques provide only local estimates of fluid motion by sampling the motion field at specific locations. The last group of techniques using the kinematic properties of the medium might be more accurate in modelling the fluid motion field, but they are strongly dependent on the correctness of their assumptions. For instance, if the content of a flowing river changes due to the contamination with natural soil residue or contaminants, then the kinematic model fails to readily absorb these deviations. Highlighting also the fact that requirements of real-world applications do not concern the fine details of the motion field but rather its abstract description, we conclude that the second class of techniques based on probabilistic modelling forms a good compromise of speed, efficiency, and accuracy for many applications. This is supported by both the qualitative and quantitative comparisons deployed in this study.

ACKNOWLEDGMENTS

This work is elaborated through an ongoing THALES project (CYBERSENSORS High Frequency Monitoring System for Integrated Water Resources Management of Rivers). The project has been co-financed by the European Union (European Social Fund - ESF) and Greek national funds through the Operational Program "Education and Lifelong Learning" of the National Strategic Reference Framework (NSRF) Research Funding Program: Thales. Investing in knowledge society through the European Social fund.

10.6 REFERENCES

1. R J Adrian. Particle-imaging techniques for experimental fluid mechanics. *Annual review of Fluid Mechanics*, 23(1):261–304, 1991.
2. R J Adrian and J Westerweel. *Particle image velocimetry*. Number 30. Cambridge University Press, 2011.
3. H Amini, E Sollier, M Masaeli, Y Xie, B Ganapathysubramanian, H A Stone, and D Di Carlo. Engineering fluid flow using sequenced microstructures. *Nature Communications*, 4:1826, 2013.

4. A H Arthington, J M Bernardo, and M Ilhéu. Temporary rivers: Linking ecohydrology, ecological quality and reconciliation ecology. *River Research and Applications*, 30(10):1209–1215, 2014.
5. T Astarita. Analysis of weighting windows for image deformation methods in PIV. *Experiments in Fluids*, 43(6):859–872, 2007.
6. K Avgerinakis, A Briassouli, and I Kompatsiaris. Real time illumination invariant motion change detection. In *Proceedings of the First ACM International Workshop on Analysis and Retrieval of Tracked Events and Motion in Imagery Streams*, pages 75–80, Firenze, Italy, 2010. ACM.
7. A B Basset. *Treatise on hydrodynamics* vol 2. *London: Deighton, Bell & Co*, 1888.
8. B Jahne and S Wass. Optical wave measurement technique for small scale water surface waves. In *Advances in Optical Instruments for Remote Sensing*, pages 147–152, 1989.
9. K Bacharidis, K Moirogiorgou, I A Sibetheros, A E Savakis, and M Zervakis. River flow estimation using video data. In *Imaging Systems and Techniques (IST), 2014 IEEE International Conference on*, pages 173–178. IEEE, 2014.
10. S Baker, R Gross, and I Matthews. Lucas–Kanade 20 years on: A unifying framework: Part 2. *International Journal of Computer Vision*, 2003.
11. S Baker, D Scharstein, J P Lewis, S Roth, M J Black, and R Szeliski. A database and evaluation methodology for optical flow. In *International Conference on Computer Vision - ICCV*, pages 1–8, 2007.
12. F Becker, B Wieneke, J Yuan, and C Schnörr. A variational approach to adaptive correlation for motion estimation in particle image velocimetry. *Pattern Recognition*, pages 335–344, 2008.
13. M Bertero, T A Poggio, and V Torre. Ill-posed problems in early vision. *Proceedings of the IEEE*, 76(8):869–889, 1988.
14. A Allen Bradley, A Kruger, E A Meselhe, and M VI Muste.. Flow measurement in streams using video imagery. *Water Resources Research*, 38(12):1315, 2002.
15. A Cavagna, I Giardina, A Orlandi, G Parisi, A Procaccini, M Viale, and V Zdravkovic. The STARFLAG handbook on collective animal behaviour: Part I, empirical methods. *arXiv preprint arXiv:0802.1668*, 2008.
16. A Cenedese. Eulerian and Lagrangian velocity measurements by means of image analysis. *Journal of Visualization*, 2(1):73–84, 1999.
17. J Chang, D Edwards, and Y Yu. Statistical estimation of fluid flow fields. In *ECCV Workshop on Statistical Methods in Video Processing*, pages 91–96, Copenhagen, 2002. Citeseer.
18. X Chen, P Zillé, L Shao, and T Corpetti. Optical flow for incompressible turbulence motion estimation. *Experiments in Fluids*, 56(1):8, 2015.
19. R Connor. *The United Nations world water development report 2015: Water for a sustainable world*, volume 1. UNESCO Publishing, 2015.
20. G Dezső-Weidinger, A Stitou, J van Beeck, and M L Riethmuller. Measurement of the turbulent mass flux with PTV in a street canyon. *Journal of Wind Engineering and Industrial Aerodynamics*, 91(9):1117–1131, 2003.
21. A Eckstein and P P Vlachos. Digital particle image velocimetry (DPIV) robust phase correlation. *Measurement Science and Technology*, 20(5):055401, 2009.
22. A C Eckstein, J Charonko, and P Vlachos. Phase correlation processing for DPIV measurements. *Experiments in Fluids*, 45(3):485–500, 2008.
23. J M Fitzpatrick. The existence of geometrical density-image transformations corresponding to object motion. *Computer Vision, Graphics, and Image Processing*,

44(2):155–174, 1988.

24. I Fujita and S Komura. Application of video image analysis for measurements of river-surface flows. *Annu. J. Hydraul. Eng*, 38:733–738, 1994.

25. A Georgakakos, P Fleming, M Dettinger, C Peters-Lidard, T.C. Richmond, K Reckhow, K White, and D Yates. Water Resources. In Terese (T.C.) Richmond Melillo, Jerry M. and Gary W. Yohe, editors, *Climate Change Impacts in the United States: The Third National Climate Assessment*, chapter 3, pages 69–112. U.S. Global Change Research Program, 2014.

26. L Gui and S T Wereley. A correlation-based continuous window-shift technique to reduce the peak-locking effect in digital PIV image evaluation. *Experiments in Fluids*, 32(4):506–517, 2002.

27. T Hadad and R Gurka. Effects of particle size, concentration and surface coating on turbulent flow properties obtained using PIV/PTV. *Experimental Thermal and Fluid Science*, 45:203–212, 2013.

28. A Hauet, A Kruger, W F Krajewski, A Bradley, M Muste, J-D Creutin, and M Wilson. Experimental system for real-time discharge estimation using an image-based method. *Journal of Hydrologic Engineering*, 13(2):105–110, 2008.

29. H W Haussecker and D J Fleet. Computing optical flow with physical models of brightness variation. *IEEE Transactions on Pattern Analysis and Machine Intelligence*, 23(6):661–673, 2001.

30. P Héas, C Herzet, and E Mémin. Bayesian inference of models and hyperparameters for robust optical-flow estimation. *IEEE Transactions on Image Processing*, 21(4):1437–1451, 2012.

31. P Héas, C Herzet, E Mémin, D Heitz, and P D Mininni. Bayesian estimation of turbulent motion. *IEEE Transactions on Pattern Analysis and Machine Intelligence*, 35(6):1343–1356, 2013.

32. P Héas, E Mémin, D Heitz, and P D Mininni. Bayesian selection of scaling laws for motion modelling in images. In *Computer Vision, 2009 IEEE 12th International Conference on*, pages 971–978. IEEE, 2009.

33. D F Hill, K V Sharp, and R J Adrian. Stereoscopic particle image velocimetry measurements of the flow around a rushton turbine. *Experiments in Fluids*, 29(5):478–485, 2000.

34. W Hongwei, H Zhan, G Jian, and X Hongliang. The optical flow method research of particle image velocimetry. *Procedia Engineering*, 99:918–924, 2015.

35. B K P Horn and B G Schunck. Determining optical flow. *Artificial Intelligence*, 17(1-3):185–203, 1981.

36. Q W Hu and Q Q Li. Image restoration based on mask technique. *Geomatics and Information Science of Wuhan University*, 29(4):317–323, 2004.

37. H T Huang, H E Fiedler, and J J Wang. Limitation and improvement of PIV: part I . Limitation of conventional techniques due to deformation of particle image patterns. *Experiments in Fluids*, 15(3):168–174, 1993.

38. K Jambunathan, X Y Ju, B N Dobbins, and S Ashforth-Frost. An improved cross correlation technique for particle image velocimetry. *Measurement Science and Technology*, 6(5):507–514, 1995.

39. M Jodeau, A Hauet, A Paquier, J Le Coz, and G Dramais. Application and evaluation of ls-piv technique for the monitoring of river surface velocities in high flow conditions. *Flow Measurement and Instrumentation*, 19(2):117–127, 2008.

40. R D Keane and R J Adrian. Theory of cross-correlation analysis of PIV images. *Applied*

Scientific Research, 49(3):191–215, 1992.

41. J-Y Kim, L-S Kim, and S-H Hwang. An advanced contrast enhancement using partially overlapped sub-block histogram equalization. *IEEE Transactions on Circuits and Systems for Video Technology*, 11(4):475–484, 2001.

42. T K Kim, J K Paik, and B S Kang. Contrast enhancement system using spatially adaptive histogram equalization with temporal filtering. *IEEE Transactions on Consumer Electronics*, 44(1):82–87, 1998.

43. K Krajsek and R Mester. A maximum likelihood estimator for choosing the regularization parameters in global optical flow methods. In *Image Processing, 2006 IEEE International Conference on*, pages 1081–1084. IEEE, 2006.

44. K Krajsek and R Mester. Bayesian model selection for optical flow estimation. In *Joint Pattern Recognition Symposium*, pages 142–151. Springer, 2007.

45. A Lecerf, B Renou, D Allano, A Boukhalfa, and M Trinité. Stereoscopic PIV: validation and application to an isotropic turbulent flow. *Experiments in Fluids*, 26(1):107–115, 1999.

46. B Lecordier, D Demare, L M J Vervisch, J Réveillon, and M Trinite. Estimation of the accuracy of PIV treatments for turbulent flow studies by direct numerical simulation of multi-phase flow. *Measurement Science and Technology*, 12(9):1382, 2001.

47. Z-C Liu, R J Adrian, C D Meinhart, and W Lai. Structure of a turbulent boundary layer using stereoscopic, large-format video-PIV. In R.J. Adrian, D.F.G. Durao, F. Durst, M.V. Heitor, M. Maeda, and J.H. Whitelaw, editors, *Developments in laser techniques and fluid mechanics*, pages 259–273. Springer, Berlin Heidelberg, 1997.

48. B D Lucas, T Kanade, et al. An iterative image registration technique with an application to stereo vision. In *Proceedings of Imaging Understanding Workshop*, pages 121–130. Vancouver, BC, Canada, 1981.

49. C D Meinhart, S T Wereley, and J G Santiago. PIV measurements of a microchannel flow. *Experiments in Fluids*, 27(5):414–419, 1999.

50. A Melling. Tracer particles and seeding for particle image velocimetry. *Measurement Science and Technology*, 8(12):1406, 1997.

51. K Moirogiorgou, et al. High frequency monitoring system for integrated water resources management of rivers. In *Improving efficiency of water systems in a changing natural and financial Environment (EWaS-MED International Conference)*, 2013.

52. M Moroni and A Cenedese. Comparison among feature tracking and more consolidated velocimetry image analysis techniques in a fully developed turbulent channel flow. *Measurement Science and Technology*, 16(11):2307, 2005.

53. H Murase. Surface shape reconstruction of a nonrigid transparent object using refraction and motion. *IEEE Transactions on Pattern Analysis and Machine Intelligence*, 14(10):1045–1052, 1992.

54. Y Nakajima, H Inomata, H Nogawa, Y Sato, S Tamura, K Okazaki, and S Torii. Physics-based flow estimation of fluids. *Pattern Recognition*, 36(5):1203–1212, 2003.

55. C L Nikias and A P Petropoulou. *Higher order spectral analysis: a nonlinear signal processing framework*, pages 313–322. Prentice Hall, New Jersey, 1993.

56. J Nogueira, A Lecuona, U Ruiz-Rivas, and P A Rodriguez. Analysis and alternatives in two-dimensional multigrid particle image velocimetry methods: application of a dedicated weighting function and symmetric direct correlation. *Measurement Science and Technology*, 13(7):963, 2002.

57. A K Prasad. Stereoscopic particle image velocimetry. *Experiments in Fluids*, 29(2):103–116, 2000.

58. A K Prasad and R J Adrian. Stereoscopic particle image velocimetry applied to liquid flows. *Experiments in Fluids*, 15(1):49–60, 1993.

59. R Pytlak. *Conjugate gradient algorithms in nonconvex optimization*, volume 89 of *Nonconvex Optimization and Its Applications*. Springer Science & Business Media, 2008.

60. M Raffel, C E Willert, J Kompenhans, et al. *Particle image velocimetry: a practical guide*. Springer Science & Business Media, 2007.

61. H Sakaino. Fluid motion estimation method based on physical properties of waves. In *Computer Vision and Pattern Recognition, 2008. CVPR 2008. IEEE Conference on*, pages 1–8. IEEE, 2008.

62. H Sakaino. Motion estimation for dynamic texture videos based on locally and globally varying models. *IEEE Transactions on Image Processing*, 24(11):3609–3623, 2015.

63. B G Schunck. Image flow fundamentals and future research. In *IEEE CPRV*, pages 560–571, San Francisco, June, 1985. IEEE.

64. L Shindler, M Moroni, and A Cenedese. Using optical flow equation for particle detection and velocity prediction in particle tracking. *Applied Mathematics and Computation*, 218(17):8684–8694, 2012.

65. A Stitou and M L Riethmuller. Extension of PIV to super resolution using PTV. *Measurement Science and Technology*, 12(9):1398, 2001.

66. M Sun. *Research on Key Technology of Automatical and Fast DOM Generation*. PhD thesis, Wuhan University, 2008.

67. K Takehara and T Etoh. A study on particle identification in PTV particle mask correlation method. *Journal of Visualization*, 1(3):313–323, 1998.

68. Z J Taylor, R Gurka, G A Kopp, and A Liberzon. Long-duration time-resolved PIV to study unsteady aerodynamics. *IEEE Transactions on Instrumentation and Measurement*, 59(12):3262–3269, 2010.

69. R Theunissen. Theoretical analysis of direct and phase-filtered cross-correlation response to a sinusoidal displacement for PIV image processing. *Measurement Science and Technology*, 23(6):065302, 2012.

70. W Thielicke. *The Flapping Flight of Birds-Analysis and Application*. PhD thesis, Rijksuniversiteit Groningen, 2014.

71. W Thielicke and E Stamhuis. PIVlab–towards user-friendly, affordable and accurate digital particle image velocimetry in MATLAB. *Journal of Open Research Software*, 2(1), 2014.

72. R Tsubaki, I Fujita, and S Tsutsumi. Measurement of the flood discharge of a small-sized river using an existing digital video recording system. *Journal of Hydro-environment Research*, 5(4):313–321, 2011.

73. O Tzoraki and N P Nikolaidis. A generalized framework for modelling the hydrologic and biogeochemical response of a Mediterranean temporary river basin. *Journal of Hydrology*, 346(3):112–121, 2007.

74. M Wang, J Pan, S Chen, and H Li. A method of removing the uneven illumination phenomenon for optical remote sensing image. In *Geoscience and Remote Sensing Symposium, 2005. IGARSS'05. Proceedings. 2005 IEEE International*, volume 5, pages 3243–3246, Korea, 2005. IEEE.

75. M P Wernet. Symmetric phase only filtering: a new paradigm for DPIV data processing. *Measurement Science and Technology*, 16(3):601, 2005.

76. J Westerweel and J van Oord. Stereoscopic PIV measurements in a turbulent boundary layer. In M. Stainslaus, J. Kompenhans, and J. Westerweel, editors, *Particle image velocimetry: Progress toward industrial application*. Kluwer, Dordrecht, 2008.

77. J Westerweel. Fundamentals of digital particle image velocimetry. *Measurement Science and Technology*, 8(12):1379, 1997.
78. J Westerweel, D Dabiri, and M Gharib. The effect of a discrete window offset on the accuracy of cross-correlation analysis of digital PIV recordings. *Experiments in Fluids*, 23(1):20–28, 1997.
79. R P Wildes, M J Amabile, A-M Lanzillotto, and T-S Leu. Recovering estimates of fluid flow from image sequence data. *Computer Vision and Image Understanding*, 80(2):246–266, 2000.

11 Computational Approaches for Incorporating Short- and Long-Term Dynamics in Smart Water Networks

KARTIKEYA BHARDWAJ, DIMITRIOS STAMOULIS, RUIZHOU DING, DIANA MARCULESCU, RADU MARCULESCU

CHAPTER HIGHLIGHTS

- We effectively investigate the correlation between carbon emissions and precipitation anomalies under a complex networks perspective. Our results yield a correlation between air pollution and precipitation anomalies which has grown stronger during recent decades.
- We present results based on complex networks that span multivariate climate dynamics across multiple temporal and spatial scales to corroborate their ability to capture long-memory temporal relationships and long-range spatial dependencies.
- We describe K-hop learning, a feature extraction method, which utilizes network information to learn upcoming river flow peaks 24-hours ahead of time.
- We effectively use these network-based features in a machine learning framework for river flow rate prediction. We incorporate the learned features in a linear model to predict 24-hours ahead flow rate for very big rivers (flow rate $\approx 10^4 \, \text{m}^3/\text{s}$), big rivers (flow rate $\approx 1000 \, \text{m}^3/\text{s}$), small rivers (flow rate $\approx 150 \, \text{m}^3/\text{s}$), and intermediate rivers (flow rate $\approx 400 \, \text{m}^3/\text{s}$).
- Finally, we present comparison results of K-hop learning against SVM regression, autoregression, random forests, and K-nearest neighbor regression. Our results clearly demonstrate how using network-based features can significantly improve the prediction accuracy over the timeseries-based models.

K Bhardwaj ✉ • D Stamoulis • R Ding• D Marculescu • R Marculescu

Department of Electrical and Computer Engineering, Carnegie Mellon University, US.

E-mail: kbhardwa@andrew.cmu.edu

11.1 INTRODUCTION AND MOTIVATION

Fresh water sources such as rivers have very important applications, ranging from a myriad of daily uses, to agriculture, all the way to hydropower generation. Nonetheless, rapidly increasing water crises call for an intelligent and sustainable use of water resources [43, 44]. To this end, several modelling approaches have been fruitfully employed in the context of smart water networks (SWN), aiming to accurately capture water dynamics of water management systems [28] or industrial SWN [33]. The need for highly accurate models is crucial especially in the context of river flow prediction and precipitation modelling, since accurately forecasting both flow rate and precipitation is critical for effective management of run-of-river hydropower and planning/preparation for natural disasters such as floods.

Indeed, many man-made as well as naturally occurring smart water systems, such as climate and terrestrial river systems, are characterized by complex networks. As a key illustrative example, Figure 11.1 shows a typical river system formed by various streams and tributaries in a river basin (the figure shows the Ohio River Basin, USA, covering more than 204,000 sq. miles). As is evident from the zoomed-in portion shown in Figure 11.1(b), the river system can be best described by a very complex network which possesses a *fractal* structure [27]. The nodes in this network are geographical locations and links are the rivers flowing between locations. Consequently, *we must explicitly account for the underlying complex network dynamics in order to optimally model such smart water networks*. This serves as the **key motivation** behind our work. In this chapter, therefore, we describe new and recently proposed computational techniques that can account for the network structure and dynamics, while studying multiscale smart water networks.

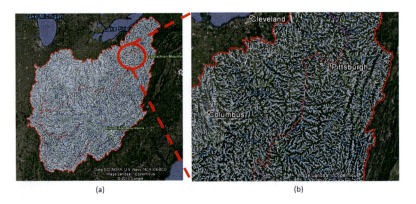

Figure 11.1: (a) River network for the Ohio River Basin, USA. (b) The zoomed-in portion of the river network clearly demonstrates its complex network characteristics such as fractal structure. Hence, the network structure must be taken into account while designing the flow rate prediction models. We plotted this figure in Google Earth [1] using Ohio Watershed KML files obtained from USGS Elevation Derivatives for National Applications (EDNA) website [2].

In recent years, achieving high prediction accuracy in precipitation and river forecast systems has become increasingly imperative, due to the rising variability and the frequency of extreme events caused by climate change [42]. Since the climate system consists of several heavily interdependent subsystems such as land, oceans, and atmosphere, it can be characterized by complex networks. These complex network characteristics are inherent in precipitation patterns and can be investigated in the context of networks of larger temporal and spatial coverage. Indeed, recent work has introduced *climate networks*, i.e., complex networks constructed from climate data, wherein the nodes are the grid points on the globe, and an edge indicates a statistically significant correlation between a pair of grid points. Existing methods use complex networks as a means of capturing large-scale meteorological and physical phenomena [6, 18, 41, 46], however their applications to climate and especially in water systems have yet to emerge [39], since these studies focus on specific physical phenomena [19, 40, 46], or limit their multivariate analysis to specific regions [14, 39, 40]. Hence, existing modelling approaches yield a trade-off between model complexity and spatiotemporal coverage. For instance, precipitation models focusing on specific rivers [5] might be agnostic to the effect of long-term and long-range climate dependencies. This observation raises an interesting research question for water networks' engineers: *given a precipitation anomaly that is recorded in an area adjacent to our model region, to what extent will we need to reassess the assumptions of our initial model?*

Addressing this problem is challenging, since it boils down to capturing the low-frequency, long-memory temporal relationships, and long-range spatial dependencies of global precipitation dynamics [15,21]. Thus, our **first key goal** is to employ an extensive analysis that spans multivariate climate dynamics across multiple temporal and spatial scales to demonstrate the ability of complex networks to capture long-memory temporal relationships and long-range spatial dependencies. In other words, we postulate the potential of modelling complex precipitation networks as a means to enhance regional, high-granularity SWN models. That way, small-scale models could be enhanced to account for long-term precipitation dynamics, enabling SWN engineers to dynamically decide whether it is necessary to revisit and re-evaluate any previous model assumptions.

Having motivated the significance of network-based computational methods in a generic context, we next demonstrate the significance of such approaches by considering concrete smart water networks at a much finer granularity, namely, the river systems. Although significant research has been reported in the literature on forecasting water resources, existing approaches use only timeseries-based models. In Figure 11.1, however, we showed that the river network possesses complex network characteristics (e.g., the fractal structure). Such complex river network characteristics give rise to the nonstationary behavior of the river flow timeseries. While river flow rate timeseries is *nonstationary* and thus characterized by the presence of many crests (or *peaks*) and troughs, river flow peaks are hard to predict. Consequently, since the existing timeseries-based models fail to account for network structure and dynamics, they suffer from *inaccurate river flow peak prediction* [7]. Nevertheless, accurately predicting such peaks is extremely important for flood planning and hydropower management.

This leads us to the **second key goal** of our work, i.e., to propose a novel machine learning approach that effectively accounts for the complex network characteristics of rivers and can, therefore, accurately predict river flow peaks. Towards this direction, we describe a new *feature learning* method that can be used by machine learning models to predict the peaks in river flow very accurately. We show that, for a given location, certain features can be *learned* from the river network which cannot otherwise be learned from timeseries data at a single location; these features contain precise information about peaks in river flow rate. We then use these learned features in machine learning models to predict river flow rate 24-hours ahead at different locations. In contrast to our *network* approach, previous machine learning models only use timeseries data at a *single* location. Finally, we demonstrate the effectiveness of our approach by extensive experimental evaluation.

11.1.1 WORK CONTRIBUTIONS

In this work, we extend the key findings of our prior work [7,11]. Towards enhancing precipitation models by accounting for multivariate and multiscale climate dynamics, we present the key contributions as discussed in [11]:

1. We effectively investigate the correlation between carbon emissions and precipitation anomalies under a complex networks perspective. Our results yield a correlation between air pollution and precipitation anomalies which has grown stronger during recent decades.
2. We present results based on complex networks that span multivariate climate dynamics across multiple temporal and spatial scales to corroborate their ability to capture long-memory temporal relationships and long-range spatial dependencies.

We further describe new computational approaches towards *improving river flow prediction* and significantly extend the analysis presented in [7]. Overall, we make the following key contributions:

3. We present K-hop learning, a feature extraction method, which utilizes network information to learn upcoming river flow peaks 24-hours ahead of time.
4. We effectively use these network-based features in a machine learning framework for river flow rate prediction. We incorporate the learned features in a linear model to predict 24-hours ahead flow rate for very big rivers (flow rate $\approx 10^4$ m^3/s), big rivers (flow rate > 1000 m^3/s), small rivers (flow rate ≈ 150 m^3/s), and intermediate rivers (flow rate ≈ 400 m^3/s).
5. Finally, we present comparison results of K-hop learning against SVM regression, autoregression, random forests, and K-nearest neighbor regression. Our results clearly demonstrate how using network-based features can significantly improve the prediction accuracy over the timeseries-based models.

Our work is organized as follows. In Section 11.2, we discuss limitations of existing methodologies towards capturing long-term precipitation dynamics or enabling accurate river flow prediction. In Section 11.3, we present our comprehensive analysis to capture multivariate climate dynamics across multiple temporal and spatial scales. That is, we discuss the methodology for constructing complex climate networks in Section 11.3.1. We then discuss the key experimental findings in Section 11.3.2. In Section 11.4, we introduce K-hop learning, our proposed method for river flow forecasting. More specifically, we introduce the datasets, we formulate the problem, and we describe the proposed method in Section 11.4.1. We then evaluate K-hop learning in Section 11.4.2. We conclude our work in Section 11.5.

11.2 RELATED WORK

11.2.1 MULTIVARIATE AND MULTISCALE CLIMATE DYNAMICS

Prior art has already motivated the need for multivariable, multiscale methods for capturing temporal relationships and long-term spatial dependencies observed in precipitation dynamics [15,21]. To this end, complex networks have emerged as a powerful modelling solution, given their effectiveness in capturing climate trends [38]. Towards this direction, the so-called *climate networks* have emerged, where each node corresponds to a location and an edge between the different nodes of the network indicates a correlation between two locations with respect to climate data. Prior art on *climate networks* can be categorized into two directions, focusing either on the network *construction* algorithms based on complex networks analytical tools, or on the *analysis* of the network properties in the context of specific climate phenomena.

First, in terms of network construction methodologies, initial attempts have mostly focused on identifying correlated signals based on statistical significance metrics. Several heuristics have been successfully investigated in this context. Gozolchiani et al. have proposed the *correlation strength* [17] as a metric able to capture the different profiles for the correlation function, and can, therefore, be used to identify strongly and intermediately correlated links with respect to temperature anomalies, and weakly correlated (WC) links where the different local maxima cannot be distinguished from noise. Similarly, the Pearson correlation has been applied directly to anomaly data [18,46]. The key disadvantage of this methodology is twofold: first, the method introduces the additional overhead of computing the correlation values for both unshuffled and shuffled data; second, the noise floor above which the correlation of two signals is considered statistically significant is graphically (manually) identified. Upon initial experimentation, we found that there was not a one-noise-floor to fit all cases based on this method, while we also observed that the quality of the results was highly sensitive to the selected noise floor. Moreover, while the case study in [18] focuses on a limited location, season, and metric of interest, we found the graphical generation of the noise floor to be the bottleneck when applying the method to arbitrary temporal and spatial scales every time. To address this suboptimality, Steinhaeuser et al. have successfully used the *p-value* as a metric which is easy to acquire, yet is effective enough to capture the statistical significance level [39], since the observed similarity of Pearson correlation and mutual information networks can be considered statistically significant [12]. In this work, we exploit the effectiveness yet simplicity of this heuristic and we use it to identify statistically significant links.

Based on the aforementioned network construction techniques, state-of-the-art approaches use complex networks to capture and understand climate anomalies and their latent patterns [38]. Nonetheless, existing work focuses on specific physical phenomena, e.g., El Niño [40,46] and North Atlantic Oscillation [19], and limit their multivariate analysis to either ocean- or continent-specific regions [14,39,40]. To this end, the key motivation behind our work is to investigate long-term global precipitation dynamics using complex networks. A related work was presented recently [39], however limiting the analysis only to ocean regions, without accounting for the correlation of precipitation dynamics with global air pollution trends. In contrast, we aim to extend our exploration to global, multifeature climate anomalies by incorporating yearly data for precipitation, temperature, and carbon emissions dating from 1950 to the present.

11.2.2 RIVER FLOW FORECASTING

Significant research has been reported in the literature on forecasting water resources. There are two types of models: conceptual models and data-driven models. Conceptual models use prior knowledge of the physical process, decompose the whole process into a series of subprocesses, and solve the problem using mathematical tools such as differential equations, etc. On the other hand, data-driven models rely less on prior knowledge and learn the models with training data. With an increasing availability of data, data-driven models have been receiving more and more attention.

Prior work on data-driven models focuses on using machine learning models to predict short-term water levels or flow rates [26, 31, 34, 36]. Nguyen et al. forecast timeseries daily river levels using machine learning models such as lasso, random forests, and SVM regression [31]. Similar evaluation of machine learning models for river flow rates has been reported in [34] and, more recently in [36]. There have also been many studies on rainfall-runoff modelling, i.e., forecasting future river flow based on historic river flow and historic rainfall. Artificial neural networks (ANNs) have also been well-studied models for this forecasting problem. In the majority of existing approaches, multilayer perception is used to model the nonlinear relationship between the input and the output. A backpropagation algorithm is used to train the neural network. Hsu et al. [22] compare an ANN, a linear model, and a conceptual model, and conclude that ANN performs better than the other two models. Recent studies on hydrological forecasting have been using variants of ANNs. For example, the emotional ANN (EANN) is proposed for river flow forecasting [32]. EANN includes simulated emotional parameters aimed at improving the network learning process.

There have also been many studies using fuzzy rule-based systems (FRBS) for water-related applications. FRBS can learn the "if-then" rules from the training data. Instead of using a 0-1 logic, FRBS models the variables using fuzzy set theory where each variable can have partial membership within different sets [37]. In many prior studies for river forecasting, genetic algorithm (GA) has been used to search for a local optimal solution of the models such as ANN or FRBS [10] [47]. Starting from an initial state, GA iteratively searches for better states with transformations inspired by the biological processes of natural selection and survival of the fittest [16]. Genetic programming (GP), which uses the framework of GA but targets on function blocks, has also been used for real-time run-off forecasting [25]. There have also been many studies using instance-based learning (IBL) for hydrological modelling [24, 35]. K-nearest-neighbors (KNN) algorithm, an example of IBL, compares the new input with all the inputs of the training set, and finds the several nearest ones, whose outputs are averaged for prediction. Karlsson et al. [24] used KNN algorithm for single-variate hydrological time series prediction. Shamseldin et al. [35] combined the KNN algorithm with the concept of perturbations from a mean value. Ensemble learning which combines different single models and makes the prediction with all the models has been a new direction for river forecasting. Abrahart et al. [3] evaluated six data fusion strategies used for hydrological forecasting for the River Ouse and the Upper River Wye. In addition to rivers, machine learning-based lake water level prediction models have also been proposed in [26].

Although valuable, this previous research uses only timeseries-based models. Indeed, like many environmental variables, river flow rate timeseries is *nonstationary* and is characterized by the presence of many crests (or *peaks*) and troughs. As emphasized in the previous section, rivers form a complex network which in turn gives rise to these nonstationarities and hard-to-predict river flow peaks. Consequently, these *single timeseries-based* machine learning models suffer from *inaccurate river flow peak prediction*. For instance, the results of Nguyen et al. show that the river flow peaks are *not* detected very accurately [31], as there is often a delay between the observed and predicted river flow peaks. Meanwhile, accurately predicting such peaks is extremely important for flood planning and hydropower management. Therefore, in this work, we focus on short-term flow rate prediction with an emphasis on very accurate peak prediction. To this end, there is a critical need for new computational approaches that can learn relevant features from the underlying network itself in order to enable highly accurate smart water network systems.

In the rest of the chapter, we first introduce such network-based computational approaches in a generic context, namely, to capture climate dynamics at arbitrary scales in space and time. Then, we address a concrete smart water networks problem at a much finer granularity, i.e., the short-term prediction in river systems.

11.3 MULTIVARIATE AND MULTISCALE CLIMATE DYNAMICS

11.3.1 METHODOLOGY

To effectively capture climate dynamics, we use complex networks constructed from climate data, i.e., *climate networks*. In climate networks, the nodes are the grids on the globe, while an edge connecting a pair of grid points indicates a significant correlation between these two points. Moreover, the weight of an edge corresponds to the magnitude of the correlation. To construct the climate networks over arbitrary temporal and spatial scales, we base our analysis on daily precipitable water and temperature data from the NCAR Reanalysis project [23]. We also use monthly carbon emission data from the CDIAC [9]. Since carbon emissions for ocean grid points are unavailable, we focus on land areas with a grid resolution of $2.5° \times 2.5°$.

Network construction: For each year y, network link weights are described by an adjacent matrix $W \in R^{long \times lat}$, where $long \times lat$ are the grid dimensions for the whole globe. To compute entries of the W matrix, we use Pearson correlation of the yearly time series data at any two grid points, as shown in [18]. That is, for each year y from 1950 to 1999, the weight between grid location i and j is:

$$w_{i,j}^{(y)} = \left| \frac{[x_i^{(y)} - \mu(x_i^{(y)})]^T \cdot [x_j^{(y)} - \mu(x_j^{(y)})]}{\sigma(x_i^{(y)}) \cdot \sigma(x_j^{(y)})} \right| \qquad (11.1)$$

where $x_i^{(y)}$ is the daily data that span the entire year y. Please note that we use the absolute value, since both positive and negative correlation coefficients indicate a strong relationship. To convert the time series of precipitable water and temperature to anomalies data, we subtract the long-term mean and divide it with the long-term standard deviation computed over all considered years (i.e., the mean of all January 1st days). This normalization is shown to effectively reduce temporal autocorrelation in the time series [39].

Finally, it is important to identify the statistically significant correlation values, such that edges with statistically insignificant weights are pruned from the network. To this end, we use the p-value of the correlation result as a significance-pruning heuristic and prune out edges when the p-value of their correlation is more that 10^{-10} [39]. To assess the correctness of the generated networks, we plot the frequency distribution (histograms) of the edge lengths of each network throughout all years in Figure 11.2. Indeed, we observe that for all three features, a left-most "peak" occurs, which corresponds to the close-proximity dependencies, while the observed "tail" on the right captures long-distance correlations. Moreover, we observe the "tail thickness" of carbon emission data that reflects the significantly higher long-range dependence across entire regions of adjacent air masses. These results are consistent with prior art [39], and especially the findings related to long-distance "connections," i.e., teleconnections [4].

11.3.2 EXPERIMENTAL RESULTS

Multivariate, Long-Memory Temporal Climate Dynamics

Having computed the edges for considered networks as discussed in the previous section, we first compute the average degree for each yearly network for all three features, i.e., the average number of connections between all nodes. Intuitively, highly-connected networks with more correlated nodes capture climate anomalies that affect increasingly wider regions. First, we compute the correlation coefficient between the

Incorporating Short- and Long-Term Dynamics in Smart Water Networks   305

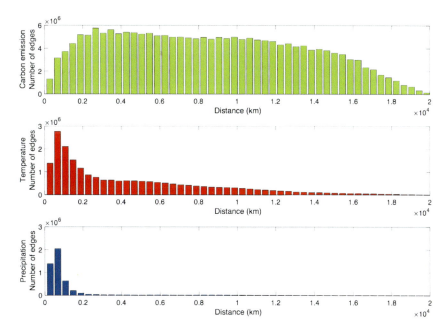

Figure 11.2: Frequency distribution (histograms) of edge lengths for each constructed network: carbon emissions (top-green), temperature (middle-red), and precipitation (bottom-blue). The networks correctly capture both proximity-based and long-range correlations. Figure is adapted with permission from [11].

acquired signals. As an interesting result, it is worth noting that the correlation between the average degree of carbon emissions and precipitation anomalies grows stronger during more recent decades. That is, while precipitation anomalies and carbon emissions have an overall correlation coefficient of 0.3091, this value increases to 0.4644 and eventually to 0.6018 the last 30 and 20 years, respectively. As expected, precipitation and temperature patterns have correlated network average degree with an overall coefficient of 0.5368.

We further highlight global climate trends by investigating the evolution of network clusters. To this end, we detect communities in all the yearly global networks for the three features, by using the Louvain algorithm [8]. Figure 11.3 shows the community detection results for precipitation anomalies (top), carbon emissions (middle), and temperature anomalies (bottom) for the years 1955, 1975, and 1995. First, similar to prior art studying precipitation and temperature patterns [39], we observe that the clusters illustrate relative stability over time. Nonetheless, we observe that carbon emission clusters increase in size and coverage, suggesting an ever-increasing trend that spans the entire globe.

Unlike prior art on complex climate anomalies networks which do not account for air pollution data, our analysis highlights several interesting observations. First, several regions, such as North America and Europe, belong to the same carbon emission

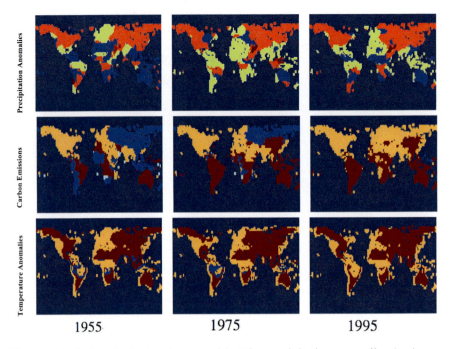

Figure 11.3: Network clusters (communities) for precipitation anomalies (top), carbon emissions (middle), and temperature anomalies (bottom) for the years 1955, 1975, and 1995. Each color corresponds to one community. Figure is adapted with permission from [11].

community over time, hence indicating that carbon data have high correlation across entire long-range continental land masses. Moreover, since carbon emission patterns are more likely to be dictated by human activity, the boundaries of carbon communities are closer to the country boundaries, such as United States, India, China, etc., compared to precipitation and temperature anomalies data. In terms of precipitation anomalies, the clusters follow a latitudinal community structure, i.e., regions in the same latitude bands are more likely to belong to the same precipitation anomaly community. Finally, the anomalies in terms of temperature exhibit a separation between land and coastal regions.

Long-Range Spatial Precipitation Anomalies

In the previous section, we exploited the ability of complex networks to capture multivariate dynamics, even under long-term and long-memory dependencies. Next, we investigate the inherent ability of precipitation networks to capture spatial dependencies. We focus on two important properties of complex networks, the degree centrality and the betweenness centrality [13]. The former metric captures the degree of each node, while the latter measures the probability that a given node appears in

Incorporating Short- and Long-Term Dynamics in Smart Water Networks 307

the shortest path between two random nodes in the network. Note that we compute betweenness centrality using unweighted networks. To highlight their evolution, the node degree and betweenness centrality metrics for four consecutive years are shown in Figure 11.4 and Figure 11.5, respectively. Please note that the values are in log scale, wherein red regions have the largest values and blue regions have the lowest ones. The presented years have been arbitrarily selected, since a detailed analysis for node degree throughout all years follows in the next section.

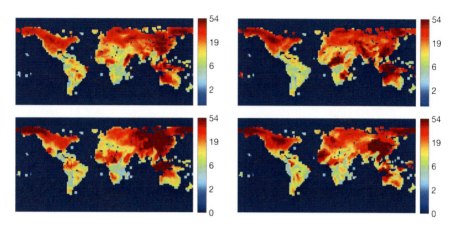

Figure 11.4: Node degree of precipitation networks, for the years 1960 (top left), 1961 (bottom left), 1962 (top right), and 1963 (bottom right). Figure is adapted with permission from [11].

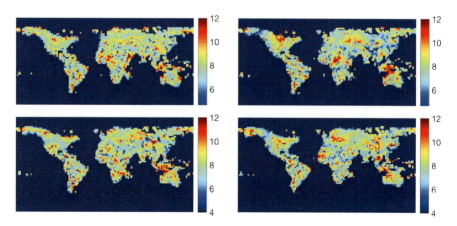

Figure 11.5: Betweenness centrality of precipitation networks, for the years 1960 (top left), 1961 (bottom left), 1962 (top right), and 1963 (bottom right). Figure is adapted with permission from [11].

By observing the centers (dark red clusters) of the centrality maps in Figure 11.5, we make certain key observations. First, the centers in the degree centrality map can be interpreted as the most influential nodes [13]. In other words, when they experience a climate anomaly, the nodes connected to them are highly likely to be affected accordingly. Most importantly, it is worth stressing that the centers are aligned in both metrics. This is to be expected, since these centers correspond to locations affected heavily by global jet streams and ocean currents. That is, the highly connected nodes simultaneously exhibit high centrality throughout all the years along the west coast of N. America, of S. America, of N. Europe, of Africa, of Australia, and the east coast of Central America and Asia.

We have shown the polar and subtropical jet stream paths and our betweenness centrality plots across all years in a supplementary video posted at `https://goo.gl/EodYBP`. The betweenness centrality video is smoothed by up-sampling the images and applying a Gaussian Kernel in order to reveal a clear pattern. As is evident, high betweenness (dark red centers) in Alaska, the USA-Canada border, and Northern Europe corresponds to the polar jet streams, while high betweenness in Mexico and Northern Africa corresponds to subtropical jets; similar considerations apply to the Southern hemishpere. From a network perspective, nodes with high betweenness centrality represent temporal and spatial *shortest paths* through which the precipitable water traverses. Since jet streams are basically very fast moving segments of atmosphere, air and, hence, water carried by jet streams will also traverse the shortest paths. Therefore, it intuitively makes sense that the shortest precipitable water traversal paths captured by our network correspond to the polar and subtropical jet stream paths. Of note, all these regions are also consistent with prior art findings related to polar jet streams [20] and ocean currents [13].

Long-Memory Multiscale Precipitation Patterns

In this section, we investigate the ability of complex networks to capture inherent multiscale dynamics between precipitation and temperature anomalies. While previously the correlation between precipitation and temperature anomalies has been reported for autumn data [11], we now focus on yearly data in North America. That way, we effectively investigate different temporal and spatial scales. Figure 11.6 shows the node degree and the average degree for precipitation and temperature anomalies networks across all the nodes for N. America. The degrees of all the nodes for each year are vectorized as a column and they are sorted by their values. The correlation coefficient is 0.56. This key observation that the average degrees of precipitation and temperature networks are strongly correlated across both scales, demonstrates the multiscale modelling capacities of complex networks. From a practical point of view, years with more temperature anomalies are also likely to witness precipitation anomalies, and vice versa, a key insight that can be used to tackle sensitivities of regional water models to dynamic climate anomalies.

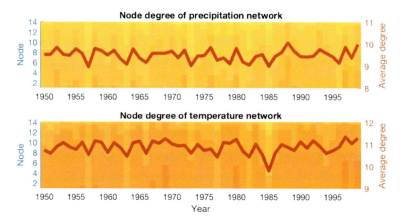

Figure 11.6: Node degree distribution and the average degree value for precipitation and temperature anomalies networks for N. America. The correlation coefficient of the two average degree curves is 0.56. Figure is adapted with permission from [11].

11.4 RIVER FLOW FORECASTING: K-HOP LEARNING

Let us now look at a specific example of how computational approaches can be used in the context of networked river systems. Below, we describe the formulation of a feature extraction technique which can significantly improve the accuracy of short-term river flow rate predictions [7].

11.4.1 PROBLEM FORMULATION

In this section, we first present the data sources and construct the river network from data. Then, we explain K-hop learning which learns features from the river network in order to improve short-term flow rate predictions.

Data Sources and Preparation

In this work, we use two major datasets: *(i)* daily river flow rate [29] and *(ii)* daily precipitation [30] for the Ohio River Basin, USA (total area: 204,000 sq. miles). Both datasets are available within NASA's North American Land Data Assimilation System 2 (NLDAS-2) database. The spatial resolution of both datasets is $0.125°$ latitude by $0.125°$ longitude. Figure 11.7(a) and (b) show the region under consideration and its gridded structure as plotted in Google Earth.

At the $0.125°$ scale, the Ohio River Basin contains a total of 3681 locations. In Table 11.1, we show a sample of flow rate data for the year 1997 ($24h \times 365 = 8760h$). Here, each node number refers to a location (i.e., a yellow placeholder shown in Figure 11.7(b)) with the given latitude and longitude. For the purpose of our experiments, we have used data for the years 1992-1997. Of note, we preprocess both datasets to generate daily timeseries from the original hourly data.

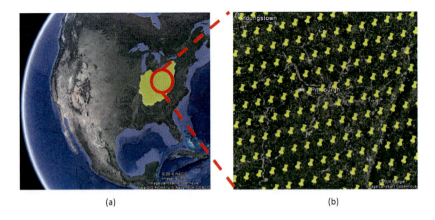

(a) (b)

Figure 11.7: (a) The region under consideration is the Ohio River Basin, USA (total area 204,000 sq. miles). (b) The datasets are gridded at 0.125° latitude/longitude resolution (11-13 km), for which the Ohio River Basin contains 3681 locations.

The River Network

Now, we must answer an important question: *Can we infer an approximate river network using just the historic river flow rate timeseries data?* To address this, suppose this network is represented by an adjacency matrix R and its nodes by Node IDs $\{1, 2, \ldots, 3681\}$ (see Table 11.1). Also, let $r_i(t)$ be the flow rate of a river at node i. The following two observations are critical to our approach:

1. Rivers combine to form larger rivers.
2. Average flow rate increases as we proceed through the network from smaller rivers towards larger rivers.

The first observation above implies that, for a given node i, the sum of flow rates of its neighboring nodes (denoted $\mathcal{N}(i)$) must be highly correlated with the flow rate at node i. Here, $\mathcal{N}(i)$ consists of 8 nearest neighbors of node i in the grid shown in Figure 11.7(b). Then, the directions in the river network can be obtained using the second observation above.

Based on these observations and assuming that no more than 3 rivers meet at a single place, we infer the river network by maximizing the correlation between flow rate at node i and sum of flow rates at its immediate neighbors. For each node i,

Incorporating Short- and Long-Term Dynamics in Smart Water Networks **311**

therefore, the problem is to:

$$\begin{aligned}
\underset{R(i,j)}{\text{maximize}} \quad & corr\left(r_i(t), \sum_{j \in \mathcal{N}(i)} r_j(t) \cdot R(i,j) \right) \\
\text{subject to} \quad & r_i^{avg} > \sum_{j \in \mathcal{N}(i)} r_j^{avg} \cdot R(i,j) \\
& \sum_{\forall j} R(i,j) \leq 3, \quad R(i,j) \in \{0,1\}, \forall i,j
\end{aligned}$$
(11.2)

where, r_i^{avg} refers to the average flow rate at node i.

The final river network is obtained by solving Eq. 11.2 for all nodes. In our approach, we connect two nodes with each other only when they exhibit a very strong correlation (e.g., when correlation is greater than 0.99). For instance, the sum of flow rates of nodes 3029 and 3185 shown in Table 11.1 would be $\approx \{1119, 1118, \ldots, 700\}$ which is nearly the same as that observed for node 3107. In fact, node 3107 represents the city of Pittsburgh where the Ohio River originates at the confluence of the Monongahela and Allegheny rivers (i.e., nodes 3029 and 3185, respectively).

Table 11.1

River flow rate data sample for the year 1997. Table is reproduced with permission from [7].

Node	Latitude	Longitude	River flow rate (in m³/s) at time			
No.	(North)	(West)	$1h$	$2h$	$\ldots$	$8760h$
3029	40.3125	−79.8125	404.97	404.34	$\ldots$	269.95
3185	40.5625	−79.8125	715.60	714.60	$\ldots$	431.63
3107	40.4375	−79.8125	1131.3	1130.1	$\ldots$	707.93

Next, in order to validate that the river network obtained from Eq. 11.2 is indeed correct, we plot a directed path from Pittsburgh in Google Earth. As expected, the path not only ends at Cairo, IL, where the Ohio River meets the Mississippi River, but also follows the exact same trajectory as the Ohio River. The trajectory of the Ohio River and that of the corresponding directed path plotted in Google Earth are shown in Figure 11.8(a) and (b), respectively. This verifies that the resulting network structure learned by solving Eq. 11.2 is correct.

K-Hop Feature Learning

Here, we address another important question: *Can we leverage the river network and the flow rate data to reveal the necessary features in order to predict the upcoming flow rate peaks 24-hours in advance?* To answer this, we describe K-hop learning [7] which utilizes the above river network to learn such hidden features.

We show the schematic of the river network in Figure 11.9(a), where nodes are

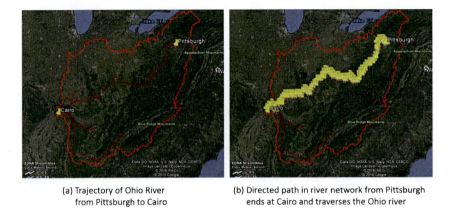

(a) Trajectory of Ohio River from Pittsburgh to Cairo

(b) Directed path in river network from Pittsburgh ends at Cairo and traverses the Ohio river

Figure 11.8: (a) Trajectory of the Ohio River from Pittsburgh to Cairo. (b) Directed path in the river network (obtained by solving Eq. 11.2) starting at Pittsburgh, traverses the Ohio River exactly (both maps were plotted in Google Earth [1]).

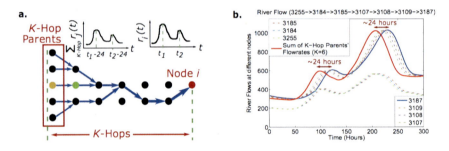

Figure 11.9: K-hop learning: (a) River network schematic — link thickness increases since the river size increases when more and more rivers combine. Upper panel (right): $r_i(t)$ plot shows the flow rate at node i with two peaks at times t_1 and t_2. Upper panel (left): the sum of K-hop parents' flow rates plot for which the corresponding peaks appear almost 24 hours earlier. (b) Real flow rate data shows that the sum of flow rates of K-hop parents can detect peaks $\approx$ 24 hours in advance. This data is for the Ohio River at node 3187 (near Pittsburgh, USA). For clarity, only one of the K-hop parents (i.e., node 3255) is shown here. Figure is reproduced with permission from [7].

locations, links show the rivers flowing between locations, and we have river flow rate data for each node. Then, the problem is to predict the river flow rate 24 hours ahead at node i as shown in Figure 11.9(a).

Imagine the flow rates at each of these nodes as a signal flowing through the river network. Next, we consider the pair of orange and green nodes shown in Figure 11.9(a). As the flow rate signal travels from the orange node (i.e., the source) towards the green node (i.e., the destination), there will be a delay as it moves from

Incorporating Short- and Long-Term Dynamics in Smart Water Networks **313**

the source to the destination. Therefore, as the signal travels through the network, *at every hop*, it will experience and accumulate some *delay*. This delay is a function of physical features of various rivers such as elevation and topographical or geographical features. Consequently, the delay will be different not only for different nodes, but also for different rivers. Since the topography, geography, and elevations of rivers do not change very quickly, we can easily learn such delays for all nodes using historic data. Hence, for every node i, we can *learn* the *optimal number of hops, K*, such that the total delay along those K hops is approximately 24 hours. In other words, as illustrated in Figure 11.9(a), if the flow rate at node i ($r_i(t)$) peaks at times t_1 and t_2, we can detect these peaks 24 hours ahead using the sum of flow rates of its parent nodes, some K-hops upstream in the network.

The above ideas can be mathematically formulated as an optimization problem. Let $\mathcal{P}_K^i$ be the K-hop parents of a given node i. Since the delay on each hop is a function of river topography/geography which does *not* change very fast, the problem is to learn such features using historic data (1992-1996). Therefore, optimal K for each node can be found by maximizing the cross-correlation between flow rate at node i and sum of flow rates of its K-hop parents:

$$\underset{K}{\text{maximize}} \quad xcorr\left(r_i(t), \sum_{j \in \mathcal{P}_K^i} r_j(t) \right)$$
$$\text{subject to} \quad 21 \leq \tau_{xcorr}^{max} \leq 25 \tag{11.3}$$
$$3 \leq K \leq 12$$

where, τ_{xcorr}^{max} is the lag which maximizes the cross-correlation function. Intuitively, τ_{xcorr}^{max} corresponds to the total delay along the K hops between node i and its K-hop parents. Therefore, the first constraint in Eq. 11.3 ensures that the K-hop parents can detect the upcoming river flow peaks at node i, 24 hours earlier. Moreover, since we can predict such peaks only approximately, the above constraint aims to keep τ_{xcorr}^{max} between 23 ± 2 hours. The second constraint, on the other hand, restricts K to be between 3 and 12 for efficiency purposes. Also, many very small rivers (average flow rate < 25 m^3/s) did not have parents more than 12 hops upstream in the network. Such smaller rivers do not have as much use as larger rivers due to their small size. Therefore, we do not focus on predictions for these rivers. The smallest rivers considered in this work have an average flow rate of ≈ 150 m^3/s.

The above optimization problem is solved for *all nodes* in the river network for five different years (1992-1996). Note that, the value of K for each node is almost the same for various years. Still, we take this small variation into account by averaging the K obtained for each node across the five years. Hence, we obtain a mean K for 844 nodes which correspond to significant rivers with flow rates ≥ 150 m^3/s. The rest 2837 nodes are very small rivers and are, therefore, insignificant. For the aforementioned 844 nodes, $K \in \{5, 6, \ldots, 11\}$.

Finding K-hop parents for every node in the network can be implemented faster than a depth first search, since for each node, $K \in \{3, 4, \ldots, 12\}$ reduces the river network to a small tree with number of nodes $n << N$, the total number of nodes.

Therefore, finding K-hop parents takes $\mathcal{O}(N \cdot n)$. Next, computing cross-correlations for each node takes $\mathcal{O}(N \cdot p)$ (where, p is the length of timeseries). Therefore, the overall complexity of the problem is $\mathcal{O}(N \cdot (n + p))$ and since $n \gg p$, this is $\approx \mathcal{O}(N \cdot p)$. Practically, it takes about 15 seconds per year in MATLAB$^{\circledR}$ with 8 parallel workers to solve this problem for all nodes in the network.

The code to find K-hop away parents in the river network (for a specific K) can be implemented in a recursive fashion (like a depth first search). For a given K, Algorithm 13 describes a pseudo-code for learning K-hop parents in the river network. Here, *parentsSet* is initialized to be an empty array, *HopCount* is initialized to 1 before calling the `Compute_KHop_Parents` routine. Also, K is the number of hops upstream in the network for which we want to find the K-hop parents of node i. The `Compute_KHop_Parents` routine is run for $K \in \{3, 4, \ldots, 12\}$ until the total delay along the K hops between node i and its K-hop parents is close to 24-hours. In other words, we run `Compute_KHop_Parents` for various K's until τ_{xcorr}^{max} satisfies the first constraint in the Eq. 11.3.

Algorithm 13 `Compute_KHop_Parents`

1: **Input:** River Network R, Node i, *parentsSet*, *HopCount*, K
2: **Output:** *parentsSet*
3: *ImmediateParents* $\leftarrow$ find$(R(:, i) == 1)$ /* Find immediate parents of node in river network i */
4: **if** $HopCount == K$ **then**
5:     *parentsSet* $\leftarrow$ *parentsSet* $\cup$ *ImmediateParents*
6:     **return**
7: **else**
8:     **for all** j in *ImmediateParents* **do**
9:         $HopCountTemp = HopCount + 1$
10:         *parentsSet* $\leftarrow$ `Compute_KHop_Parents`$(R, j, parentsSet, \ldots$
11:                                 $\ldots, HopCountTemp, K)$
12:     **end for**
13: **end if**

Finally, we demonstrate the power of K-hop learning in detecting river flow peaks 24 hours ahead of time. As an example, we analyze the effectiveness of K-hop parents for detecting flow rate peaks of node 3187. Figure 11.9(b) shows the flow rate of node 3187 (thick blue line) and sum of flow rates of its K-hop parents (thick red line, $K = 6$). Note that, the river hierarchically flows from node 3255 to node 3184 to 3185, all the way to 3187. Looking closely at the river flow plots for each of the locations reveals that there is some delay between the peaks for each successive hop. Hence, going sufficiently upstream in the network (e.g., $K = 6$ here) can enable us to detect the peaks at 3187, 24 hours earlier. As evident, the peaks at node 3187 can be detected 24 hours ahead by looking at its K-hop parents (node 3255 shown in Figure 11.9(b) is only one of the K-hop parents). Therefore, K-hop learning can significantly improve the prediction accuracy.

Short-Term Prediction Model and Experimental Setup

In the last section, we saw how to learn relevant features from the network that can accurately capture river flow peaks 24 hours ahead. Here, we use these K-hop parents-based features in a machine learning model to predict the short-term flow rates for various rivers. Specifically, to predict the flow rate 24-hours (i.e., 1-day) ahead at each node i, we use a simple linear model consisting of two features:

1. Cumulative 24-hours precipitation at node i.
2. Sum of daily flow rates of its K-hop parents.

In order to compare our model with other timeseries-based machine learning models, we replace the *second* feature above by timeseries of daily flow rate at node i itself; we refer to these models as *timeseries-based machine learning models* throughout the chapter. Of note, thus far, we have used only the historic data (1992-1996) to create the network and to learn the K-hop parents. On the other hand, to perform all prediction tasks, we train and test all the models on 1997 data, while using the K-hop features based on historic 1992-1996 data. This further demonstrates that our K-hop learning method is robust across the years.

We account for nonstationary behavior of flow rate data using an established technique called *rolling window*, often used for nonstationary financial timeseries prediction [45]. In a rolling window method, the training period of a model is "rolled" or moved forward, as soon as the next observation becomes available. Such techniques are especially important for one-step ahead predictions. Therefore, we train our model for 20 days, test on the next day, then move the training set forward by one day, test on the day after that, and so on. For river applications, a similar technique has also been used in [31].

We conduct the prediction experiments at several locations in the Ohio River Basin, USA. Specifically, we show results for four different sized rivers – small (average flow rate ≈ 150 m^3/s), intermediate (flow rate ≈ 400 m^3/s), large (flow rate > 1000 m^3/s), and very large (flow rates of the order of 10^4 m^3/s). The locations for which we predict the 24-hours ahead river flow rate, along with their corresponding sizes are shown in Figure 11.10.

11.4.2 EXPERIMENTAL RESULTS

We now present the flow rate prediction results using K-hop learning and compare them to four timeseries-based machine learning models: *(i)* autoregression (AR), *(ii)* random forests (RF), *(iii)* SVM regression (SVM-R), and *(iv)* K-nearest neighbors regression (KNN-R).

Figure 11.11 shows the flow rate prediction results for six different locations (two rivers from each category – large (a,d), intermediate (b,e), and small (c,f)) in the Ohio River Basin. Each figure shows three lines: *(i)* observed flow rate (blue), *(ii)* predicted flow rate using K-hop features (red), and *(iii)* predicted flow rate using SVM-R based on timeseries data at the given node (green). Here, we have shown only the results corresponding to SVM-R because, as we shall see shortly, it performs

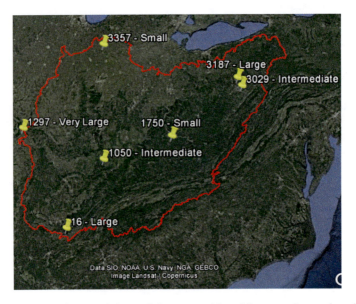

Figure 11.10: Locations and sizes of rivers considered for experimental evaluation.

the best (in most cases) among the timeseries-based models we have considered for comparison. Clearly, the prediction based on our proposed K-hop learning outperforms the traditional timeseries-based SVM-R and, therefore, all other methods for all six rivers (see Figure 11.11(a)-(f)).

Further, our model predicts the river flow rate peaks much more accurately than timeseries-based SVM-R. The peaks predicted by SVM-R and other models are generally one or two days after they actually occur. Our K-hop learning method, on the other hand, predicts these peaks on the correct day. Accurate prediction of these river flow peaks is critical for flood planning and run-of-river (ROR) hydropower management and, therefore, K-hop learning can play a fundamental role in this problem space. Below, we explain this in more detail using a concrete example.

We perform 24-hours ahead flow rate prediction for a very large river (flow rates of the order of 10^4 m^3/s) at node 1297. Node 1297 is located towards the end of the Ohio River Basin where the Ohio River ultimately meets the Mississippi River. Fig. 11.12 illustrates the prediction results with K-hop learning and SVM-R. The inset in Figure 11.12 shows a zoomed-in portion of the results (between the 112th–116th days) which reveals that the flow rates predicted by K-hop learning (red) are very close to the observed flow rates (blue), while SVM-R (green) still predicts the peaks and troughs one day late. This can have a significant impact on run-of-river hydropower management. For instance, consider the observed and predicted flow rates for the 115th day (see $\{X,Y\}$ measurements shown as blue, red, and green rectangles in Figure 11.12 inset). Clearly, SVM-R overestimates the flow rate by 460 m^3/s while K-hop learning's prediction is only 21 m^3/s below the true observed value. As a result, if we were to depend on SVM-R for 1-day ahead predictions, we

Incorporating Short- and Long-Term Dynamics in Smart Water Networks 317

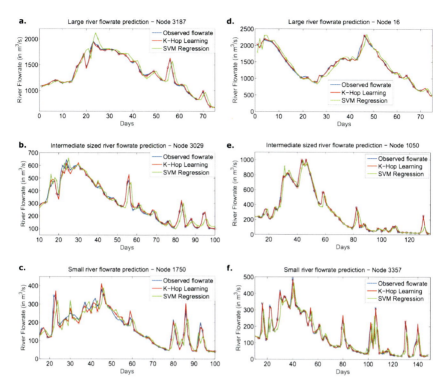

Figure 11.11: River flow rate prediction: (a,d) Large rivers (average flow rates > 1000 m^3/s), (b,e) Intermediate sized rivers (average flow rates ≈ 400 m^3/s), and (c,f) Small rivers (average flow rates ≈ 150 m^3/s). Predictions based on the proposed K-hop feature learning outperform the best timeseries-based model – SVM-R. Moreover, our approach can detect the river flow peaks very accurately. Please note that the results above generalize the analysis discussed [7] to multiple considered cases. Figure is extended with permission from [7].

would not be able to prepare for this sudden hydropower deficit observed on the 115th day. Therefore, K-hop learning can enable a significantly more effective run-of-river hydropower management.

Next, in Table 11.2, we report the root mean square error (RMSE) results for the flow rate prediction experiments shown in Figure 11.11 and Figure 11.12. The best results are indicated in bold. The table clearly demonstrates that K-hop learning *significantly* reduces the RMSE for rivers of all sizes, with a net reduction of 57 – 84%. More importantly, we still improve the prediction RMSE for the very large river case by around 75%. This shows that the performance of our proposed model does not degrade with increasing river sizes. In contrast, error for traditional models increases very rapidly as the river sizes increase. However, for K-hop learning, the error indeed increases as the river size increases from a small river to a very large

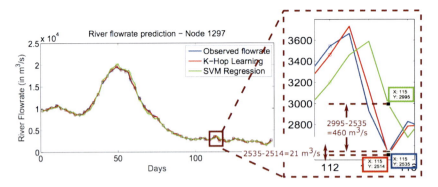

Figure 11.12: River flow rate prediction for a very large river with flow rates of the order of 10^4 m^3/s. Inset: Observed and predicted flow rates zoomed-in between 112th-116th days. Clearly, the trough predicted on 115th day by K-hop learning is much more accurate than the one predicted by SVM-R (with error of $21 m^3/s$ for K-hop learning vs. $460 m^3/s$ for SVM-R).

Table 11.2
River flow rate prediction RMSE. Table is extended with permission from [7].

Node & Size*	RMSE (in m^3/s)					% Gain
	AR	RF	SVM-R	KNN-R	K-hop	
1750-S	33.28	39.69	33.27	39.49	**14.09**	57.64
3357-S	55.44	60.18	50.71	56.59	**9.00**	82.25
3029-M	57.65	61.21	48.20	60.86	**14.78**	69.33
1050-M	55.16	85.21	51.85	76.36	**8.17**	84.21
3187-L	95.80	132.88	85.61	117.9	**14.95**	82.53
16-L	88.71	180.24	86.48	143.70	**32.71**	62.17
1297-VL	341.95	1335.79	373.65	929.26	**86.16**	74.80**

*S– Small, M– Medium or Intermediate, L– Large, VL– Very Large River.

** % Gain reported with respect to AR since AR performs the best among all timeseries models for this case.

river, but not nearly as rapidly as that observed for timeseries-based machine learning models.

Table 11.3 shows the results for mean absolute errors (MAE) for all locations and river sizes. Note that, except for the very large river, the SVM-R performs the best among all other timeseries-based machine learning models. For the very large river, AR performs slightly better than SVM-R (in both MAE and RMSE). Table 11.3 clearly demonstrates that K-hop learning performs the best among all other models

Table 11.3

River flow rate Prediction Mean Absolute Error (MAE)

| Node & | MAE (in m^3/s) | | | | | % |
Size*	AR	RF	SVM-R	KNN-R	K-hop	Gain
1750-S	19.98	25.61	19.24	23.71	**7.83**	59.30
3357-S	34.39	41.25	30.24	36.37	**5.65**	81.31
3029-M	35.62	45.53	29.72	41.51	**9.65**	67.53
1050-M	36.15	58.70	31.99	47.38	**5.38**	83.18
3187-L	67.98	101.86	59.74	87.66	**9.62**	83.89
16-L	62.82	131.75	59.76	94.01	**14.78**	75.26
1297-VL	253.10	904.29	274.97	555.89	**62.99**	75.11**

*$S-$ Small, $M-$ Medium or Intermediate, $L-$ Large, $VL-$ Very Large River.

** % Gain reported with respect to AR since AR performs the best among all timeseries models for this case.

and yields MAE improvements ranging from 59% to 83%. Also, we note that the MAE values for K-hop learning also do not scale up quickly as the river size increases. We also ran experiments on more rivers in the network and observed similar results.

Finally, in Figure 11.13, we analyze how the performance of various models varies with different training set sizes. As is evident, among the timeseries-based machine learning models, SVM-R performs the best (magenta line in Figure 11.13). However, K-hop learning not only significantly reduces the RMSE but is also robust to varying training set sizes. Moreover, here we have compared a linear model using K-hop features with advanced machine learning models using timeseries data at a given node, thereby showing the power of K-hop feature learning. Note that, K-hop features automatically take network characteristics (i.e., the characteristics responsible for complex river peaks and troughs) into account. This possibly explains why K-hop learning performs better even with a linear model than advanced machine learning models. When K-hop features were used in conjunction with advanced models such as SVM-R, the RMSE results obtained were very similar. Hence, like in many machine learning tasks, the performance improvement is governed by the quality of features. This further highlights how using network-based features can significantly improve the prediction accuracy as opposed to using local features (e.g., features based on a single location).

11.5 CONCLUDING REMARKS AND FUTURE DIRECTIONS

In this work, we have comprehensively evaluated the modelling capacity of complex networks to capture long-memory temporal, large-scale spatial, and multiscale cli-

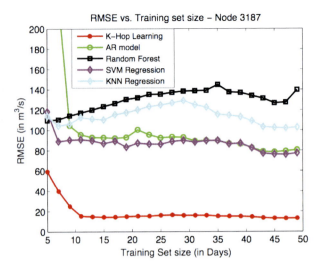

Figure 11.13: Impact of varying training set size on RMSE for the Ohio River at node 3187. Clearly, K-hop learning outperforms all other timeseries-based machine learning models. Also, SVM regression performs best among other models. Figure is reproduced with permission from [7].

mate dynamics by investigating their properties through the respective case studies. Towards this direction, we were the first to capture the correlation between carbon emissions and precipitation anomalies using complex networks. Moreover, through our key findings, we postulated the potential of complex network approaches as a complementary means to dynamically evaluate the need to reassess regional water models, and to eventually enhance their accuracy.

More importantly, we have proposed K-hop learning, a novel feature extraction method, which utilizes network information to learn upcoming river flow peaks 24-hours in advance. We have used these learned features to forecast the river flowate 24-hours ahead at various locations in the Ohio River Basin, USA. Finally, we have compared our approach to timeseries-based machine learning models such as SVM regression, autoregression, random forests, and K-nearest neighbor regression. Our results have clearly demonstrated that the predictions using K-hop learning are far superior to the above models. This ultimately illustrates the significant advantage of using network information in the field of computational sustainability.

As a future work, we first plan to compare our correlations-based river network to real river connectivity data. We will also incorporate long-term precipitation dynamics in our current framework to evaluate its sensitivity to anthropogenic climate change [11]. In addition, our work has motivated that these key results can be flexibly incorporated into fine-grained smart water networks. An orthogonal step to our analysis is to train a model on specific water-systems (e.g., a river model such as the one used in the K-hop learning method [7]) throughout multiple decades, and to

Incorporating Short- and Long-Term Dynamics in Smart Water Networks **321**

investigate the correlation between the model's coefficients and the characteristics of a complex network spanning a wider region around the point of interest.

ACKNOWLEDGMENTS

This work was supported in part by the US National Science Foundation (NSF) under CyberSEES Grant CCF-1331804.

11.6 REFERENCES

1. Google Earth. `https://www.google.com/earth/`.
2. Ohio Watershed Google Earth KML Files - USGS Elevation Derivatives for National Applications. `https://edna.usgs.gov/watersheds/kml_index.htm`.
3. R J Abrahart and L See. Multi-model data fusion for river flow forecasting: an evaluation of six alternative methods based on two contrasting catchments. *Hydrology and Earth System Sciences Discussions*, 6(4):655–670, 2002.
4. M A Alexander, I Bladé, M Newman, J R Lanzante, N-C Lau, and J D Scott. The atmospheric bridge: The influence of enso teleconnections on air–sea interaction over the global oceans. *Journal of Climate*, 15(16):2205–2231, 2002.
5. E Artinyan, B Vincendon, K Kroumova, N Nedkov, P Tsarev, S Balabanova, and G Koshinchanov. Flood forecasting and alert system for arda river basin. *Journal of Hydrology*, 541:457–470, 2016.
6. K Bhardwaj and R Marculescu. Network-based modelling and analysis of cloud fraction and precipitation: A case study for the ohio river basin. In *Proceedings of the 1st ACM International Workshop on Cyber-Physical Systems for Smart Water Networks*, page 12. ACM, 2015.
7. K Bhardwaj and R Marculescu. K-hop learning: a network-based feature extraction for improved river flow prediction. In *Proceedings of the 3rd International Workshop on Cyber-Physical Systems for Smart Water Networks*, pages 15–18. ACM, 2017.
8. V D Blondel, J-L Guillaume, R Lambiotte, and E Lefebvre. Fast unfolding of communities in large networks. *Journal of statistical mechanics: theory and experiment*, 2008(10):P10008, 2008.
9. T A Boden, G Marland, and R J Andres. Global, regional, and national fossil-fuel co2 emissions. Carbon Dioxide Information Analysis Center, Oak Ridge National Laboratory, U.S. Department of Energy, Oak Ridge, Tenn., U.S.A., 2013.
10. O Cordón, F Gomide, F Herrera, F Hoffmann, and L Magdalena. Ten years of genetic fuzzy systems: current framework and new trends. *Fuzzy Sets and Systems*, 141(1):5–31, 2004.
11. R Ding, D Stamoulis, K Bhardwaj, D Marculescu, and R Marculescu. Enhancing precipitation models by capturing multivariate and multiscale climate dynamics. In *Proceedings of The 3rd International Workshop on Cyber-Physical Systems for Smart Water Networks (CySWATER 2017)*, pages 1 4, April 2017.
12. J F Donges, Y Zou, N Marwan, and J Kurths. Complex networks in climate dynamics. *The European Physical Journal Special Topics*, 174(1):157–179, 2009.
13. J F Donges, Y Zou, N Marwan, and J Kurths. The backbone of the climate network. *EPL (Europhysics Letters)*, 87(4):48007, 2009.

14. I Ebert-Uphoff and Y Deng. A new type of climate network based on probabilistic graphical models: Results of boreal winter versus summer. *Geophysical Research Letters*, 39(19), 2012.
15. L Goddard, S J Mason, S E Zebiak, C F Ropelewski, R Basher, and M A Cane. Current approaches to seasonal to interannual climate predictions. *International Journal of Climatology*, 21(9):1111–1152, 2001.
16. D E Goldberg and J H Holland. Genetic algorithms and machine earning. *Machine Learning*, 3(2):95–99, 1988.
17. A Gozolchiani, K Yamasaki, O Gazit, and S Havlin. Pattern of climate network blinking links follows el nino events. *EPL (Europhysics Letters)*, 83(2):28005, 2008.
18. O Guez, A Gozolchiani, Y Berezin, S Brenner, and S Havlin. Climate network structure evolves with north atlantic oscillation phases. *EPL (Europhysics Letters)*, 98(3):38006, 2012.
19. O Guez, A Gozolchiani, Y Berezin, Y Wang, and S Havlin. Global climate network evolves with north atlantic oscillation phases: Coupling to southern pacific ocean. *EPL*, 103(6):68006, 2013.
20. R Hall, R Erdélyi, E Hanna, J M Jones, and A A Scaife. Drivers of north atlantic polar front jet stream variability. *International Journal of Climatology*, 35(8):1697–1720, 2015.
21. M P Hoerling, A Kumar, and T Xu. Robustness of the nonlinear climate response to ensos extreme phases. *Journal of Climate*, 14(6):1277–1293, 2001.
22. K-l Hsu, H V Gupta, and S Sorooshian. Artificial neural network modelling of the rainfall-runoff process. *Water Resources Research*, 31(10):2517–2530, 1995.
23. E Kalnay, M Kanamitsu, R Kistler, W Collins, D Deaven, L Gandin, M Iredell, S Saha, G White, J Woollen, et al. The ncep/ncar 40-year reanalysis project. *Bulletin of the American Meteorological Society*, 77(3):437–471, 1996.
24. M Karlsson and S Yakowitz. Nearest-neighbor methods for nonparametric rainfall-runoff forecasting. *Water Resources Research*, 23(7):1300–1308, 1987.
25. S T Khu, S-Y Liong, V Babovic, H Madsen, and N Muttil. Genetic programming and its application in real-time runoff forecasting. *JAWRA Journal of the American Water Resources Association*, 37(2):439–451, 2001.
26. B Li, G Yang, R Wan, X Dai, and Y Zhang. Comparison of random forests and other statistical methods for the prediction of lake water level: a case study of the Poyang Lake in China. *Hydrology Research*, in press, pages 1–15, 2016.
27. B B Mandelbrot. *Fractals*. Wiley Online Library, 1977.
28. L Montestruque and M D Lemmon. Globally coordinated distributed storm water management system. In *Proceedings of the 1st ACM International Workshop on Cyber-Physical Systems for Smart Water Networks*, page 10. ACM, 2015.
29. NASA. NLDAS-2 Noah Streamflow Dataset. *Accessed: 2016.* *http://ldas.gsfc.nasa.gov/nldas/NLDASnews.php*, 1979-2008.
30. NASA. NLDAS-2 Forcing File A Dataset. *Accessed: 2016.* *http://ldas.gsfc.nasa.gov/nldas/NLDAS2forcing.php*, 1979-2016.
31. T-T Nguyen, Q N Huu, and M J Li. Forecasting time series water levels on Mekong river using machine learning models. *Int'l Conf. on Knowledge and Systems Engineering*, pages 292–297, 2015.
32. V Nourani. An emotional ann (eann) approach to modelling rainfall-runoff process. *Journal of Hydrology*, 544:267–277, 2017.
33. A Panousopoulou and P Tsakalides. On graph-based feature selection for multi-hop

performance characterization in industrial smart water networks. In *Cyber-Physical Systems for Smart Water Networks (CySWater), 2016 International Workshop on*, pages 19–24. IEEE, 2016.

34. M S Shahriar, M Kamruzzaman, and S Beecham. Multiple resolution river flow time series modelling using machine learning methods. *Proc. Machine Learning for Sensory Data Analysis*, pages 62–66, 2014.

35. A Y Shamseldin and K M O'Connor. A nearest neighbour linear perturbation model for river flow forecasting. *Journal of Hydrology*, 179(1-4):353–375, 1996.

36. J E Shortridge, S D Guikema, and B F Zaitchik. Machine learning methods for empirical streamflow simulation: A comparison of model accuracy, interpretability, and uncertainty in seasonal watersheds. *Hydrol. Earth Syst. Sci.*, 20:2611–2628, 2016.

37. D Solomatine, L M See, and R J Abrahart. Data-driven modelling: concepts, approaches and experiences. In *Practical Hydroinformatics*, pages 17–30. Springer, 2009.

38. K Steinhaeuser and N V Chawla. Identifying and evaluating community structure in complex networks. *Pattern Recognition Letters*, 31(5):413–421, 2010.

39. K Steinhaeuser, A R Ganguly, and N V Chawla. Multivariate and multiscale dependence in the global climate system revealed through complex networks. *Climate Dynamics*, 39(3-4):889–895, 2012.

40. A A Tsonis and K L Swanson. Topology and predictability of el nino and la nina networks. *Physical Review Letters*, 100(22):228502, 2008.

41. A A Tsonis, K L Swanson, and P J Roebber. What do networks have to do with climate? *Bulletin of the American Meteorological Society*, 87(5):585–595, 2006.

42. M Vliet et al. Vulnerability of US and European electricity supply to climate change. *Nature Climate Change*, 2:676–681, 2012.

43. C J Vorosmarty, A Y Hoekstra, S E Bunn, D Conway, and J Gupta. What scale for water governance? *Science*, 349(6247):478–479, 2015.

44. C J Vorosmarty, P B McIntyre, et al. Global threats to human water security and river biodiversity. *Nature*, 467:555–561, Sept. 2010.

45. I Welch and A Goyal. A comprehensive look at the empirical performance of equity premium prediction. *The Review of Financial Studies*, 21(4):1455–1508, 2008.

46. K Yamasaki, A Gozolchiani, and S Havlin. Climate networks around the globe are significantly affected by el niño. *Phys. Rev. Lett.*, 100:228501, Jun 2008.

47. X Yao and Y Liu. A new evolutionary system for evolving artificial neural networks. *IEEE Transactions on Neural Networks*, 8(3):694–713, 1997.

Biographical Notes of Contributors

Anastasia Aidini. Anastasia Aidini received her B.Sc. degree from the Department of Mathematics of the University of Crete in 2014. In 2016 she received her M.Sc. degree in mathematics and its applications, with specialization in cryptography and coding theory, in mathematical foundations of informatics from the Department of Mathematics and Applied Mathematics of University of Crete. Currently, she is a graduate research assistant at the Institute of Computer Science (FORTH-ICS), in Signal Processing Laboratory and she is pursuing her Ph.D. degree at the Computer Science Department in the University of Crete, under the supervision of Prof. Panagiotis Tsakalides. Her main research interests lie in the fields of matrix and tensor completion, compressed sensing and coding.

Daniel Alonso Román. Daniel Alonso Román received the B.S. in telecommunications engineering from the Universidad Politcnica de Madrid, Spain, in 1994, and the M.Sc. in electrical engineering from the Universidad de Valencia, in 2011 (Master Special Award). He has worked as project and business manager in different companies in the field of communications and information systems. He worked as a research assistant at University of Valencia for three years, participating in several national and international projects. He is currently pursuing the Ph.D. degree at University of Agder, Norway, in the areas of distributed signal processing, communication, and networking for wireless sensor networks.

César Asensio Marco. César Asensio Marco received the B.S. degree in computer science and in telecommunications from the University of Valencia, Spain, in 2006, and the M.Sc. in electrical engineering in 2008. He worked as a research assistant at University of Valencia for six years, participating in several national and international projects. He got his Ph.D. degree at the University of Agder, Norway, in 2017. He is currently a postdoctoral researcher at the WISENET Lab at the University of Agder working in the areas of distributed signal processing, communication, and networking for wireless sensor networks.

Konstantinos Bacharidis. Konstantinos Bacharidis completed his B.Sc. and M.Sc. in electrical and computer engineering (ECE) at the Technical University of Crete (TUC) in 2014 and 2016, respectively. He is currently pursuing his Ph.D. in computer science, at the University of Crete (UoC) in association with the Computer Vision and Robotics Laboratory (CVRL), at the Foundation for Research and Technology — Hellas (FORTH). His current research focuses on fine-grained action detection and recognition. His main research interests involve object detection and tracking, motion estimation and stereoscopic vision.

Despina Bakogianni. Despina Bakogianni is a chemical engineer, holding a master's degree in environment and development. Currently, she is a Ph.D. candidate in the Unit of Environmental Science and Technology of the National Technical University of Athens, Greece. Her postgraduate research focuses on wastewater management, resource efficiency, circular economy, and waste valorization. She specializes also in combining waste management methods and pollution prevention with geographical information and remote sensing systems. Her work experience focuses on the implementation and monitoring of EU-funded international projects considering wastewater management, especially leachate valorization and brine management.

Baltasar Beferull-Lozano. Prof. Baltasar Beferull-Lozano received the Ph.D. degree in electrical engineering from University of Southern California (USC), Los Angeles, in 2002. In October 2002, he joined EPFL as a research associate. In December 2005, he joined the School of Engineering at University of Valencia as Associate Professor. Since August 2014, he has been a professor at the Department of Information and Communication Technology, University of Agder, leading the Lab Intelligent Signal Processing and Wireless Networks (WISENET). He has served as a member of the technical program committees for several ACM and IEEE international conferences. His research interests are in the areas of distributed signal processing and communications for wireless sensor and communication networks, cross-disciplinary tools, data analytics, and machine learning. At USC, Dr. Beferull-Lozano received several awards including the Best Ph.D. Thesis Paper Award in 2002 and the Outstanding Academic Achievement Award in 1999. He also received the Best Paper Award at the IEEE DCOSS 2012 international conference and is a recipient of a TOPPFORSK Grant Award, from the Research Council of Norway, 2016. He is a senior member of the IEEE.

Evaggelina Belia. Dr. Evaggelina Belia is the co-founder and principal of Primodal Inc. She is a process engineer with over 20 years of experience in wastewater treatment. She specializes in combining mathematical modelling with her experience in operations to solve optimization problems around the globe. Her recent projects include the use of real-time data quality methods and dynamic modelling for real-time plant control. In addition, for the past nine years, she has been working in the field of model-based uncertainty evaluation as the chair of the joint IWA/WEF Task Group on Design and Operational Uncertainty (DOUT). Her focus is the integration of dynamic simulation, explicit uncertainty evaluations and static design guidelines into a cohesive approach for the design of water resource recovery facilities.

Kartikeya Bhardwaj. Kartikeya Bhardwaj received his B.E. (Hons.) degree in electrical and electronics engineering from BITS Pilani Goa Campus, India, in 2014. He is working toward the Ph.D. degree in the Department of Electrical and Computer Engineering, Carnegie Mellon University, Pittsburgh. He is the recipient of the DAAD-WISE Scholarship for a research internship in the Karlsruhe Institute of Technology, Germany (2013). He contributed several research papers in reputed international journals, conferences, and workshops. His research interests are artificial

Biographical Notes of Contributors

intelligence, machine learning, and network science. He also focuses on designing new machine learning and network science techniques for applications in sustainability, climate, social, and biological systems. He is a student member of the IEEE and the IEEE Computer Society.

Rachel Cardell-Oliver. Rachel Cardell-Oliver is an associate professor in the School of Computer Science and Software Engineering at the University of Western Australia. Her research interests are in wireless sensor networks for applications in water, buildings, transport and natural environments, analytics for sensor data, sensor network protocols and computer science education. She holds a Ph.D. in protocol verification from the Computer Laboratory of the University of Cambridge, and an M.Sc. in distributed systems software from the University of Western Australia.

Andrea Castelletti. Andrea Castelletti received an M.S. degree in environmental engineering and a Ph.D. in information engineering from Politecnico di Milano, Italy, in 1999 and 2005. Since 2006 he has been Assistant Professor of Management of Natural Resources in the same university (Como campus). He was visiting scholar at IDSIA (Lugano), Lancaster University (UK), and University of Western Australia. Since 2007 he has also been Honorary Research Fellow at the Centre for Water Research of the University of Western Australia. His main research interests focus on participatory and integrated modelling and control of environmental systems, namely water resource systems, and Decision Support System design. He is co-author of two international books on integrated water resource management and a number of papers in international journals and conference proceedings. Since 2008 he has been chair of the IFAC Technical Committee on Modelling and Control of Environmental Systems (TC 8.3) and since 2011 member of the ASCE/EWRI Environmental & Water Resources Systems Technical Committee. He is a member of the editorial board of *Environmental Modelling & Software* and associate editor of *Acta Geophysica*. In 2010 he was awarded an Early Career Excellence Award by the International Environmental Modelling and Software Society.

Andrea Cominola. Andrea Cominola received a BS.c. cum laude in environmental engineering at Politecnico di Milano in 2010 and a double-degree M.Sc. cum laude at Politecnico di Milano and Politecnico di Torino in 2013. Since Fall 2013 he has been a Ph.D. candidate in information technology in the Natural Resources Management group of Professor Andrea Castelletti at the Department of Information, Electronics and Bioengineering at Politecnico di Milano. He was formerly a visiting Ph.D. candidate at the Watershed Sciences Center of UC Davis (CA - USA), visiting researcher at Korea University (Seoul, Korea), and visiting student at Penn State University (PA - USA) and Norwegian University of Science and Technology (Trondheim, Norway). He was selected to join the double-degree M.Sc. program of Alta Scuola Politecnica (Politecnico di Milano, Politecnico di Torino). His research interests include urban water consumers behavioural modelling, water and energy demands modelling and management and smart metering.

John B. Copp. Dr. John B. Copp is a principal at Primodal Inc. and has over 20 years applied wastewater treatment process modelling experience. John specializes in custom model development, the development of nutrient removal process models and improving data quality for clients around the globe. Dr. Copp received his Ph.D. from McMaster University in Hamilton, Canada, before two postdoc appointments in the Netherlands and South Korea investigating online model-based control. John has co-authored several books on the subject, as well as a chapter in MOP31. He is an active member of several IWA and WEF specialty groups including being the chair of WEFs OWWL Task Force and the vice-chair of WEFs MEGA sub-committee.

Amitava Datta. Amitava Datta is a professor in the School of Computer Science and Software Engineering at the University of Western Australia. His research interests are in data mining, high performance computing, and bioinformatics. He has published many articles in reputed international journals and conference proceedings.

Ruizhou Ding. Ruizhou Ding received the B.S. degree in electronic and information science and technology from Peking University, Beijing, China, in 2015. He is currently pursuing the Ph.D. degree in electrical and computer engineering with Carnegie Mellon University, Pittsburgh, PA, USA. His current research interests include applied machine learning, time series forecasting, and energy-aware deep neural networks. He was a recipient of the Best Paper Award for his research at GLSVLSI 2017.

Rong Du. Rong Du is currently a Ph.D. student at KTH Royal Institute of Technology, Electrical Engineering and ACCESS Linnaeus Center, Stockholm, Sweden. He received the master's degree (2014) and the bachelor's degree (2011) from Shanghai Jiao Tong University, China. His research interests include optimization with applications to wireless sensor networks for smart cities, vehicular sensor networks, and wireless energy transfer for sensor networks. He has publications in journals such as *IEEE Journal of Selected Areas in Communications, IEEE Transactions on Control of Network Systems,* and *IEEE Transactions on Vehicular Technology,* and in conferences such as IEEE Infocom, ICDCS, ICC, and Globecom. He has received several awards and scholarships, including Excellent Master's Thesis of Shanghai, China (2015), and National Scholarship for Graduate Students, China.

Alessandro Facchini. Alessandro Facchini is a logician, working since 2015 as a researcher at the Swiss AI Lab IDSIA. After finishing a Ph.D. in 2010 under the co-supervision of Jacques Duparc (University of Lausanne) and Igor Walukiewicz (LaBRI-University of Bordeaux 2), he has been a visiting researcher at the Database Group of the UCSC, and a post-doc at the Information and Language Processing Systems group of the University of Amsterdam. Before joining IDSIA, he was an assistant professor at the Informatics Institute of the University of Warsaw, where he was leading a two-year Homing Plus project on fixpoint logics, co-financed by the European Union ("Grants for Innovation"). His work has been published in journals such as *Logical Methods in Computer Science,* the *ACM Transactions on Computational Logic,* the *Journal of Symbolic Logic* and *Physical Review Letters A,* and at conferences such as LICS, MFCS and CSL. He is the recipient of the 2011 Paul Bernays Award.

Biographical Notes of Contributors

Carlo Fischione. Dr. Carlo Fischione is currently a tenured associate professor at KTH Royal Institute of Technology, Electrical Engineering and ACCESS Linnaeus Center, Stockholm, Sweden. He received the Ph.D. degree in electrical and information engineering (3/3 years) in May 2005 from University of L'Aquila, Italy, and the Laurea degree in electronic engineering (laurea, summa cum laude, 5/5 years) in April 2001 from the same University. He has held research positions at Massachusetts Institute of Technology, Cambridge, MA (2015, visiting professor); Harvard University, Cambridge, MA (2015, associate); and University of California at Berkeley, CA (2004-2005, visiting scholar, and 2007-2008, research associate. His research interests include optimization with applications to networks, wireless and sensor networks, networked control systems, security and privacy. He has co-authored over 150 publications, including a book, book chapters, international journals and conferences, and international patents. He has received or co-received a number of awards. He is editor of *IEEE Transactions on Communications* and associated editor of *IFAC Automatica*. He has chaired or served as a technical member of program committees of several international conferences and is serving as referee for technical journals.

Konstantina Fotiadou. Konstantina Fotiadou is currently pursuing the Ph.D. degree in computer science from the Computer Science Department of the University of Crete (UoC), Greece, under the supervision of Prof. Panagiotis Tsakalides. She received her M.Sc. degree in computer science from the Computer Science Department of the University of Crete, and B.Sc. degree in applied mathematics from the Department of Applied Mathematics (UoC), in 2014 and 2011, respectively. Since 2012 she has been affiliated with the Signal Processing Laboratory at FORTH-ICS as a research assistant. Her main research interests involve machine learning techniques for computational photography applications, with special emphasis in hyperspectral image enhancement techniques. She is currently involved with the H2020 PHySIS and DEDALE European projects, wherein she is developing optimization algorithms for terrestrial and remote sensing hyperspectral applications.

Piero Fraternali. Piero Fraternali received a Laurea Degree in electrical engineering (cum laude) in 1989 and a Ph.D. in computer science from the Politecnico di Milano in 1994. He is a full professor in the Department of Electronics, Information and Bioengineering of Politecnico di Milano and Head of the Degree Course in Computer Engineering of the Como Campus of Politecnico di Milano. His main research interests concern database integrity, active databases, software engineering, and methodologies and tools for WEB application development. He was the technical manager of the W3I3 Project "Web-Based Intelligent Information Infrastructures" (1998-2000). IIc is co-inventor of WebMI (http://www.webml.org) a model for the conceptual design of Web applications (US Patent 6,591,271, July 2003) and co-founder of Web Models (http://www.webratio.com), a start-up of the Politecnico di Milano focused on the commercialization of an innovative tool suite called WebRatio for the Model-Driven Development of Web applications. He was Program

Chair of the International Conference on Web Engineering in 2004 and Vice President of the Software Engineering Track of the WWW conference in 2005.

Michalis Giannopoulos. Michalis received his M.Sc. degree in signal processing for communications and multimedia from the Department of Informatics and Telecommunications of the National and Kapodistrian University of Athens, and his B.Sc. degree in informatics and telecommunications from the Department of Informatics and Telecommunications of the National and Kapodistrian University of Athens, in 2016 and 2012, respectively. Currently, he is pursuing his Ph.D. degree in computer science in the Computer Science Department of the University of Crete under the supervision of Prof. Panagiotis Tsakalides. Since September 2016, he has been affiliated with the Institute of Computer Science of the Foundation for Research and Technology Hellas (FORTH-ICS), Heraklion, Crete, as a graduate research assistant in the Signal Processing Laboratory (SPL). His main research interests lie in the fields of low-rank modelling, deep learning, and tensor decompositions and factorizations for Big Data signal processing.

Matteo Giuliani. Matteo Giuliani received his M.Sc. degree in environmental and land planning engineering from Politecnico di Milano and Politecnico di Torino in 2010, and obtained the Diploma of the Alta Scuola Politecnica V cycle in 2011. He received a Ph.D. in information technology from Politecnico di Milano in 2014, after a visiting period at Penn State University (PA, USA). He was post-doctoral research fellow in the research group on planning and management of environmental systems. Since November 2016, he has been an assistant professor at DEIB. His research interests include the integrated management of water resources in complex systems characterized by the presence of multiple decision makers and many conflicting stakeholders through the development and the applications of multi-objective optimization and control algorithms, decision-making under uncertainty, multi-agent systems. He is currently a member of the IFAC Technical Committee on Modelling and Control of Environmental Systems and of the ASCE/EWRI Environmental & Water Resources Systems Technical Committee.

Katherine-Joanne Haralambous. Katherine-Joanne Haralambous is a professor at the School of Chemical Engineering of the National Technical University of Athens, in General Chemistry Laboratory/Unit of Environmental Science and Technology. She has 20 years of experience in environmental issues and is an expert in the development of methods and systems for the treatment of industrial wastewater and in the development of guidelines for clean technologies and best available techniques. Her research activities involve environmental pollution issues emphasizing the chemical processes of metals in solid and aquatic environments. She has more than 90 publications in scientific journals and international conferences and she has also participated in 35 national and international research projects regarding waste and wastewater management.

Sergio Herrera. Sergio Herrera is an assistant researcher in the Dipartimento di Elettronica e Informazione at Politecnico di Milano. He received his M.Sc. in computer science in 2016 with a focus on data engineering and machine learning. His main

Biographical Notes of Contributors

research interests include web architectures and ICT for behavioral change; other interests include data analysis and knowledge discovery.

Sokratis Kartakis. Sokratis Kartakis received the B.Sc. degree (with honors) and M.Sc. in computer science from the University of Crete, Greece, in 2006 and 2008, respectively, and the Ph.D. degree in computer science from Imperial College London, Department of Computing, UK, in 2017. Between July 2003 and October 2008, he was an undergraduate and graduate fellow at HCI Laboratory of ICS-FORTH. The next two years, 2009-2011, he worked as Lecturer and Lab Coordinator in the Science Department of Technological Institute of Crete, Greece, and as Senior Software Engineer and Researcher in the project of Ambient Intelligence at HCI Laboratory of ICS-FORTH. In 2011, he moved to Yale University and the ENALAB Laboratory of Electrical Engineering Department as Visiting Assistant in Research in the domain of cyber physical systems. From January 2013 to July 2015, he cooperated with Seldera LLC of Ameresco, Framingham, MA, US, as Senior Software Engineer and Researcher designing algorithms and sensor networks for real-time monitoring and control of energy consumption for commercial building. Since March 2017, he has been an IoT research scientist at Intel Labs Europe, UK. His research interests include signal processing, data reduction, edge decision making, and communication optimization for Internet of Things (IoT) and cyber-physical systems.

George Livanos. George Livanos holds a master's and a bachelor's degree from the Technical University of Crete, Department of Electronics and Computer Engineering. His diploma thesis was regarding wavelet analysis for CT and MRI imaging while his master's concerned processing of immunohistochemical images for evaluation of the HER2/neu membrane overexpression. He is currently a Ph.D. candidate at the same department, assistant personnel of the Digital Image and Signal Processing Laboratory (DISPLAY) and is involved in research on image histogram modelling and segmentation for biomedical applications, polarimetric imaging and statistical analysis for measuring material properties, optimization tools and spectrum deconvolution approaches for analyzing spectroscopy data, machine vision and 3D image reconstruction approaches for environmental surveillance and inspection of industrial infrastructure, navigation of remotely operated vehicles via optical recognition. He has been involved in two book chapters and 26 paper publications/announcements in international journals and conferences in related areas of image/signal processing, biomedical engineering, 3D reconstruction from stereo and machine vision, chemometrics and spectrum analysis. He is an experienced researcher, having participated in several projects and has served as a reviewer in several journals and conferences in the above-mentioned scientific areas.

Maria Loizidou. Maria Loizidou is a professor at the School of Chemical Engineering of the National Technical University of Athens and the Head of the Unit of Environmental Science and Technology (UEST) that belongs to the same university. She has 35 years of technical and scientific experience on environmental issues. She has been the leader and scientist responsible for over 150 national, European and international projects related to environmental issues (available at:

http://www.uest.gr/projects.html). Prof. Loizidou was the scientific coordinator of the SOL-BRINE project, which was selected as the best Life Environment project out of 4,306 LIFE projects implemented over the past 25 years (Green Award). She has over 450 publications in scientific journals and international conferences.

Diana Marculescu. Diana Marculescu is a Professor of electrical and computer engineering at Carnegie Mellon University. She received the Dipl. Ing. degree in computer science from the Polytechnic University of Bucharest, Bucharest, Romania, and the Ph.D. degree in computer engineering from the University of Southern California, Los Angeles, CA, in 1991 and 1998, respectively. Her research interests include energy- and reliability-aware computing, and CAD for non-silicon applications, including e-textiles, computational biology, and sustainability. Diana was a recipient of the National Science Foundation Faculty Career Award (2000-2004), the ACM SIGDA Technical Leadership Award in 2003, the Carnegie Institute of Technology George Tallman Ladd Research Award in 2004, and several best paper awards. She was an IEEE Circuits and Systems Society Distinguished Lecturer (2004-2005) and the Chair of the Association for Computing Machinery (ACM) Special Interest Group on Design Automation from 2005 to 2009. Diana chaired several conferences and symposia in her area and is currently an associate editor for *IEEE Transactions on Computers*. She was selected as an ELATE Fellow (2013-2014), and is the recipient of an Australian Research Council Future Fellowship (2013-2017) and the Marie R. Pistilli Women in EDA Achievement Award (2014). Diana is an IEEE Fellow and an ACM Distinguished Scientist.

Radu Marculescu. Radu Marculescu is a professor in the Department of Electrical and Computer Engineering at Carnegie Mellon University. He received his Ph.D. in electrical engineering from the University of Southern California (1998). He has received multiple best paper awards, NSF Career Award, Outstanding Research Award from the College of Engineering (2013). He has been involved in organizing several international symposia, conferences, workshops, and tutorials, as well as guest editor of special issues in archival journals and magazines. His research focuses on design methodologies for embedded and cyber-physical systems, biological systems, and social networks. Radu Marculescu is an IEEE Fellow.

Julie A. McCann. Julie McCann is a professor of computer systems at Imperial College. Her work focuses on decentralized algorithms and protocols for low-powered sensing and control systems that dynamically adapt to their environments. She has over 100 publications, authored *Autonomic Computing: Principles, Design and Implementation* (Springer, 2013), and has accumulated over 10M GBP in research funding. McCann is currently PI for the Intel Collaborative Institute on Sustainable Cities, and Co-I for both the NEC and FP7 WISDOM (619795) smart water initiatives and the NERC FUSE: Floodplain Underground Sensing Network projects.

Mark Melenhorst. Mark Melenhorst is a senior researcher at the European Institute for Participatory Media. He has a master's in communication science and has received a Ph.D. in human-computer interaction (University of Twente, the Netherlands). He has a profound knowledge of research methodology, requirements elicitation, and ICT-mediated behavioral change processes. Furthermore, he is interested in user aspects of recommender systems. He has participated in various FP7 projects (Smart H2O, PHENICX, CrowdRec, CUbRIK), leading and contributing to the requirements, formative evaluation, and summative evaluation parts of these projects.

Isabel Micheel-Mester. Isabel Micheel-Mester is researcher at the European Institute for Participatory Media in Berlin. She studied media and information technologies at University of Bremen, and Aalborg University Copenhagen. Her research focuses on the areas of interaction design, user experience and user-centered design. Recent projects include the application of Web2.0-technologies, gamification, information visualization, and social networks in the areas of green ICT, citizen services, crowdsourcing and social interfaces.

Konstantia Moirogiorgou. Konstantia Moirogiorgou completed her B.Sc. and M.Sc. in electronic and computer engineering (School of ECE) at the Technical University of Crete (TUC), Greece. Currently, she is a Ph.D. candidate and a research and teaching assistant in the Digital Signal & Image Processing Laboratory (Display Lab) at the School of ECE in TUC. She has more than 10 years of (educational) laboratory experience in digital signal processing and a multiyear research experience in digital image processing (from 2005 until today). She is the author of several scientific publications in international journals and international conference proceedings. Her research interests mainly focus on digital signal and image processing, machine vision, pattern recognition, statistical analysis, design and implementation of real-time video processing systems, DSP-based imaging systems, video transmission over network, mainly in the research areas of environmental monitoring, early detection systems and surveillance. She has been involved in many EU funded R&D projects from 2005 until today.

Luis A. Montestruque. Luis A. Montestruque received the Ph.D. degree in control systems from the electrical engineering department at the University of Notre Dame in 2004. His research on model-based networked control systems concentrates on the use of real-time data and models to control networked infrastructure. He founded EmNet in 2004 to focus on the application of IoT, Big Data, and artificial intelligence to solve problems in stormwater and wastewater sewers. Its first pilot in South Bend, Indiana, deployed a wireless sensor network of over 150 sensors, making it the most densely monitored sewer system in the country. The South Bend project eliminated over a billion gallons of combined sewer overflows per year and decreased E. coli concentrations by half in the Saint Joseph River. Today EmNet is helping over 20 cities across the U.S. reduce flooding and overflows through smart sewer applications.

Jasminko Novak. Jasminko Novak is Managing Chairman of the European Institute for Participatory Media (EIPCM) and Professor of Information Systems at the University of Applied Sciences Stralsund. He was also guest lecturer at various European universities (University of Zurich, University of Applied Sciences in Basel). Prof.

Novak is an internationally acknowledged expert with 12 years of scientific and professional experience in user-centered design and evaluation of socio-technical systems, human-computer interaction, knowledge visualization and participatory media. He has extensive experience in the participation in and leadership of large-scale research projects (EU-IST, German Federal Ministry of Education and Research) as well as in innovation management and the transfer of research results to business exploitation. Prof. Novak has advised the EU DG-INFSO as member of Expert Groups on User-centric Media, Future Internet and Social Networks. He also advises companies and public institutions in IT, media, financial services, culture, and creative industries.

Athanasia Panousopoulou. Athanasia Panousopoulou has been a research scientist with FORTH since 2013 and a visiting lecturer at the Computer Science Department at the University of Crete since 2015. She received her Diploma and Ph.D. in electrical and computer engineering from the University of Patras, Greece in 2004 and 2009, respectively. Athanasia has 10 years of expertise on the design and deployment of wireless sensor networks and CPS in hostile and RF-harsh environments, and she was involved in several large- and small-scale research projects funded by the European Union, Greece, and the United Kingdom, including the FP7 STREP Hydrobionets Project on smart water treatment. Her research interests include wireless sensor/actuator networks, distributed networked architectures for machine learning and signal processing, and she has co-authored more than 30 peer-reviewed conference papers, including journal articles and book chapters. In 2015 she established the International Workshop series on Cyber-Physical Systems for Smart Water Networks, and is a member of the organizing committee since then. She is a chartered engineer from the Technical Chamber of Greece since 2004, a member of IEEE, and she is serving as a reviewer on numerous research forums, edited by IEEE and Springer.

Chiara Pasini. Chiara Pasini received the M.Sc. degree in computer engineering from Politecnico di Milano, Milan, Italy, in 2010, where she is currently working as assistant researcher in the Department of Electronics, Information and Bioengineering. Her research interests include web architectures, with particular focus on the development of web applications for environmental sustainability.

Andrea-Emilio Rizzoli. Andrea-Emilio Rizzoli received the Laurea (master's) degree in control engineering in 1989 and the Ph.D. in control engineering and informatics in 1993 from Politecnico di Milano (Italy). He spent a one-year post-doc at CSIRO, Division of Water Resources, where he worked on the design and implementation of the first core of the Integrated Catchment Modelling Toolkit, a software toolkit for modelling and simulation of water management processes at the catchment scale, aimed at supporting catchment managers in the decision making process. After this experience, he went back to Italy where he collaborated with the Department of Electronics of the Politecnico di Milano in the design and development of TwoLe, a decision support system for the management of multipurpose reservoirs. TwoLe has been applied as a participatory decision support tool in the management of Lake Maggiore. Presently, he is senior research scientist at IDSIA (Istituto Dalle

Molle di Studi sull'Intelligenza Artificiale) a research institute of the University of Applied Sciences of Southern Switzerland (SUPSI) and of the University of Lugano (USI). His current research interests have expanded the application of model reuse and encapsulation to different domains, such as transportation and logistics systems. He also holds the title of Professor at SUPSI. In 2001 he has been awarded the "Early Career Research Excellence" award by the Modelling and Simulation Society of Australia and New Zealand. He is a founding member and a fellow of the International Environmental Modelling and Software Society, whose aim is to develop and use environmental modelling and software tools to advance the science and improve decision making with respect to resource and environmental issues.

Cristina Rottondi. Cristina Rottondi received both her bachelor's and master's degrees cum laude in telecommunications engineering and a Ph.D. in information engineering from Politecnico di Milano in 2008, 2010 and 2014, respectively. From 2014 to 2015 she was postdoctoral researcher at the Department of Electronics, Information and Bioengineering in Politecnico di Milano. Currently, she is researcher at the Dalle Molle Institute for Artificial Intelligence (IDSIA) in Lugano, Switzerland. Her research focuses on data privacy and security in the automatic metering infrastructure of smart grids. She is co-author of more than 50 publications in scientific journals and peer-reviewed conference proceedings.

Andreas Savakis. Andreas Savakis is Professor of Computer Engineering at Rochester Institute of Technology (RIT), where he served as department head of Computer Engineering from 2000 to 2011. He received the B.S. with highest honors and M.S. degrees in electrical engineering from Old Dominion University in Virginia, and the Ph.D. in electrical and computer engineering with mathematics minor from North Carolina State University at Raleigh, NC. He was Senior Research Scientist with the Kodak Research Labs before joining RIT. His research interests include object tracking, expression and activity recognition, domain adaptation, scene understanding, change detection, deep learning and computer vision applications. Prof. Savakis has co-authored over 100 publications and holds 12 U.S. patents. He received the NYSTAR Technology Transfer Award for Economic Impact in 2006, the IEEE Region 1 Award for Outstanding Teaching in 2011, and the RIT Trustees Scholarship Award in 2015. He is Associate Editor of the *IEEE Transactions on Circuits and Systems for Video Technology* and the *Journal for Electronic Imaging*. He co-organized the first International Workshop on Extreme Imaging at ICCV 2015, and is one of the guest editors at the *IEEE Transactions on Computational Imaging* for a 2017 *Special Issue on Extreme Imaging*.

Ary Shiddiqi. Ary Shiddiqi is a lecturer at Institut Teknologi Sepuluh Nopember, Indonesia. He holds a master's in computer science from Monash University and is currently studying for his Ph.D. at the University of Western Australia. His research interests are wireless sensor networks, data mining and computer networks.

Dimitrios Stamoulis. Dimitrios Stamoulis received a diploma from the Department of Electrical and Computer Engineering, National Technical University of Athens, Greece in 2013. He received a master's of engineering from the

Department of Electrical and Computer Engineering, McGill University, Montreal, 2015. He is currently pursuing a Ph.D. degree in electrical and computer engineering at Carnegie Mellon University, Pittsburgh, USA. His research interests include performance optimization of heterogeneous systems under power and thermal constraints.

Grigorios Tsagkatakis. Dr. Grigorios Tsagkatakis (g. male) (http://spl.edu.gr/index.php/people/tsagkatakis/) received his diploma and M.Sc. degrees in electronics and computer engineering from the Technical University of Crete, Greece in 2005 and 2007, respectively, and his Ph.D. in imaging science from the Rochester Institute of Technology, New York, in 2011. During the period 2011–2013, he was a Marie Curie post-doctoral fellow at the Institute of Computer Science (ICS), FORTH and he is currently working as a research associate with the Signal Processing Laboratory at FORTH-ICS. He has been involved in various European projects, including H2020 PHySIS and DEDALE, FP7 CS-ORION & HYDROBIONETS, FP6 OPTAG, and in industry-funded projects by leading companies including Kodak and Cisco. He has co-authored more than 30 peer-reviewed conferences, journals, and book chapters in the areas of signal and image processing, computer vision and machine learning. He was awarded the best paper award in the Western New York Image Processing Workshop in 2010 for his paper, "A Framework for Object Class Recognition with No Visual Examples."

Panagiotis Tsakalides. Panagiotis Tsakalides received a diploma in Electrical Engineering from Aristotle University of Thessaloniki, Greece, in 1990. The same year, he moved to Los Angeles for graduate studies with scholarships from the Fulbright and Bodossakis Foundations. He received the M.Sc. degree in computer engineering and the Ph.D. degree in electrical engineering from the University of Southern California (USC) in 1991 and 1995, respectively. From 1996 to 1999, he was a research assistant professor with the Signal & Image Processing Institute at USC and he consulted for the US Navy and Air Force. From 1999 to 2002 he was with the Department of Electrical Engineering, University of Patras, Greece. In September 2002, he joined the Computer Science Department at the University of Crete as an associate professor and FORTH-ICS as a researcher. Currently, he is a professor of computer science, the Vice Rector for Finance at the University of Crete, and the Head of the Signal Processing Laboratory at FORTH-ICS. His research interests lie in the field of statistical signal processing with emphasis in non-Gaussian estimation and detection theory, sparse representations, and applications in sensor networks, audio, imaging, and multimedia systems. He has co-authored over 150 technical publications in these areas, including 35+ journal papers. Prof. Tsakalides has extended experience in transferring research and interacting with the industry. During the last 10 years, he has been the project coordinator on 7 European Commission and 9 national projects totaling more than 4 M Euros in actual funding for FORTH-ICS and the University of Crete.

Dimitris Xevgenos. Dr. Dimitris Xevgenos is the managing director of SEALEAU B.V. He has a strong engineering background, having obtained his bachelor's degree from the School of Mechanical Engineering at the National Technical University of Athens (2008), M.Sc. in electrical engineering (2010) and Ph.D. degree in chemical engineering (2016) from the same university, while he continued as a post doc in civil engineering at TU Delft. Throughout his studies, he obtained 8 scholarships, being in the top 2% of his class. His research activities involve the design of innovative processes on (waste)water and waste management, while his Ph.D. was focused on the design of an innovative system for the treatment of seawater desalination brine and resource recovery. This research was carried out in the framework of the EU project called SOL-BRINE (BestLife project). He has published 6 high-quality research papers in international peer-reviewed journals and has presented his work at 13 international, scientific conferences. He has been involved in 11 European research projects, while he has raised more than 15 million Euros by writing five successful EU proposals.

Michalis Zervakis. Michalis Zervakis holds a Ph.D degree from the University of Toronto, Department of Electrical Engineering, since 1990. He joined the Technical University of Crete in January 1995, where he is serving as Professor in the Department of Electronic and Computer Engineering. He was an assistant professor with the University of Minnesota-Duluth, USA, from September 1990 to December 1994. Prof. Zervakis is the director of the Digital Image and Signal Processing Laboratory (DISPLAY) at the Technical University of Crete. His research interests include modern aspects of signal and image processing, data mining and pattern recognition, imaging systems and integrated automation systems. Under his direction, the DISPLAY lab www.display.tuc.gr covers a wide range of design, analysis, implementation and validation areas. Applications include bioinformatics, biosignal analysis and medical imaging, biomarker selection from mass genomic data, time-frequency biosignal and EEG characterization, modelling of disease state and progression, as well as cancer research on diagnosis and prognosis. He is co-author of more than 280 scientific papers in international journals and conference proceedings. He has participated in several national and European research projects and has participated in the scientific committees of several IEEE conferences. He served as Associate Editor for the *IEEE Transactions on Signal Processing* from 1994 to 1996. He currently serves as Associate Editor in *IEEE Transactions on Biomedical and Health Informatics*, as well as the Journal of *Biomedical Signal Processing and Control.*

Index

A

Adaptive application-independent data aggregation (AIDA), 88

Adaptive thresholding-based real-time anomaly detection, 98–100

Advanced Simulation, Forecasting, and Optimization Platform, 236

Agent-based storm water management system, 151–165, *See also* CSOnet; Storm water management

Air pollution and precipitation anomalies correlation, 297, 300, 304, 305–306

Approximation ratio, 10

Artificial neural networks (ANNs), 135–136, 302

Autoregression (AR) machine learning model, 315–319

B

Bartle's player taxonomy, 176*f*

Basis pursuit denoising, 110

Bayesian optical fluid flow estimation models, 268–271

Behavioral water management interventions, 169–171

  big data analytics and end use disaggregation, 173–174

  gamified approaches, 169, 170–171, 175–196

  persuasive technologies, 174–175

  resource efficiency improvement applications, 187–191

  SmartH20 project, 171

  smart water meters, 170, 171, 172–173

  some commercial products, 186–187

  user modeling, 174

  visualization interfaces, 175, 191–193

  See also Gamified water management approaches

BentoBox, 78–86

Big data analytics

  offline graph-based burst localization algorithm, 101–103

  stormwater management system, 171, 173–174

Biofilm detection and state estimation, 41–45

  autonomous control, 50

  biofilm formation and evolution, 43–48

  future research directions, 68

  numerical simulation results, 64–68

  reverse osmosis membrane systems, 42*f*, 43

  wireless sensor networks, 43, 44, 48–50

    consensus-based distributed detection and tracking, 53–64

    optimal detection, 50–52

    optimal state estimation, 52–53

Block matching particle tracking velocimetry approaches, 265, 275

Brine management

  brine sources and natures, 204–206

  concentrate management options, 206–208

  industrial water use and, 216–220

  Mediterranean-area applications, 220–227

  mineral and metal recovery, 204, 215–216

  need for, 203–204

  seawater brine effluent valorization, 215–216, 227

See also Desalination
Burst localization algorithm, 101–103
bzip2, 90

C

CANDECOMP/PARAFAC
decomposition, 116–117
Carbon emissions and precipitation
anomalies correlation, 297,
300, 304, 305–306
Cellular communication systems, 157
Chasqui mote, 154–155
Classification algorithms, machine
learning for leak
quantification, 130, 133–134
Clean Water Act, 152
Climate change, 255–256
Climate modeling, 298–299, *See also*
Climate networks; River flow
rate prediction
Climate networks, 299, 301–303
long-memory multiscale
precipitation patterns, 308
long-range spatial precipitation
anomalies, 306–308
multivariate, long-memory
temporal dynamics, 304–306
precipitation and temperature
anomalies, 306, 308
Combined sewer overflow (CSO)
mitigation, 151–153, *See also*
CSOnet; Storm water
management
Communication optimization, 71–74
compressive sensing, 27–28, 73,
87–91, 92–96
edge adaptive compression, 96–98
edge analytics, 71, 73–74, 98
adaptive thresholding-based
real-time anomaly detection,
98–100
offline graph-based burst
localization algorithm,
101–103

lossless compression, 73, 87–91,
96–98
low power wide area (LPWA)
technologies, 73, 74–76
evaluation platform, 78–80
evaluation results, 82–87
experimental methodology,
80–82
spread spectrum and
ultra-narrow band, 76–77
See also Compressive sensing;
Data compression; Signal
representation; Wireless
sensor networks
Complex network dynamics, 298–308,
See also Climate networks
Compressive sensing, 27–28, 73, 87–98,
109, 120–121, *See also* Data
compression; Signal
representation
Cosine similarity (CS) function, 143
CSOnet, 151, 153–154
communication subsystem,
156–157
distributed control subsystem,
158–163
implementation and data analytics,
163–166
monitoring hardware, 154–156
performance and future directions,
166–167
See also Storm water management
Current-flow centrality, 140–141

D

Data aggregation techniques, 88
Data compression, 73
adaptive thresholding-based
real-time anomaly detection,
98–100
edge adaptive compression, 96–98
lossless compression, 73, 87–91
lossless compression and
compressive sensing

combination, 96–98
lossy compression, 91–92
recovery from compressed
measurements, 120–121
See also Compressive sensing;
Signal representation
Data fusion, 88
Deep well injection (DWI), 207
Demand-side water management
interventions, *See* Behavioral
water management
interventions
Desalination, 203–204, 225*t*
brine concentrate management
options, 206–208
brine effluents, 203–206
capacity by region and technology,
213*f*
energy considerations, 212
global capacity, 205, 206*f*
industrial water use and, 216–220
Mediterranean-area applications,
220–227
mineral and metal recovery, 204
renewable-energy desalination,
213–215
reverse osmosis membrane
systems, 211–212
seawater brine effluent valorization,
215–216
technologies, 42*f*, 43, 208–212
industrial water use and,
221–222
Mediterranean-area, 221–223
See also Biofilm detection and state
estimation; Reverse osmosis
(RO) membranes
Dictionary learning, 110–111
Differential optical fluid flow estimation
approaches, 265–267,
271–274
comparing different algorithmic
schemes, 279–288
Directed acyclic graph (DAG), 140–143

District metered area (DMA) leak
quantification, 143–145
Drop! board game, 184–186

E
Edge adaptive compression, 96–98
Edge analytics, 71, 73–74, 98
adaptive thresholding-based
real-time anomaly detection,
98–100
offline graph-based burst
localization algorithm,
101–103
Electrodialysis (ED), 211
End use disaggregation, 173–174
End user (demand-side) water
management interventions,
See Behavioral water
management interventions
Energy beamforming, 33–34
Environmental Protection Agency
(EPA), 152
EPANET, 7, 24, 132–133
European Union's Seventh Framework
Program, 178–179, 183
Eutectic freeze crystallization (EFC),
210
Evaporation ponds, 207

F
Flash flood monitoring, 256, 257, *See
also* Optical flow estimation
methods; River flow rate
prediction
Flood prediction, 256, 290, *See also*
Optical flow estimation
methods; River flow rate
prediction
Flow estimation methods, *See* Optical
flow estimation methods;
River flow rate prediction
Fluid mechanics-based optical flow
models, 272–273
Fractal structure, 298

Fuzzy rule-based systems (FRBSs), 302–303

G

Gamified water management approaches, 169–171, 175–178
 Bartle's player taxonomy, 176*f*
 Drop! board game, 184–186
 future directions, 195–196
 IBM Intelligent Water, 186–187
 Oracle Utilities, 187
 resource efficiency improvement applications, 187–191
 SmartH20, 171, 183–186, 191–195
 summary table, 189–190*t*
 Urban Water, 178–179
 Water Mansion, 179
 Waternomics, 180–183
 WatERP, 179–180
Genetic algorithm (GA), 134, 139–140, 303
Genetic programming, 303
Global climate change, 255–256
Global Water Intelligence, 204
Grand Rapids water resource recovery facility, 238–250
Graph-based burst localization algorithm, 101–103
Greedy algorithms, 110
Green infrastructure, 152

H

Horn and Schunck method, 277–285

I

IBM Advanced Simulation, Forecasting, and Optimization Platform, 236
IBM Intelligent Water, 186–187
Illumination invariance correction, 288–289
Image deformation particle image velocimetry, 261

inControl Solutions Monitoring and Control Platform (M&CP), 236–237, 240
Industrial brine effluents, 204–205
Industrial water use and desalination, 216–220
Instance-based learning, 303
Integrated Computer Control System (IC^2S), 235
Intelligent Water, 186–187
Internet of things (IoT), 71, 73, 236

J

Jacard Index (JI), 143

K

Kalman filter, 44, 52, 55, 60, 64, 66, 100
Khatri-Rao products, 115–116
K-hop learning, 297, 300, 309–319
K-nearest-neighbors (KNN) algorithm, 303, 315–319
Kronecker vector product, 115–116
K-SVD algorithm, 111
Ky Fan r-norm, 114

L

LASSO, 110
Leak detection and location, 72–73, 129–130
 leakage characteristics and consequences, 4
 PipeNet system, 7
 See also Leak quantification techniques
Leak quantification techniques, 129–146
 current-flow centrality, 140–141
 directed acyclic graph, 140–143
 future research directions, 146
 large-scale water distribution systems, 143–145
 leak signature database, 141
 machine learning algorithms, 130, 133–137

mathematical modeling, 130, 131–132

multiple leak quantification, 135

optimization and classification algorithms, 130, 133–134, 137

sensor placement strategies, 130, 137–143

small-leak quantification, 142–143

water balance law, 130

water distribution system simulation, 132–133

See also Leak detection and location

Levenberg-Marquardt methods, 131

Local area networks (LANs), smart metering technologies, 172–173

LoRa, 78–86

Lossless compression, 73, 87–91, 96–98

Lossy compression, 91–92

Louvain algorithm, 305

Low power wide area (LPWA) communication technologies, 73, 74–76

  adaptive thresholding-based real-time anomaly detection, 98–100

  evaluation platform, 78–80

  evaluation results, 82–87

  experimental methodology, 80–82

  spread spectrum and ultra-narrow band, 76–77

Low rank modeling, 108, 111–115

  applications, 120

  recovery from noisy measurements, 122

Lucas–Kanade method, 277–285

LZ4, 90

LZMA, 90

M

Machine learning

  artificial neural networks, 135–136, 302

  genetic algorithm, 134, 139–140, 303

  leak quantification, 130, 133–137

  leak quantification sensor location strategy, 139–140

  optimization and classification algorithms, 130, 133–134, 137

  river flow rate prediction applications, 300, 302–303, 309–319

  rule learning, 136–137

  support vector machine, 135

  timeseries-based models, 315–319

  See also K-hop learning

Matrix low rank modeling, 108, 111–115

Matrix unfolding (tensor decomposition), 115–118, 122

  tensor recovery, 118–120

Mediterranean-area desalination and integrated brine management, 220–227

Mica2 mote, 154–155

Mineral and metal recovery from desalination brine, 204, 215–216

MiniLZO, 90–91, 94, 99

Minimum test cover (MTC), 138–139

Missing measurements, 108

  matrix recovery or completion, 113–115

  tensor completion, 118–120, 121–122

Mobile sensor nodes, 6, 37

  deployment problems, 21–25

Model-assisted, real-time monitoring and control, 231, 236

  benefits of using, 231–232, 234

  case study, 238–250

    ABAC-SRT controller, 244–250

    data quality of online sensors, 242–243

    process model development, 241

  development and implementation

challenges, 234
driving forces in, 232–233
dynamic process models, 234, 241
essential functionalities, 233–234
examples
   inControl Solutions Monitoring
   and Control Platform,
   236–237, 240
   Integrated Computer Control
   System, 235
   Star Utility Solutions, 235–236
Monod equation, 45
Multilayer perceptron model, 302
Multiple-effect evaporation/distillation
   (MED) desalination, 209
Multistage flash evaporation/distillation
   (MSF) desalination, 209

N

Network dynamics approach, *See*
   Climate networks; Complex
   network dynamics
Newton-Raphson methods, 131
Neyman-Pearson detector, 44, 51, 57–58
Noisy measurements, 108
   recovery applications, 121–122
Nuclear norm, 114, 118
nWave, 78–82

O

Ohio River Basin, 298, 309
Optical flow estimation methods,
   255–257, 289–290
   comparing different algorithmic
   schemes, 274–288
   differential approaches, 265–267,
   271–274
      comparing different algorithmic
      schemes, 279–288
      fluid mechanics-based, 272–273
      particle tracking velocimetry,
      265–267
      wave properties, 273–274
   experimental results, 276–286

illumination invariance correction,
   288–289
integrated water flow monitoring
   system, 256
local-based versus global
   approaches, 285–286
particle image velocimetry (PIV),
   257–264
particle tracking velocimetry
   (PTV), 257–259, 264–267
probabilistic and Bayesian
   approaches, 267–271
real-world applications, 286, 290
See also River flow rate prediction
Optimization algorithms, machine
   learning for leak
   quantification, 130, 133–134,
   137
Oracle Utilities, 187
Orthogonal matching pursuit (OMP)
   algorithm, 110

P

Particle-based optical fluid flow
   estimation methods
   comparing different algorithmic
   schemes, 274–288
   particle image velocimetry (PIV),
   257–264
   particle tracking velocimetry
   (PTV), 257–259, 264–267
Petroleum refinery water use, 218–219
Phase cross-correlation particle image
   velocimetry, 261–263
PipeNet project, 6, 7
PipeProbe, 6
Polar jet streams, 308
Power industry water use, 219
Power sources for sensor nodes,
   154–155
PPMd, 90
Precipitation and temperature
   anomalies, 306, 308
Precipitation anomalies and air pollution

INDEX

345

correlation, 297, 300, 304,
305–306
Precipitation modeling, 298–299, *See
also* River flow rate prediction
Predictive applications, *See* Optical flow
estimation methods; River
flow rate prediction
Principal components analysis, 114, 122
Privacy issues, smart metering
technologies, 172
Probabilistic optical fluid flow
estimation approaches,
267–271
comparing different algorithmic
schemes, 278–288
Process models, 232, 234, 241, *See also*
Model-assisted, real-time
monitoring and control
PRODES, 214
Programmable logic controllers (PLCs),
232

Q

Quartile Digest, 88

R

Random forests (RF) machine learning
model, 315–319
Real-time anomaly detection exploiting
compression rates, 98–100
Real-time monitoring and control, *See*
Model-assisted, real-time
monitoring and control
Refinery water use, 218–219
Relaxation techniques, 110
Renewable-energy desalination
(RES-D), 213–215
Reverse osmosis (RO) membranes, 42*f*,
43, 211–212
biofilm detection, *See* Biofilm
detection and state estimation
industrial desalination applications,
219–220
Mediterranean-area desalination

applications, 221–222
RIPPER, 136–137
River flow monitoring, 256–257, 286,
290, *See also* Optical flow
estimation methods; River
flow rate prediction
River flow rate prediction, 297–301
carbon emissions and precipitation
anomalies correlation, 297,
300, 304, 305–306
climate dynamics, 303–308
complex network dynamics,
298–299
inaccuracies of timeseries-based
models, 299, 303
K-hop learning, 297, 300, 309–319
long-memory multiscale
precipitation patterns, 308
long-range spatial precipitation
anomalies, 306–308
Ohio River Basin, 298, 309
previous approaches and
limitations, 302–303
timeseries-based machine learning
models, 315–319
See also Optical flow estimation
methods; Storm water
management
Robust principal components analysis
(RPCA), 114, 122
Rolling window technique, 315
Rule-based classification, 136–137

S

Schatten l-norm, 114
Scheduling wireless sensor
transmissions, *See*
Transmission scheduling of
wireless sensor networks
Seawater desalination, *See* Desalination
Seawater desalination brine valuation,
215–216, 227
Security and privacy issues, smart
metering technologies, 172

Sensor nodes, 4–6, 155
    cooperation schemes, 43–44
    CSOnet agent-based stormwater management system, 154–156
    mobile nodes, 6, 21–25, 37
    power sources, 154–155
    static network deployment, 10–21
    transmission scheduling, 25–31
    See also Wireless sensor networks
Sensor placement strategies for leak quantification, 130, 137–143
Sensor signal representation, *See* Data compression; Signal representation
Separate sewer overflows (SSOs), 152
Serious games, 175, 179, *See also* Gamified water management approaches
Sewer discharge of brine effluents, 207
Sewer system overflow mitigation, *See* Storm water management
SIGFOX, 77
Signal representation, 107–108
    low rank modeling of matrices, 108, 111–115
    matrix recovery or completion, 113–115
    recovery from noisy measurements, 122
    missing measurements and, 108, 118, 121–122
    recovery from compressed measurements, 120–121
    recovery from noisy measurements, 121–122
    sparse signal coding, 108–111
    dictionary learning, 110–111
    recovery from compressed measurements, 120–121
    recovery from noisy measurements, 121–122
    tensor modeling, 115, 122–123
    decompositions or matrix unfolding, 115–118

tensor completion, 118–120
tensor completion applications, 121–122
See also Compressive sensing; Data compression
Simulation models
    biofilm detection and state estimation, 64–68
    EPANET TOOLS, 7, 24, 132–133
    IBM Advanced Simulation, Forecasting, and Optimization Platform, 236
    leak quantification, 132–133
    mobile nodes, 24–25
Singapore's Wireless Water Sentinel project, 6, 8
Singular value decomposition (SVD), 112–113
    K-SVD algorithm, 111
    Tucker decomposition, 117–118
Sleep/awake scheduling of wireless sensor networks, 27–31
S-LZW-MC, 90–91
SmartBall, 6
SmartH20 project, 171, 183–186, 191–195
Smart water grid communication optimization, *See* Communication optimization
Smart water meters, 170, 171, 172–173
Solar-powered renewable-energy desalination applications, 214
Sparse signal coding, 108–111
    recovery from compressed measurements, 120–121
    recovery from noisy measurements, 121–122
Spatio-temporal autoregressive (STAR) model, 268
Spray irrigation, 207
Spread-spectrum technologies, 76–77, 155
Star Utility Solutions, 235–236
State place search, 137–138

INDEX

Stereoscopic particle image velocimetry (SPIV), 263–264
Storm water management, 151–165
    combined sewer overflow (CSO) mitigation, 151–153
    CSOnet agent-based system, 153–154
        communication subsystem, 156–157
        distributed control subsystem, 158–163
        implementation and data analytics, 163–166
        monitoring hardware, 154–156
        results and future directions, 166–167
    green infrastructure, 152
    separate sewer overflows, 152
    See also River flow rate prediction
Subtropical jet streams, 308
Superior Tuning and Reporting Utility Solutions™ (STAR), 235–236
Supervised learning, 133–134
Supervisory control and data acquisition (SCADA) systems, 232, 238, *See also* Model-assisted, real-time monitoring and control
Support vector machine (SVM), 135
Surface water discharge of brine effluents, 207
SVM regression (SVM-R) machine learning model, 315–319
Synopsis Diffusion Framework, 88

T
Temporal coherency-aware in-network aggregation (TiNA), 88
Tensor modeling, 115, 122–123
    applications, 120
    decompositions, 115–118
    missing measurements and tensor completion, 118–120, 121–122

Timeseries-based machine learning models, 315–319
Timeseries-based river flow forecasting models, 299, 303
Transmission scheduling of wireless sensor networks, 25
    routing, 26–27
    sleep/aware scheduling, 27–31
    wireless energy transfer, 31–36, 37
Triopus, 6
Tucker decomposition, 117–118

U
Ultra-narrow band (UNB) communication technologies, 76–77
United Nations World Water Development Report, 256
Urban Water project, 178–179
User modeling, 174

V
Vapor compression (VC) desalination, 209–210
Visualization interfaces, 175, 191–193

W
Wastewater treatment plants (WWTPs), real-time monitoring and control, 231–251, *See also* Model-assisted, real-time monitoring and control
Wastewater treatment plants (WWTPs), storm water management, 151–165, *See also* Storm water management
Water balance law, 130
Water distribution system simulation, *See* Simulation models
Water flow estimation methods, *See* Optical flow estimation methods
Water Framework Directive, 256
Water management behavioral

interventions, *See* Behavioral water management interventions; Gamified water management approaches

Water Mansion, 179

Waternomics, 180–183

WatERP, 179–180

Water quality preservation, 256

Water resource recovery facilities (WRRFs), real-time monitoring and control, 231–251, *See also* Model-assisted, real-time monitoring and control

Water sustainability, 256

Wave properties-based optical fluid flow models, 273–274

Weightless, 77

Wide area network (WAN)
- CSOnet agent-based stormwater management system, 154, 156–157
- smart metering technologies, 172–173

Window-based particle image velocimetry, 261

Wireless energy transfer (WET), 31–36, 37

Wireless sensor networks (WSNs), 4, 72, 120
- cooperation schemes, 43–44
- CSOnet agent-based stormwater management system, 154–156, *See also* CSOnet
- current water grid monitoring systems, 7–8
- sensor nodes, 4–6, 155
- signal coding and representation, *See* Data compression; Signal representation

sparse coding applications, 120–121
See also Sensor nodes

Wireless sensor networks (WSNs), biofilm detection and state estimation system, 43, 44, 48–50
- consensus-based distributed detection and tracking, 53–64
- numerical simulation results, 64–68
- optimal detection, 50–52
- optimal state estimation, 52–53

Wireless sensor networks (WSNs), communication optimization, *See* Communication optimization

Wireless sensor networks (WSNs), node deployment and scheduling, 3, 8–10
- approximation ratio, 10
- future research directions, 37
- mobile nodes, 21–25, 37
  - maximum coverage, 23–24
  - simulation, 24–25
- static networks, 10
  - maximum coverage, 14–21
  - minimum number of nodes, 10–14
- transmission scheduling, 25
  - routing, 26–27
  - sleep/aware scheduling, 27–31
  - wireless energy transfer, 31–36, 37
See also Communication optimization

Wireless Water Sentinel project, 6, 8

X

Xbee868, 78–86

Z

Zero liquid discharge (ZLD), 208, 219